T0329574

Silica Optical Fiber Technology for Devices and Components

EY SERIES IN MICROWAVE AND OPTICAL ENGINEERING

CHANG, Editor
A&M University

mplete list of the titles in this series appears at the end of this volume.

Silica Optical Fiber Technology for Devices and Components

Design, Fabrication, and International Standards

KYUNGHWAN OH

UN-CHUL PAEK

A JOHN WILEY & SONS, INC., PUBLICATION

Published by John Wiley & Sons, Inc., Hoboken, New Jersey
Published simultaneously in Canada

For general information on our other products and services or for technical support, please contact
our Customer Care Department within the United States at (800) 762-2974, outside the
United States at (317) 572-3993 or fax (317) 572-4002.

Wiley also publishes its books in a variety of electronic formats. Some content that appears in print
may not be available in electronic formats. For more information about Wiley products, visit our
web site at www.wiley.com.

Library of Congress Cataloging-in-Publication Data:

Oh, Kyunghwan
    Silica optical fiber technology for optical devices and components : design,
fabrication, and international standards / Kyunghwan Oh, Un-Chul Paek – 1st
ed.
        p. cm. – (Wiley series in microwave and optical engineering ; 158)
    ISBN 978-0-471-45558-5 (hardback)
    1.   Optical fibers. 2.   Fiber optic cables–Materials. 3.   Optical fiber
communication–Equipment and supplies–Design and construction. 4.
Optoelectronic devices–Materials.   I. Oh, Kyunghwan.   II. Title.
    TK5103.592.F52P34 2011
    621.382'75–dc23
                                                        2011024790

10 9 8 7 6 5 4 3 2 1

Contents

Preface   **ix**

Acknowledgment   **xiii**

1 Introduction   **1**

   1.1 Brief Historical Review of Silica Optical Fibers   1

   1.2 International Standards for Silica Optical Fibers   8

   1.3 Classifications of Silica Optical Fibers   12

   References   16

2 Review on Single-Mode Fiber Design and International Standards   **18**

   2.1 Optical Modes in Cylindrical Waveguides   18

   2.2 Material Dispersion in Optical Fibers   37

   2.3 Optical Attributes for Single-Mode Fiber Characterization
       and Classification   48

   2.4 International Standards for Single-Mode Fibers   64

   References   79

3 Preform Fabrication and Optical Fiber Drawing Process   **83**

   3.1 Preform Fabrication Based on Chemical Vapor Deposition Process   83

   3.2 Postprocesses for Geometrical Modification of Preform   101

   3.3 Optical Fiber Drawing   104

   References   126

4 Dispersion-Managed Single-Mode Fibers for Wavelength Division Multiplexing   **131**

    4.1 Wavelength Allocations in Single-Mode Fibers for WDM Applications   131

    4.2 Optimization of Waveguide Parameters for Dispersion Control   137

    4.3 Refractive Index Profile Analysis for Dispersion-Shifted Fibers   150

    4.4 Dispersion-Compensating Fibers Using the Fundamental Mode   157

    4.5 Dispersion Compensation Using High-Order Modes   168

    References   181

5 Multimode Fibers for Large-Bandwidth Applications   **185**

    5.1 History and Recent Application Trends of Multimode Optical Fibers   185

    5.2 Principle of Multimode Optical Fiber Design   190

    5.3 Impacts of Nonideal α-Refractive Index Profile on Transmission Bandwidth   197

    5.4 Main Attributes of GI-MMFs-Bandwidth and Differential Modal Delay   203

    5.5 Multimode Optical Fiber Standards   213

    References   221

6 Optical Nonlinearity Control in Optical Fibers   **224**

    6.1 Historical Review of Optical Nonlinearity in Optical Fibers   224

    6.2 Origin of Optical Nonlinearities in Optical Fibers   226

    6.3 Specifications of Nonlinear Optical Processes in Optical Fibers   231

    6.4 Comparison of Raman and Brillouin Scattering in Single-Mode Optical Fibers   234

    6.5 Control of Raman Scattering in Silica Optical Fibers   236

    6.6 Brief Review on Raman Amplifiers and Lasers   245

    6.7 Control of Brillouin Scattering in Silica Optical Fibers   253

    6.8 Review on Fiber Brillouin Sensors and Recent Novel Applications   267

    References   273

7   Birefringence Control in Optical Fibers   **280**

7.1 Physical Parameters for the Polarization Characterization
     in Optical Fibers   280

7.2 Representation of the State of Polarization in Optical
     Fiber Using Poincaré Sphere   286

7.3 Classifications of Linear Polarization Maintaining Fibers   290

7.4 Fabrication Methods for High Birefringence Fibers   294

7.5 Control of Birefringence by Waveguide Design
     in Birefringent Fibers   296

7.6 Single-Polarization Single-Mode Fibers   304

7.7 Low Linear Birefringence Fibers   310

     References   316

8   Optical Fibers Based on Air–Silica Guiding Structure   **321**

8.1 Review of Air–Silica Guidance in Optical Fibers   321

8.2 Fabrication Technique—Stack and Draw Method   322

8.3 Effective Index Guiding Air–Silica Optical Fibers   324

8.4 Large Mode Area and Bending Loss Based on Effective
     Index Guiding Air–Silica Holey Fibers   331

8.5 Dispersion Control in Effective Index Guiding Air–Silica
     Holey Fibers   335

8.6 Optical Loss in Effective Index Guiding Air–Silica
     Holey Fibers   342

8.7 Kerr Nonlinearity in Effective Index Guiding Air–Silica
     Holey Fibers   347

8.8 Birefringence Control in Effective Index Guiding
     Air–Silica Holey Fibers   351

8.9 Hollow Optical Fiber and its Applications   355

     References   368

9   Fiber Mode Analysis Using OFACAD   **378**

9.1 Theoretical Review on Cascaded Boundary Matrix Method   378

9.2 Algorithm for CBM to Find Optical Properties of
     Guided Modes   384

9.3   Mode Analysis Example Using OFACAD   386

References   399

Appendix A: OFACAD Installation/Operation Manual   401

Appendix B: Operation Manual of OFACAD   415

Index   447

Preface

The proposal of high capacity optical communication using silica optical fiber initiated by Dr. Charles Kao in 1966 and subsequent announcement of a silica-based low-loss fiber fabrication by the Corning in 1970 stimulated intensive global research and development efforts in optical fibers and fiber optical communication. Rapidly growing optical fiber technologies successfully shifted toward commercial communication applications in the massive scale worldwide, which in turn has generated profound impacts over information technology revolution in the late 20'st centuries. The huge success and contributions in optical fiber technology in shaping the present information based society have been recognized by scientific communities to have Dr. Charles Kao nominated as a Nobel laureate in Physics 2009.

Today in 21'st century, silica optical fiber is regarded as a "de facto transmission medium" to transport high bandwidth information including voice, picture, video and data. Since the advent of internet and internet based services, demands on flexible connectivity and wide bandwidth have been soaring in recent years and optical communication systems based on optical fibers have proven that fiber optic technology is the only viable options to cope with those ever-increasing demands.

Optical fibers have successfully served their initial roles as transmission media for optical signals, providing both low attenuation and large bandwidth. However, it should be also emphasized that optical fibers can be used not only as a mere transmission medium but also as a functional building block for valuable active and passive optical devices. Recently silica optical fibers have evolved to provide variety of applications as discrete devices or components with versatile functions such as amplification, wavelength shift, filtering, and modulation of optical signals. Especially recent air silica holey fibers (ASHFs) or generally known as photonic crystal fibers have introduced new concept to control the guiding properties by air hole arrangements across the fiber cross section.

Prior index guiding principles in solid core/solid clad optical fibers were still kept in ASHF yet the air holes could provide unprecedented ranges of waveguide properties. Nonlinear fiber ASHF of a few tens of meters can serve as either an amplifier or a frequency shifter with proper pumping mechanisms. A few meters of ASHFs can provide sufficient isolation of a specific polarization of the fundamental mode. Furthermore a few tens of centimeters of ASHFs can serve as highly efficient liquid or gas sensors. With these rapid developments of both conventional and holey silica optical fibers, thorough knowledge of design and fabrication of optical fibers in terms of material, modal analysis, linear and nonlinear optical behavior became essential for integrating novel functionalities into optical fibers.

This book aims to provide readers fundamental and basic perspectives of silica optical fibers as the transmission medium and functional building blocks for optical devices. The authors also added the latest international standards for conventional transmission optical fibers so that readers can be prepared for practical applications such as optical fiber and component manufacturing, optical system engineering, and novel integrated optical device research. There have been varieties of optical fiber related texts and monographs but systematic efforts to combine conventional optical fibers with air silica holey fibers have been very scarce. Systematic review on material properties and fabrication processes in this book would also serve as a good introductory text book for the readers who just enter into the field.

A user-friendly software program (Optical Fiber Analysis Computer Aided Design, OFCAD) is also included for the full vectorial modal analysis of optical fiber waveguides. The program calculates effective index, group delay, chromatic dispersion, field/intensity profiles, mode field diameter, effective area, optical loss, and splice loss for a selected guided mode in a given step index profile. The program will provide readers with hands-on experiences of optical fiber design and parametric analysis on the subjects that are dealt in this book. This is a first attempt to provide a fiber design tool in a book covers optical fiber waveguide analysis. The level of the book is set to the first or second year graduate students in both science and engineering departments.

In Chapter 1-Introduction, we start with brief historical review on silica optical fibers along with discussions on principal fiber properties such as attenuation, dispersion, birefringence and modal area for optical communications applications. International standards on transmission fibers are summarized in tables. Classification of silica fibers is then introduced for further discussions in the following chapters.

In Chapter 2-Review on single mode fiber design and international standards, we explain optical modal analysis for cylindrical waveguide. Material dispersion of silica glass and conventional dopants are also discussed. Optical attributes of optical fibers used in international standards are then explained along with discussion on their characterization methods. Latest international standards in ITU-T and IEC are compared and discussed in detail for transmission single mode optical fibers.

In Chapter 3-Preform fabrication and optical fiber drawing process, we explained the principles of chemical vapor deposition process, which are being shared by major optical fiber manufacturers worldwide. Post processes to enlarge preforms or to modify the geometrical structures are then discussed. Detailed review on optical fiber drawing process is presented along with thermo-mechanical analysis of fiber pulling, cooling, and acrylate polymer coating.

In Chapter 4-Dispersion managed single mode fibers for wavelength division multiplexing, we review the waveguide designs for dispersion shifted fibers, non-zero dispersion shifted fibers, dispersion compensating fibers in the fundamental mode, dispersion compensating fibers in the higher order modes. The frequency and wavelength standards are summarized for wavelength division multiplexing systems.

In Chapter 5-Multimode fibers for large bandwidth applications, historical review on multimode fiber development and recent applications in large bandwidth data communications are explained. Refractive index control in multimode fibers and differential modal delays are discussed for wide bandwidth applications. Latest international standards on multimode fibers are summarized.

In Chapter 6-Optical nonlinearity control in optical fibers, historical review on nonlinearity issues in optical fiber is made and the discussion on the origin of fiber nonlinearity is followed. Scope of the book is set to scattering processes: Raman and Brillouin scattering and their principal characteristics are compared. For Raman scattering in single mode fibers, Raman gain and its dependence on material and waveguide parameters are discussed along with amplifier and laser applications. Impacts of the single mode fibers parameters on Brillouin scattering gain are then explained along with discussion on Brillouin optical sensors.

In Chapter 7-Birefringence control in optical fibers, basic polarization descriptions in optical fibers are reviewed along with classification of polarization maintaining fibers. Detailed fabrication processes are then discussed and waveguide designs for high birefringent fibers, single mode single polarization fibers, and low linear birefringent fibers are explained.

In Chapter 8-Optical fibers based on air-silica guiding structure, we primarily discuss on air silica holey fibers (ASHFs), or photonic crystal fibers, in terms of fabrication process, effective mode area, and chromatic dispersion. Among ASHFs the scope of discussion is focused on effective index guiding structures and controls of optical nonlinearity, birefringence are explained. Another type of air-silica guiding structure, hollow optical fiber, is then discussed along with various device applications.

In Chapter 9-Fiber mode analysis using OFACAD, we provide a simulation tool, optical fiber analysis computer aided design (OFACAD) that can calculated the effective index, chromatic dispersion, mode field area, and modal intensity distribution. Basic principles of the mode calculation using cascaded boundary matrix method are explained and examples of mode analyses are then explained.

The readers will understand the basic principles of silica optical fibers and learn the related applications of optical fibers in both high capacity transmission media and novel photonic devices. The readers will also be familiar with the latest international standards in optical fibers so that they could apply their knowledge in practical fields of optical communications.

The simulation program referenced in this book may be found at the following ftp site: ftp://ftp.wiley.com/public/sci_tech_med/optical_fiber

Acknowledgment

The authors would like to express their sincere thankfulness to their beloved family, Un-Chul Paek's wife and his daughter: Jae Hack, and Audrey, and Kyunghwan Oh's wife and his daughters: Hyeyon, Seiyon, and Seijung. Without their kind support, continuous encouragement and sincere understanding, completion of this book would have not been possible.

The authors also acknowledge numerous helps from former students: Hong Seok Seo, Woojin Shin, Soan Kim, Yongmin Jung, Jun Ki Kim, at Gwangju Institute of Science and Technology and graduate students at Yonsei University, Sejin Lee, Woosung Ha, Sehun Kang, Jongki Kim, Jiyoung Park, Marzier Pournoury. Especially Dr. Hong Seok Seo's effort and dedication in the software program for optical fiber design was essential in this book.

I would have never expected to have this book published without bright smile and bit hug from the co-author, Prof. Un-Chul Paek, who has suddenly passed away in May 3'rd 2011. As an inventor of fast optical fiber drawing process and dispersion shifted optical fiber, Prof. Paek contributed to change optical fiber from a laboratory specimen to an industrial necessity in the information era. In addition to technical achievement, he served as a source of professional inspiration and as a role model for generation of young students and researchers. It was Prof. Paek's ardor and gusto that made this book started. As a co-author, I sincerely hope that his passion toward fiber optics would be carried to the readers of this book.

On behalf of the co-author, the Late Un-Chul Paek,
Kyunghwan Oh

CHAPTER ONE

Introduction

1.1 BRIEF HISTORICAL REVIEW OF SILICA OPTICAL FIBERS

The invention of the low-loss silica glass optical fiber using chemical vapor deposition in 1970 [1] has opened new horizon in communication technology, and led to the possibility of ultrahigh capacity information flow. Worldwide competitive research and development efforts followed that invention and brought about astounding progress in photonics technology. Two seminal inventions made almost simultaneously in 1970, silica-based glass fiber [1] and continuous wave semiconductor laser operating at room temperature [2], marked the inception of the photonics era. Owing to excellent transmission properties of low loss and high bandwidth as predicted by Kao and Hockham [3], silica optical fiber is the status quo broadband telecommunications medium. Real-time audio, video, and data applications for advanced broadband systems prompted optical fiber manufacturers to develop high volume mass production technologies such as large perform fabrication [4] and high speed drawing [5] and consequently lower the cost of the fiber. As a result of these efforts, fiber technology has provided various types of optical fibers and fiber devices that enabled all-fiber platform in telecommunications, sensing, and metrology. All of these fiber optic innovations and contributions to today's information technologies have led to recent recognition among scientific community such that Dr. Charles Kao won the Nobel prize in physics in 2009 [6].

Major development efforts in optical fiber began with low-loss fibers to extend the optical link distance. After the demonstration of fiber with a loss of 20 dB/km in 1970, the fabrication process advanced rapidly adopting chemical vapor deposition (CVD) technologies. By 1973, the loss of less than 5 dB/km was reported at 850 nm [7], approaching the intrinsic scattering limit at the wavelength. Longer wavelengths were then explored to take advantage of lower intrinsic scattering loss. In 1976, an optical

Silica Optical Fiber Technology for Devices and Components: Design, Fabrication, and International Standards, First Edition. By Kyunghwan Oh and Un-Chul Paek.
© 2012 John Wiley & Sons, Inc. Published 2012 by John Wiley & Sons, Inc.

1

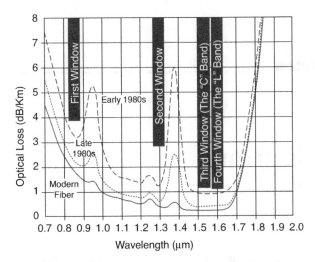

FIGURE 1.1 Optical loss spectrum of silica glass single-mode fiber [11].

fiber loss of 0.47 dB/km at 1200 nm was reported [8]. Within 3 years, the fiber attenuation reached 0.2 dB/km at 1550 nm [9,10], close to the theoretical limit of 0.15 dB/km. The evolution of optical loss reduction efforts is summarized in Figure 1.1.

In 1970s, the available semiconductor light sources were only near ∼850 nm, which has named the corresponding spectral range as the first window in the optical fiber. With the advancement of both compound semiconductor technology and fiber fabrication process, the second window near 1300 nm was flourished in 1980s. As the ultimate low loss was achieved near 1550 nm region, the communication widow shifted to the third window in order to take fullest advantage of longer link distances. The third window is also called conventional band (C band). With further development of bending loss reduction technology, the third window expanded into longer wavelength near 1620 nm, which is called long band (L band). In recent international standards the optical communication bands have been further classified as O, E, S, C, L, and U bands, which will be discussed in more detail in the following section.

Along with the efforts to reduce the optical loss, there have been intensive researches to control the waveguide properties of optical fibers in order to manipulate the modal dispersion and the chromatic dispersion, which are directly related to bandwidth and data rate in optical communications. Two types of fibers have been developed almost in parallel: multimode fiber (MMF) and single-mode fiber (SMF). Li and Nolan [12] have made a detailed review on optical fiber development history, and schematic routes of development for multimode and single-mode optical fibers are summarized in Figures 1.2 and 1.3, respectively.

Starting from a step-index fiber, MMFs quickly evolved to graded-index profile MMF to increase the bandwidth for 850 nm light-emitting diode (LED) light sources. Graded-index MMF diameter started from 50 μm, and then increased to 62.5 μm to

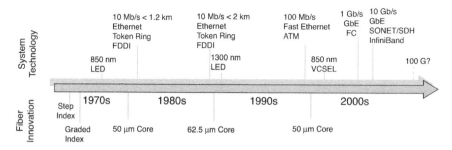

FIGURE 1.2 Multimode fiber technology evolution [12]. FDDI, fiber distributed data interface; ATM, asynchronous transfer mode; GbE, gigabit Ethernet; FC, fiber channel; SONET, synchronous optical networking; SDH, synchronous digital hierarchy.

accommodate 1300 nm LED light source. As high speed narrow linewidth light sources such as vertical cavity surface-emitting lasers (VCSELs) and laser diodes (LDs) replaced LEDs, graded-index MMF diameter once again reduced to 50 µm with further tightened specifications in the refractive index especially near the center in order to secure high bandwidths in data communications. Presently 100 Gbit/s signals can be transmitted over several hundred meters length, using graded-index 50 µm core MMFs, and further increase in both data rate and link distance is being explored.

In the case of single-mode optical fibers the evolution routes have been more dynamic since the introduction of wavelength division multiplexing (WDM) based on optical amplifiers [13] such as erbium-doped fiber amplifier (EDFA) [14] and Raman fiber amplifier (RFA) [15]. In WDM SMF links, multitudes of channels are carried within a certain wavelength range where the channels are amplified by an optical amplifier to restore the attenuated intensity. However, the chromatic dispersion linearly accumulates along the SMF to result in signal pulses spread out in the time domain and bit errors in the data streams. Therefore, it was, and still is, imperative that

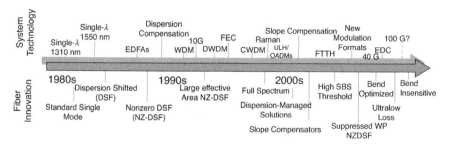

FIGURE 1.3 Single-mode fiber technology evolution [12]. DWDM, dense wavelength division multiplexing; CWDM, coarse wavelength division multiplexing; ULH, ultralong haul; OADM, optical add drop multiplexer; FTTH, fiber to the home; SBS, stimulated Brillouin scattering; EDC, electric dispersion compensation; WP, water peak.

chromatic dispersion should be properly managed or compensated in long-haul WDM systems. This important demand has initiated development of various dispersion-controlled fibers such as dispersion-shifted fiber (DSF), dispersion-flattened fiber (DFF), nonzero dispersion-shifted fiber (NZ-DSF), and dispersion-compensating fiber (DCF).

Dense WDM systems have also introduced new concepts in fiber design to manage optical nonlinearities in optical fiber caused by the high level of total optical power in WDM channels. This brought international standardization of effective mode areas and nonlinear coefficients. Nonlinear effects in optical fibers were known to be significantly suppressed if the modal area is large enough, and large mode area (LMA) SMFs have been developed with novel refractive index profiles [16].

Massive deployment of optical fiber networks into the customer premises such as in fiber to the home (FTTH) networks has raised more stringent optical nonlinearity issues especially for the case of passive optical network, where high power signals are distributed to multitude of customer premises using optical power splitters. Especially stimulated Brillouin scattering (SBS) has become a critical issue because the incident optical power is transferred to backward-propagating Stokes shifts [17]. Spatial distribution of compensating dopants such as Al and Ge across the core has been proposed to suppress SBS [18]. FTTH applications have also raised another fundamental issue in optical fiber design: macrobending loss caused by fiber installation in FTTH environments where tight bending of optical fiber is unavoidable. Bend-insensitive fibers have been recently developed to overcome the macrobending loss [19].

The optical loss due to the overtone of hydroxyl (OH) bonds near 1380 nm has been recently removed in a mass production level by establishing gas-phase dehydration process [20] and expanded the optical communication bands into S and E bands. These suppressed water peak SMFs are being widely deployed in optical networks that look forward to expansion of optical communications into S bands from existing C and L bands.

The communication bandwidth and channel speed have been continuously increasing with new developments in optical device and system technologies, which subsequently demand modification of SMF structures for optimal dispersion management within the new available bands. Especially the optical waveguide design in SMF and the optical amplifier spectral bandwidth are strongly correlated and they are continuously evolving to further accommodate future optical communications.

In this book we will focus on dispersion, birefringence, and nonlinear properties in optical fibers and describe various types of optical fiber design, fabrication, and related international standards.

Dispersion in optical fibers can be categorized into three types depending on their origins: (1) modal dispersion in MMFs, (2) chromatic dispersion in SMFs, and (3) polarization mode dispersion in SMFs.

Modal dispersion in MMF is schematically illustrated in Figure 1.4. The number of guided modes in MMF is in the order of 10^3–10^4 and each mode has different group velocity. As a result of the distribution in group velocities in step-index MMF, the effective modal delay is of a few nanoseconds, 10^{-9} s [21]. The effect of modal

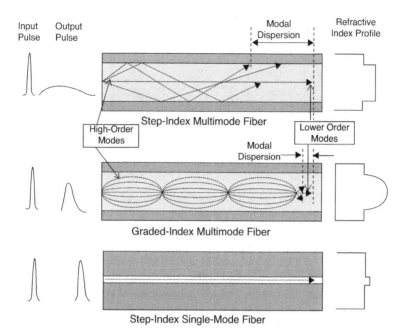

FIGURE 1.4   Impacts of modal delay in multimode fibers (MMF) over digital pulse spreads in comparison with single-mode fibers (SMF).

dispersion in MMFs is schematically illustrated in Figure 1.4. The magnitude of the modal delay in step-index MMFs is comparable to the bit period of a few 100 Mbit/s, approximately a few 10^{-9} s, which can result in severe overlap and spread of input pulses. Therefore, the maximum data rate along a step-index MMF cannot exceed 100 Mbit/s for 1 km length.

The pulse spread due to modal dispersion in MMFs can be significantly reduced by changing the refractive index distribution in the core from the step profile to a graded one, which allocates a higher refractive index to lower order modes propagating in the shorter optical paths and a lower refractive index to higher order modes propagating in the longer optical paths to effectively synchronize them in the output. This concept is schematically illustrated in Figure 1.4. The issues in MMF designs are discussed in Chapter 5.

Despite the elaborated graded-index core design in MMFs, there still exists a finite modal delay and reducing the number of modes drastically was sought after as a fundamental solution to overcome the modal dispersion. SMF has achieved an efficient guidance of only the fundamental mode in the second and third windows of communications, 1300–1600 nm, and the higher order modes were cut off in the spectral range over ~1200 nm.

In an ideal cylindrically symmetric SMF, the fundamental LP_{01} mode holds two degenerate polarization modes with the identical effective indices and they behave as a single mode. Even though the modal delay in MMF disappears in SMFs, there exists

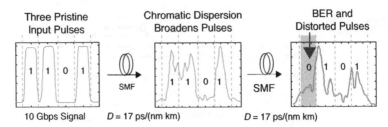

FIGURE 1.5 Impacts of chromatic dispersion (D) in single-mode fiber (SMF) over the optical signals at $\lambda = 1550$ nm. Gbps, gigabit per second.

a chromatic dispersion that linearly accumulates over the propagation length. The chromatic dispersion originates from the fact that the group velocities vary with the wavelength components, or the "color" of the light signal, which results in fast and slow components. The effect of chromatic dispersion in SMF is schematically illustrated in Figure 1.5. Optical pulse trains at $\lambda = 1550$ nm experience the chromatic dispersion of $\sim$17 ps/(nm km) in typical SMFs to result in pulse broadening in the time domain. The chromatic dispersion linearly accumulates along the fiber length and further broadening of pulses will result in distortion of signals, which is called intersymbol interference (ISI) [22]. The ISI causes bit errors, as indicated by an arrow in Figure 1.5, which is referred as the chromatic dispersion penalty. The chromatic dispersions penalties in SMFs have been overcome by waveguide designs with novel refractive index profiles, which significantly modify the waveguide dispersion to result in appropriate dispersion value and slope within the optical bands of interests. We will discuss these waveguide design issues for dispersion managements in Chapters 4 and 9.

Chromatic dispersion spectra of some of dispersion-controlled fibers are shown in Figure 1.6. Conventional SMF has the zero dispersion near 1310 nm and its value monotonically increases to $\sim$17 ps/(nm km) at 1550 nm with its slope of 0.58 ps/ $(nm^2$ km). In order to shift the zero-dispersion wavelength to the lowest loss spectral position near 1550 nm, DSFs have been developed for single-wavelength signal transmission [23]. In order to utilize wider optical communication window, DFF developments were followed, where zero dispersion occurred near both 1310 and 1550 nm [24]. In WDM systems, however, having the zero-dispersion wavelengths within the WDM channels was found to be critically detrimental by four-wave mixing (FWM) penalties [25]. A small but finite chromatic dispersion was found to be optimal and optical fibers with nonzero chromatic dispersion within the band of interests and so-called NZ-DSFs with optimal dispersion slopes have been playing major roles in long-haul applications [26]. As an alternative way to control the CD of SMF, fibers with a negative dispersion and a negative slope near 1550 nm were developed as DCFs [27]. Detailed discussions of these dispersion-controlled fibers will be given in Chapter 4.

Nominal chromatic dispersion value of DSFs is a few ps/(nm km) in the wavelength region of interests, which corresponds to an order of picosecond ($\sim$10^{-12} s) group delay for 10 km fiber with a 0.1 nm linewidth ($\Delta\lambda$) light source. This delay is negligible

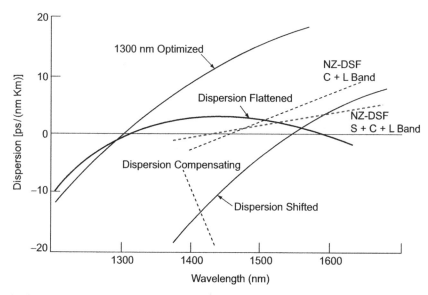

FIGURE 1.6 Chromatic dispersion spectra of SMFs with various dispersion optimizations. NZ-DSF, nonzero dispersion-shifted fiber. S, C, and L bands are short wavelength, conventional, and long wavelength bands, respectively, and their definitions are summarized in Table 1.2.

in comparison to bit period of 10 Gbps digital signals, which is an order of 10^{-10} s, and the chromatic dispersion can be overcome. However, as the data rate further increases to 40 Gbps and higher whose bit period is in the order of 10^{-11} s, another type of dispersion in optical fiber becomes the major limitation in optical transmission. By slight random deviation of the optical fiber from the perfect cylindrical symmetry, the fundamental LP_{01} mode is no longer degenerate and two modes in orthogonal polarizations will have different effective indices to result in a differential group delay as schematically shown in Figure 1.7. This dispersion is called polarization mode dispersion (PMD).

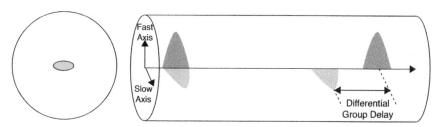

FIGURE 1.7 Impact of polarization mode dispersion in SMF voer the time delay between two orthogonal polarization modes along the slow and fast axes in an elliptical core optical fiber.

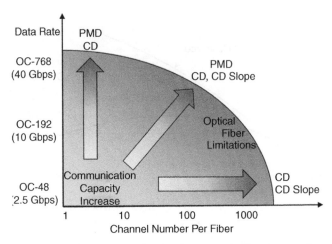

FIGURE 1.8 High speed large capacity WDM network directions and accompanying limitations. PMD, polarization mode dispersion; CD, chromatic dispersion [28].

The group delay between two orthogonal polarization modes could exceed a few picoseconds for SMF link of a few tens of kilometers, which is comparable to bit period of high data rate digital signals. In order to reduce PMD in optical fiber, various fabrication processes have been developed in both perform manufacturing and optical fiber drawing. Presently the PMD nominal values are less than $0.2\,\mathrm{ps}/\sqrt{\mathrm{km}}$, which allow 40 Gbps transmission in long-haul systems.

The future technology trends in optical communications are summarized in Figure 1.8, where the horizontal axis is the number of WDM channels per optical fiber and the vertical axis is the data rate for each channel. It is noted that the system capacity upgrade plans should face different challenges in dispersion managements depending on the choice of the data rate and number of channels. For example, if the number of channels are increased with a fixed low data rate, say OC-48 (2.5 Gbit/s), then transmission fibers should resolve the issues of compensation of chromatic dispersion (CD) and its slope within the communication band of interests. If the number of channels are fixed to a relatively small number (<10), and only the data rate is increased over OC-768 (40 Gbps), then the major problem in optical fiber is PMD and CD. If the direction of capacity expansion is to increase both data rate and channel counts, then the fiber designer should resolve the combined issues. As the optical communication system evolves to cope with ever-increasing communication capacity, so does optical fiber design dealing with fundamental issues in chromatic dispersion, polarization mode dispersion, and optical nonlinearity within optical gain bands, which are also expanding.

1.2 INTERNATIONAL STANDARDS FOR SILICA OPTICAL FIBERS

With rapid deployment of optical fiber communication system worldwide, international standards and agreements on key characteristics, measurements, and tests of optical fibers have started from the middle of 1970s right after the first installation of fiber optic link. International standards in optical fibers have played pivotal roles to draw consensus on performances of optical fibers and their future development directions [29]. Today, standards for optical fibers and cables are being studied by international organizations such as International Electrotechnical Commission (IEC), International Organization for Standardization (ISO), International Telecommunication Union—Telecommunication (ITU-T), European Telecommunications Standards Institute (ETSI), and Bell Communications Research (Bellcore) to name a few. One of the main organizations on standardization for fiber optic telecommunication technique is ITU-T, where standardization efforts are published in "Recommendations" by 13 study groups. Among the recommendations, the following series are closely related to optical fibers and cables: series G (transmission systems and media, digital systems and networks), series K (protection against interference), and series L (construction, installation, and protection of cables and other elements of outside plant).

IEC also plays equivalently important roles in fiber optic international standard as ITU-T, and in most of optical fiber standards both ITU-T and IEC share common references. IEC has been widely used in tests and measurements, while ITU-T has been widely cited for optical communication system applications. In the following, we list standards in both IEC and ITU-T that are important in optical fiber design, fabrication, tests, and installation.

IEC standards for optical fiber specifications are as follows:

- IEC 60793-1-1   Optical fibre—Part 1: Generic specification—Section 1: General
- IEC 60793-2   Optical fibre—Part 1: Generic specification—Section 2: Product specification
- IEC 60794-1-1   Optical fibre cables—Part 1: Generic specification—General
- IEC 60794-2   Optical fibre cables—Part 2: Indoor cables
- IEC 60794-3   Optical fibre cables—Part 3: Duct, buried and aerial cables

Corresponding ITU-T recommendations are as follows:

- G.651   Characteristics of a 50/125 μm multimode graded index optical fiber cable for optical access network
- G.652   Characteristics of a single-mode optical fiber cable
- G.653   Characteristics of a dispersion-shifted single-mode optical fiber cable
- G.654   Characteristics of a cut-off shifted single-mode optical fiber and cable
- G.655   Characteristics of a non-zero dispersion shifted single-mode optical fiber cable
- G.656   Characteristics of a fiber and cable with non-zero dispersion for wideband optical transport
- G.657   Characteristics of a bending loss insensitive single mode optical fibre and cable for the access network

TABLE 1.1 Optical Fibers Specified in ITU-T and Corresponding IEC Standards

ITU-T		IEC	
Recommendation	Fiber Category	Standard	Fiber Category
G.651	50/125 μm multimode gradient index optical fiber	IEC 60793-2-10	A1 multimode fiber
G.652	Single-mode optical fiber	IEC 60793-2-50	B1.1, B1.3 single-mode fiber
G.653	Dispersion-shifted single-mode optical fiber	IEC 60793-2-50	B2 single-mode fiber
G.654	Cutoff shifted single-mode optical fiber	IEC 60793-2-50	B1.2 single-mode fiber
G.655	Nonzero dispersion-shifted single-mode optical fiber	IEC 60793-2-50	B4 single-mode fiber
G.656	Nonzero dispersion-shifted single-mode optical fiber for wideband optical transport	IEC 60793-2-50	B5 single-mode fiber
G.657	Bending loss insensitive single-mode optical fiber and cable for the access network	IEC 60793-2-50	B6 single-mode fiber

Types of MMFs and SMFs specified in ITU-T and IEC are listed in Table 1.1. ITU-T also defined communication bands in optical fibers, O, E, S, C, L, and U bands, and the corresponding fiber recommendations are listed in Table 1.2.

The standards for test and measurements of optical fiber properties are listed in Table 1.3.

TABLE 1.2 Communication Bands and Corresponding Fiber Recommendations in ITU-T

ITU-T Optical Bands	Wavelength (nm)	Fiber Recommendations
O	1260–1360	G.652
E	1360–1460	G.652.C, D
S	1460–1530	G.652, G.654, G.655
C	1530–1560	G.652, G.653, G.654, G.655, G.656, G.657
L	1560–1625	G.652, G.654, G.655, G.656
U	1625–1675	Under study

O: original; E: extended; S: short; C: conventional; L: long; U: ultralong.

TABLE 1.3   Test and Measurement Standards for Optical Fibers

IEC		ITU-T	
Tests and Measurements	*Document*	*Tests and Measurements*	*Recommendation*
Fiber geometry	IEC 60793-1-20	Cladding diameter, core concentricity error, cladding noncircularity	G.650.1-5.2
Coating geometry	IEC 60793-1-21	None	
Fiber length	IEC 60793-1-22	None	
Fiber proof test	IEC 60793-1-30	Proof testing	G.650.1-5.6
Tensile strength	IEC 60793-1-31	None	
Coating strippability	IEC 60793-1-32	None	
Stress corrosion susceptibility	IEC 60793-1-33	None	
Fiber curl	IEC 60793-1-34	None	
Attenuation	IEC 60793-1-40	Attenuation	G.650.1-5.4
Chromatic dispersion	IEC 60793-1-42	Chromatic dispersion	G.650.1-5.5
Cutoff wavelength	IEC 60793-1-44	Cutoff wavelength	G.650.1-5.3
Mode field diameter	IEC 60793-1-45	Mode field diameter	G.650.1-5.1
None		Chromatic dispersion uniformity	G.6650.1-Appendix II
Macrobending loss	IEC 60793-1-47	None	
Polarization mode dispersion	IEC 60793-1-48	Polarization mode dispersion	G.650.2-5.1
Damp heat	IEC 60793-1-50	None	
Dry heat	IEC 60793-1-51	None	
Temperature cycle	IEC 60793-1-52	None	
Water immersion	IEC 60793-1-53	None	
Gamma-ray irradiation	IEC 60793-1-54	None	
Attenuation uniformity	IEC/TS 62033	None	
Microbending sensitivity	IEC 62221	None	
Nuclear radiance	IEC/TR62283	None	
Effective area (A_{eff})	IEC/TR 62284	Effective area (A_{eff})	G.650.2-Appendix III
Nonlinear coefficient (n_2/A_{eff})	IEC/TR 62284	Nonlinear coefficient (n_2/A_{eff})	G.650.2-Appendix II
Raman gain efficiency	IEC/TR 62324	None	
None		Stimulated Brillouin scattering	G.650.2-Appendix II

In this book, we will discuss the relevant international standards for various optical fibers in the following chapters.

1.3 CLASSIFICATIONS OF SILICA OPTICAL FIBERS

Silica optical fibers can be grouped in several categories depending on points of view. In this section, we will categorize optical fibers as in Table 1.4. These categories of optical fibers will be briefly reviewed with their key specifications.

1.3.1 Solid Core/Clad Fiber

1.3.1.1 Modal Delay Optimized MMFs   These fibers correspond to ITU-T G.651 and IEC 60793-2-10 standards and the refractive index profiles are optimized to reduce modal delay enabling high data rate transmission.

GbE-MMF   Gigabit Ethernet (GbE) systems are in general based on MMFs with specific characteristics. Several standardization bodies and forums (IEEE, OIF, TIA, IEC, etc.) stipulate that low-cost, short-distance network systems should combine MMFs with an 850 nm VCSEL or LD sources to connect buildings within a distance of 600 m. In these applications, two types of MMFs are being used. One is a fiber with a 62.5 μm core diameter, and the other with a 50 μm core diameter. The larger core makes it easier to couple a laser to the fiber. The number of guided modes in a fiber is

TABLE 1.4   Classification of Optical Fibers

Solid core/clad fiber		
MMF	Modal delay optimized MMF	Graded-index MMF
SMF	Dispersion tailored SMF	DSF, DCF, DFF, NZ-DSF
	Birefringence controlled SMF	PMF, SMSPF
	Nonlinear SMF	Raman fiber, Brillouin fiber
	high Kerr effect fiber	
	Photonic device SMF	Rare earth–doped SMF, photosensitive fiber, attenuation fiber
Air–silica holey fiber		
MMF	High numerical aperture (N.A.) MMF	High N.A. large core fiber for laser delivery
SMF	Dispersion tailored SMF	DSF, DCF, DFF, NZ-DSF
	Birefringence controlled SMF	PMF, SMSPF
	Nonlinear SMF	Raman fiber, Brillouin fiber, FWM, OPA, supercontinuum
	Photonic device SMF	Rare earth–doped SMF, photosensitive Fiber

SMF: single-mode fiber; MMF: multimode fiber; DSF: dispersion-shifted fiber; DCF: dispersion-compensating fiber; DFF: dispersion-flattened fiber; NZ-DSF: nonzero dispersion-shifted fiber; PMF: polarization maintaining fiber; SMSPF: single-mode single-polarization fiber; FWM: four-wave mixing; OPA: optical parametric amplification.

roughly proportional to square of the core radius and the number of modes will be about 50% more in the 62.5 μm fiber than the 50 μm fiber. The greater the mode number, the more bandwidth degradation will occur due to the modal delay. For example, differential modal delay (DMD) values at $\lambda = 850$ nm are 2 and 4 ns/km, respectively, for 50 and 62.5 μm fibers [30]. As shown in Figure 1.2 recent MMFs for high data rate communications are based on 50 μm core diameter fibers.

1.3.1.2 Dispersion Tailored SMF   This fiber includes DSF, DFF, DCF, and NZ-DSF, and their typical chromatic dispersion spectra are shown in Figure 1.6. We will discuss detailed optical properties in Chapter 4.

Dispersion-Shifted Fiber   A single-mode optical fiber whose zero-dispersion wavelength is near $\lambda = 1550$ nm is classified as DSF [23]. Chromatic dispersion in an SMF has contributions from the material and the waveguide dispersion, both in units of ps/(km nm). The former has a positive value beyond 1280 nm for pure SiO_2 glass and depends only on the material composition. The latter maintains a negative value in the communication windows, and is a function of fiber waveguide parameters in the refractive index profile. Therefore, flexible shift of the zero-dispersion wavelength is obtainable when one can find a set of the optimum values of fiber waveguide parameters making the sum of the waveguide and the material dispersion equal to zero at a desired wavelength. To have zero dispersion near 1550 nm region, it is necessary to make the relative index difference (Δ) between the core and cladding higher than that of conventional SMF. Typically Δ ranges from 0.6 to 1.0% in DSFs, which is larger than ∼0.4% of standard SMFs. Another parameter is the core diameter. The magnitude of the waveguide dispersion increases as the fiber core size decreases and the optimum core diameter in DSF is in the range of 4–6 μm, which is about one-half of ∼10 μm core of standard SMFs.

Nonzero Dispersion-Shifted Fiber   DSF is optimized for the single-channel transmission at $\lambda = 1550$ nm but its output is highly affected by optical nonlinearities due to its high index difference and small core diameter. It can take advantage of both zero dispersion and lowest attenuation at $\lambda = 1550$ nm only with moderate power levels of single-channel signal. Multiplexing dense channels, however, induce nonlinear effects such as an FWM to limit the transmission capacity [25]. NZ-DSF was invented to overcome this problem to allow nonzero finite dispersion within the spectral bands of interests [26]. Optimization of waveguide dispersion was achieved by modifying the refractive index profiles of DSF such as adding a high index pedestal in the core or high index ring in the outer cladding. Typical dispersion value of NZ-DSFs is ∼4 ps/(km nm) for 10 Gb/s system and ∼8 ps/(km nm) for 40 Gb/s system within the band of interests.

Dispersion-Compensating Fiber   Reduction of total accumulated dispersion along standard SMF can be achieved by adding a certain length of a negative dispersion fiber segment. This negative dispersion fiber is called DCF [27]. A large negative dispersion can be obtained by controlling the waveguide dispersion as in the case of

DSF. Most of DCFs have a depressed trench in their refractive index profiles so the propagation constants become much more sensitive to the wavelength than SMF [27]. DCF requires even higher Δ of $>1.5\%$ and smaller core diameter $\sim 3\,\mu m$ than DSFs. Because of these waveguide parameters, DCF has a higher attenuation than standard SMF and shorter length of DCF is desirable to minimize the loss budget allocated in the transmission system. Usually ~ 1 km of DCF can compensate the accumulated chromatic dispersion of about ~ 5 km conventional SMF.

1.3.1.3 Birefringence Controlled SMF

Polarization Maintaining Fibers The fundamental design issue of polarization maintaining fibers PMFs is creating anisotropy in the optical fiber by breaking the cylindrical symmetry of conventional SMF [31]. This anisotropy can be introduced either geometrically with elliptic waveguides or elasto-optically with asymmetric stress-applying parts (SAPs). The degeneracy in the x- and y-components of the linear polarization of the fundamental mode is removed by the anisotropy so that two orthogonal polarization modes will have different propagation constants. The magnitude of the difference in the propagation constants is represented by a physical measurand, birefringence; high-birefringence fibers are widely used to maintain the polarization states of the propagating light signal. Using these special transmission characteristics of PMFs, many applications have been demonstrated such as sensors, polarization controllers, polarizers, optical soliton, and coherent optical communications. In a particular case of PMF design, the loss of one polarization mode may become much larger than that of the other polarization to effectively provide a single-polarization single-mode (SPSM) guidance. PMFs are classified according to the origins of anisotropy, for example, PANDA, bow-tie, elliptic core, and elliptic cladding.

1.3.1.4 Nonlinear SMF
Nonlinearities in optical fibers can be generally categorized into two classes [32]:

(1) Scattering effects, the frequency-dependent optical power transfer, such as SBS and stimulated Raman scattering (SRS);
(2) Kerr effect, the intensity-dependent refractive index–related phenomena, such as self-phase modulation (SPM), cross-phase modulation (XPM), modulation instability (MI), and FWM.

In contrast to the conventional viewpoint to suppress these nonlinearities for telecom applications, the opposite viewpoint is recently growing to enhance nonlinearities and utilize them for optical sensing, optical signal processing, as well as light sources and amplifiers.

Raman Fiber Stolen et al. first observed the stimulated Raman emission in a single-mode glass optical fiber in 1972 [33] and feasibility of a broadband fiber Raman amplifier (FRA) has been demonstrated in WDM systems [15]. Raman scattering is a nonlinear optical effect related to optical phonons and its cross-section is highly dependent on the glass compositions and the waveguide design of an optical fiber. Stimulated Raman scattering process provides an efficient gain and FRAs have been demonstrated as power amplifiers, preamplifiers, and distributed amplifiers. The gain spectrum of an FRA can be flexibly controlled using the multiwavelength pumping technique. Raman gain in optical fibers also provided a new type of fiber laser where the pump photon can be downconverted to Stokes shifts.

Brillouin Fibers   The light and acoustic wave can interact within an optical fiber and, as a result of this interaction, a portion of the incident optical energy is transferred to the backward-propagating light whose frequency shifted from the original light. This effect was first observed by French physicist Brillouin around 1920 and named after him. SBS is observed only at a light intensity level above the Brillouin threshold level [17]. From the telecommunication transmission standpoint, this phenomenon is undesirable and is to be avoided. However, it was later realized that this effect can be utilized not only in fiber lasers and amplifiers but also in distributed optical sensing for both temperature and strain [34].

Air–Silica Holey Fiber   Since Knight et al. [35] first reported a unique optical fiber structure composed of periodic air hole arrays in 1996, varieties of innovative air–silica structures in optical fibers have been explored along with their novel applications that were not possible in prior fibers made of solid glass core/cladding structures. There are various names for these air–silica guiding optical fibers such as photonic crystal fibers (PCF), microstructured optical fiber (MOF), and holey fiber (HF). In this book, we will use a term air–silica holey fiber (ASHF) to take a neutral position and to be more inclusive in classification of similar types of fibers. According to Russell [36], ASHFs have provided unprecedented and unique optical properties: (1) a high refractive index difference, Δ, which enabled flexible controls in the group velocity dispersion, birefringence, nonlinearity, and bending performance; (2) hollow core guidance that effectively enhanced light–matter interaction with an optimal overlap. There are excellent review papers [35–40] and books [41,42] on ASHFs, and in this book we will confine the discussions over the four types of ASHF as shown in Figure 1.9.

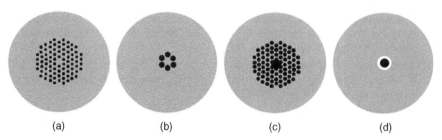

(a)   (b)   (c)   (d)

FIGURE 1.9   Four types of ASHF to be discussed in this chapter: (a) effective index guiding (EIG) fiber; (b) air hole cladding (AHC) fiber; (c) photonic band gap (PBG) fiber; (d) hollow ring core (HLC) fiber. Here black regions indicate air and gray parts are pure silica; brighter gray in (d) is GeO_2-doped silica.

REFERENCES

[1] F. P. Kapron, D. B. Keck, and R. D. Maurer, "Radiation losses in glass optical waveguides," Applied Physics Letters, vol. 17, pp. 423–425, 1970.

[2] I. Hayashi, M. B. Panish, P. W. Foy, and S. Sumski, "Junction lasers which operate continuously at room temperature," Applied Physics Letters, vol. 17, pp. 109–111, 1970.

[3] K. C. Kao and G. A. Hockham, "Dielectric-fiber surface waveguides for optical frequencies," Proceedings of the IEE, vol. 133, pp. 1151–1158, 1966.

[4] J. B. MacChesney and D. J. DiGiovanni, "Materials development of optical fiber," Journal of American Ceramic Society, vol. 73, no. 12, pp. 3537–3556, 1990.

[5] U. C. Paek, "Free drawing and polymer coating of silica glass optical fibers," Journal of Heat Transfer, vol. 121, no. 4, pp. 774–788, 1999.

[6] http://nobelprize.org/nobel_prizes/physics/laureates/2009/.

[7] D. B. Keck, R. D. Maurer, and P. C. Schultz, "On the ultimate lower limit of attenuation in glass optical waveguides," Applied Physics Letters, vol. 22, no. 7, pp. 307–309, 1973.

[8] M. Horiguchi and H. Osanai, "Spectral losses of low-OH-content optical fibres," Electronics Letters, vol. 12, no. 12, pp. 310–312, 1976.

[9] T. Miya, Y. Terunuma, T. Hosaka, and T. Moyashita, "Ultimate low-loss single-mode fibre at 1.55 μm," Electronics Letters, vol. 15, no. 9, pp. 106–108, 1979.

[10] M. Ohashi, K. Shiraki, and K. Tajima, "Optical loss property of silica based single mode fibers," Journal of Lightwave Technology, vol. 10, no. 5, pp. 539–543, 1992.

[11] http://www.fiber-optics.info/history.

[12] M.-J. Li and D. A. Nolan, "Optical transmission fiber design evolution," Journal of Lightwave Technology, vol. 26, no. 9, pp. 1079–1092, 2008.

[13] Multiwavelength Optical Technology and Networks, Journal of Lightwave Technology, vol. 14, Special Issue, 1996.

[14] E. Desurvire, "Erbium doped fiber amplifiers: principles and applications," John Wiley & Sons, Inc., 1994.

[15] J. Bromage, "Raman amplification for fiber communication systems," Journal of Lightwave Technology, vol. 22, no. 1, pp. 79–93, 2004.

[16] Y.-M. Liu and A. J. Antos, "Dispersion-shifted large-effective-area fiber for amplified high-capacity long-distance systems," Conference on Optical Fiber Communication, OFC97, Technical Digest, pp. 69–70, 1997.

[17] Y. Koyamada, S. Sato, S. Nakamura, H. Sotobayashi, and W. Chujo, "Simulating and designing Brillouin gain spectrum in single-mode fibers," Journal of Lightwave Technology, vol. 22, no. 2, pp. 631–639, 2004.

[18] A. Kobyakov, S. Kumar, D. Chowdhury, A. B.Ruffin, M. Sauer, S. Bickham, and R. Mishra, "Design concept for optical fibers with enhanced SBS threshold," Optics Express, vol. 13, no. 14, pp. 5338–5346, 2005.

[19] K. Himeno, S. Matsuo, N. Guan, and A. Wada, "Low-bending-loss single-mode fibers for fiber-to-the-home," Journal of Lightwave Technology, vol. 23, no. 1, pp. 3494–3499, 2005.

[20] G. A. Thomas, B. I. Shraiman, P. F. Glodis, and M. J. Stephen, "Toward the clarity limit in optical fibre," Nature, vol. 404, pp. 262–264, 2000.

[21] R. Olshansky and D. D. Keck, "Pulse broadening in graded-index optical fibers," Applied Optics, vol. 15, no. 2, pp. 483–491, 1976.

[22] J. Senior, "Optical fiber communications: principles & practice," 3rd edition,Prentice Hall, 2008.

[23] U. D. Paek, G. E. Peterson, and A. Carnevale, "Dispersionless single-mode lightguides with a index profiles," Bell System Technical Journal, vol. 60, pp. 583–598, 1981.

[24] D. Leers, H. Wehr, D. U. Wiechert, J. A. Van Steenwijk, D. L. A. Tjaden, and E. R. Wehrhahn, "Dispersion-flattened single-mode fibers prepared with PCVD: performance, limitations, design optimization," Journal of Lightwave Technology, vol. 4, no. 7, pp. 858–863, 1986.

[25] A. R. Chraplyvy, "Limitations on lightwave communications imposed by optical-fiber nonlinearities," Journal of Lightwave Technologies, vol. 8, no. 10, pp. 1548–1557, 1990.

[26] S. Yin, K.-W. Chung, H. Liu, P. Kurtz, and K. Reichard, "A new design for non-zero dispersion-shifted fiber (NZ-DSF) with a large effective area over 100 μm^2 and low bending and splice loss," Optics Communications, vol. 177, no. 1–6, pp. 225–232, 2000.

[27] A. J. Antos and D. K. Smith, "Design and characterization of dispersion compensating fiber based on the LP01 mode," Journal of Lightwave Technology, vol. 12, no. 10, pp. 1739–1745, 1994.

[28] IEC, "Light without limits: taming dispersion in tomorrow's high-speed networks," White paper.

[29] V. Katok and V. Krivutsa, "Fundamental trends of optical fiber cables standardization," International Conference on Transparent Optical Networks (ICTON) 2005, paper Th. B1.3, pp. 217–220, 2005.

[30] P. F. Kolesar and D. J. Mazzarese, "Understanding multimode bandwidth and differential mode delay measurements and their applications," Proceedings of the 51st International Wire and Cable Symposium of IWCS Inc., Lake Buena Vista, FL, pp. 453–460 2002.

[31] D. N. Payne, A. J. Barlow, and J. J. R. Hansen, "Development of low- and high-birefringence optical fibers," IEEE Journal of Quantum Electronics, vol. QE-18, no. 4, pp. 477–488, 1982.

[32] T. Schneider, "Nonlinear optics in telecommunications," Springer, 2004.

[33] R. H. Stolen, E. P. Ippen, and A. R. Tynes, "Raman oscillations in optical waveguides," Applied Physics Letters, vol. 20, pp. 62–64, 1972.

[34] M. Niklès, L. Thévenaz, and P. A. Robert, "Simple distributed fiber sensor based on Brillouin gain spectrum analysis," Optics Letters, vol. 21, no. 10, pp. 758–760, 1996.

[35] J. C. Knight, T. A. Birks, P. St. J. Russell, and D. M. Atkin, "All-silica single-mode optical fiber with photonic cladding," Optics Letters, vol. 21, pp. 1547–1549, 1996.

[36] P. St. J. Russell, "Photonic crystal fibers," Journal of Lightwave Technology, vol. 24, no. 12, pp. 4729–4749, 2006.

[37] J. C. Knight, "Photonic crystal fibres," Nature, vol. 424, no. 6950, pp. 847–851, 2003.

[38] P. St. J. Russell, "Photonic crystal fibers," Science, vol. 299, no. 5605, pp. 358–362, 2003.

[39] T. M. Monro and D. Richardson, "Holey optical fibres: fundamental properties and device applications," Comptes Rendus Physique, vol. 4, pp. 175–186, 2003.

[40] J. Broeng, D. Mogilevstev, S. E. Barkou, and A. Bjarklev, "Photonic crystal fibers: a new class of optical waveguides," Optical Fiber Technology, vol. 5, pp. 305–330, 1990.

[41] A. Bjaklev, J. Broeng, and A. S. Bjarklev, "Photonic crystal fibres," Springer Science, New York, 2003.

[42] F. Zolla, G. Renversez, A. Nicolet, B. Kuhlmey, S. Guenneau, and D. Felbacq, "Foundations of photonic crystal fibres," Imperial College Press, London, 2005.

CHAPTER TWO

Review on Single-Mode Fiber Design and International Standards

2.1 OPTICAL MODES IN CYLINDRICAL WAVEGUIDES

Optical fiber carries information over electromagnetic wave and it is one of dielectric waveguides. In optical fiber act, we are dealing with electromagnetic wave in the infrared range, say from 0.8 to 1.6 μm for optical communications, and we are concerned with its propagation along a cylindrical dielectric waveguide comprised of silica glass core and cladding. In order to understand the propagation properties of light over an optical fiber, it is, therefore, necessary to review the electromagnetic wave theory in a dielectric medium.

First, we will review Maxwell's equations in an infinite dielectric medium, and then try to find the guided electromagnetic waves defined by a cylindrical dielectric waveguide with a set of characteristic propagation properties, which are called modes.

Maxwell's equations 2.1–2.4 in an infinite medium are coupled partial differential equations of four physical quantities, $\vec{E}$, $\vec{B}$, $\vec{D}$, and $\vec{H}$, for given source terms ρ and $\vec{J}$. In MKS unit those fields and sources have the following units as shown in Table 2.1:

$$\vec{\nabla} \cdot \vec{D} = \rho \tag{2.1}$$

$$\vec{\nabla} \cdot \vec{B} = 0 \tag{2.2}$$

Silica Optical Fiber Technology for Devices and Components: Design, Fabrication, and International Standards, First Edition. By Kyunghwan Oh and Un-Chul Paek.
© 2012 John Wiley & Sons, Inc. Published 2012 by John Wiley & Sons, Inc.

TABLE 2.1   Units and Names for Electromagnetic Fields

$\vec{E}$: electric field (V/m)
$\vec{D}$: electric flux density (C/m^2)
$\vec{H}$: magnetic field (A/m)
$\vec{B}$: magnetic flux density (T)
ρ: electric charge density (C/m^3)
$\vec{J}$: electric current density (A/m^2)

$$\vec{\nabla} \times \vec{E} = -\frac{\partial \vec{B}}{\partial t} \tag{2.3}$$

$$\vec{\nabla} \times \vec{H} = \vec{J} + \frac{\partial \vec{D}}{\partial t} \tag{2.4}$$

In addition to these equations, the medium relates the electromagnetic fields by constitutive relations, which are characterized by its material properties, such as dielectric, magnetic, or conducting medium:

$$\vec{D} = \vec{D}(\vec{E}, \vec{B}) \tag{2.5}$$

$$\vec{H} = \vec{H}(\vec{E}, \vec{B}) \tag{2.6}$$

$$\vec{J} = \vec{J}(\vec{E}, \vec{B}) \tag{2.7}$$

In our discussion, we will make two general assumptions on the medium and, consequently, the constitutive relations. First, we assume that the medium is isotropic such that the medium does not show any directional dependence. And second, we are assuming that the medium is linear such that the constitutive relation is linear in electromagnetic field. Under these assumptions, we have the following relations:

$$\vec{D} = \varepsilon \vec{E} \tag{2.8}$$

$$\vec{B} = \mu \vec{H} \tag{2.9}$$

$$\vec{J} = \sigma \vec{E} \tag{2.10}$$

Here ε and μ are electric permittivity and magnetic permeability, respectively. σ stands for electric conductivity. Note that those coefficients in the constitutive relations are scalar under our assumption of an isotropic and linear medium. In general cases, they are expressed in tensors such that they show different values depending on the direction of electromagnetic fields.

In optical fiber, which is mainly composed of silica glass and therefore nonmagnetic, nonconducting dielectric, we can further simplify the constitutive relations by setting those coefficients as follows:

$$\mu = \mu_0 \tag{2.11}$$

$$\sigma = 0 \tag{2.12}$$

$$\varepsilon = n^2 \varepsilon_0 \tag{2.13}$$

where $\mu_0 = 4\pi \times 10^{-7}$ H/m, $\varepsilon_0 = 8.854 \times 10^{-12}$ N/m, and n is refractive index of the medium. And we get the simplified constitutive relation in an isotropic, linear dielectric medium:

$$\vec{D} = \varepsilon_0 n^2 \vec{E} \tag{2.14}$$

$$\vec{B} = \mu_0 \vec{H} \tag{2.15}$$

In optical waveguide, we assume that there are no sources within, such that both charge and current density are zero: $\rho = 0$ and $\vec{J} = 0$. By these assumptions, the Maxwell's equations lead us to

$$\vec{\nabla} \cdot \vec{D} = 0 \tag{2.16}$$

$$\vec{\nabla} \cdot \vec{B} = 0 \tag{2.17}$$

$$\vec{\nabla} \times \vec{E} = -\frac{\partial \vec{B}}{\partial t} \tag{2.18}$$

$$\vec{\nabla} \times \vec{H} = \frac{\partial \vec{D}}{\partial t} \tag{2.19}$$

The above equations are coupled time-dependent differential equation for electromagnetic fields. In order to solve the equation, we need to apply the constitutive relations 2.14 and 2.15 to reduce the equation to the wave equation.

In the following, we will derive wave equations for $\vec{E}$ and $\vec{H}$. First, let us take the curl of equation 2.18 along with equation 2.15:

$$\vec{\nabla} \times (\vec{\nabla} \times \vec{E}) = -\frac{\partial}{\partial t}(\vec{\nabla} \times (\mu_0 \vec{H})) \tag{2.20}$$

By utilizing vector calculus, $\vec{\nabla} \times (\vec{\nabla} \times \vec{V}) = -\nabla^2 \vec{V} + \vec{\nabla}(\vec{\nabla} \cdot \vec{V})$, and equation 2.19, equation 2.20 can be arranged into partial differential equation in terms of electric field, $\vec{E}$:

$$\nabla^2 \vec{E} - \varepsilon_0 \mu_0 n^2 \frac{\partial^2 \vec{E}}{\partial t^2} = \vec{\nabla}(\vec{\nabla} \cdot \vec{E}) \tag{2.21}$$

Note that the right-hand side of equation 2.21 is not zero in general and it is further arranged by equations 2.14 and 2.16.

Using equation 2.16, we find

$$\vec{\nabla} \cdot \vec{D} = \vec{\nabla} \cdot (\varepsilon_0 n^2 \vec{E}) = \varepsilon_0 n^2 \vec{\nabla} \cdot \vec{E} + \varepsilon_0 \vec{E} \cdot \vec{\nabla}(n^2) = 0 \tag{2.22}$$

Therefore, we have differential equation for the electric field

$$\nabla^2 \vec{E} - \varepsilon_0 \mu_0 n^2 \frac{\partial^2 \vec{E}}{\partial t^2} = -\vec{\nabla}\left(\frac{\vec{E} \cdot \vec{\nabla}(n^2)}{n^2}\right) \tag{2.23}$$

In the case of magnetic field, $\vec{H}$, we take curl of equation 2.19:

$$\vec{\nabla} \times (\vec{\nabla} \times \vec{H}) = \frac{\partial}{\partial t}(\vec{\nabla} \times (\varepsilon_0 n^2 \vec{E})) \tag{2.24}$$

By the vector calculus and equation 2.15, we get

$$\nabla^2 \vec{H} - \varepsilon_0 \mu_0 n^2 \frac{\partial^2 \vec{H}}{\partial t^2} = -\frac{\vec{\nabla}(n^2)}{n^2} \times (\vec{\nabla} \times \vec{H}) \tag{2.25}$$

Equations 2.23 and 2.25 are general wave equations and they are not homogeneous such that there are nonzero terms on the right-hand sides. The refractive index, n, is in general a function of spatial position such that $\vec{\nabla}n(x, y, z) \neq 0$. For example, graded-index fiber will have a radial distribution of refractive index and exact solution of wave should include the nonzero terms in equations 2.23 and 2.25.

Now we make the final assumption in the analysis that the medium is homogeneous or, in other words, the refractive index is not a function of spatial position and it is uniform. Then we have homogeneous wave equations, which are as follows:

$$\boxed{\nabla^2 \vec{E} - \varepsilon_0 \mu_0 n^2 \frac{\partial^2 \vec{E}}{\partial t^2} = 0} \tag{2.26}$$

$$\boxed{\nabla^2 \vec{H} - \varepsilon_0 \mu_0 n^2 \frac{\partial^2 \vec{H}}{\partial t^2} = 0} \tag{2.27}$$

Note that equations 2.26 and 2.27 are valid for isotropic, linear, and homogeneous dielectric and we can apply these to both the core and cladding of step-index optical fibers and our analysis on optical fibers will be based on these equations.

The assumptions for the homogeneous wave equations 2.26 and 2.27 are summarized in the following.

The dielectric medium is assumed to be

- Free of external sources, such that $\rho = 0$ and $\vec{J} = 0$
- Isotropic, such that there is no directional dependence
- Linear, such that constitutive relations are linear
- Homogeneous, such that physical properties do not depend on spatial position
- Nonmagnetic, such that $\mu = \mu_0$
- Nonconducting, such that $\sigma = 0$

Now from the wave equations we will consider electromagnetic solution in a cylindrical waveguide as follows.

The fiber has the core with a radius of a and refractive index n_1. The cladding has the radius of b and refractive index of n_2. The fiber is composed of dielectric and provides cylindrical symmetry, as shown in Figure 2.1.

Due to its symmetry, the wave equation should be solved in the cylindrical coordinate (r, θ, z). Equations 2.26 and 2.27 are vector equations and we need to solve six scalar components, E_r, E_θ, E_z, H_r, H_θ, and H_z. We assume that the electromagnetic wave of frequency ω is propagating along $+z$ direction with a propagation constant β and harmonic time dependence such that the fields are expressed as follows:

$$\vec{E}(\vec{r}, t) = \vec{E}^0 \, e^{j(\omega t - \beta z)} = (E_r^0 \hat{r} + E_\theta^0 \hat{\theta} + E_z^0 \hat{z}) e^{j(\omega t - \beta z)} \qquad (2.28)$$

$$\vec{H}(\vec{r}, t) = \vec{H}^0 \, e^{j(\omega t - \beta z)} = (H_r^0 \hat{r} + H_\theta^0 \hat{\theta} + H_z^0 \hat{z}) e^{j(\omega t - \beta z)} \qquad (2.29)$$

Here $\vec{E}^0$ and $\vec{H}^0$ are the complex amplitudes for the electric field and the magnetic field, respectively.

Before we solve the wave equation, we need to take a note on the relationship between longitudinal components and transverse components of electromagnetic fields.

Under the assumption of solution given in equations 2.28 and 2.29, we go back to Maxwell's equations 2.16–2.19, and we take into account the harmonic time dependence for time derivatives to get

$$\vec{\nabla} \times \vec{E} = -j\omega\mu_0 \vec{H} \qquad (2.30)$$

$$\vec{\nabla} \times \vec{H} = j\omega\mu_0\varepsilon_0 n^2 \vec{E} \qquad (2.31)$$

These equations relate the vector components of electromagnetic field as follows

$$\frac{1}{r}\frac{\partial E_z^0}{\partial \theta} + j\beta E_\theta^0 = -j\omega\mu_0 H_r^0 \qquad (2.32)$$

$$\frac{\partial E_z^0}{\partial r} + j\beta E_r = j\omega\mu H_\theta^0 \tag{2.33}$$

$$\frac{E_\theta^0}{r} + \frac{\partial E_\theta^0}{\partial r} - \frac{1}{r}\frac{\partial E_r^0}{\partial \theta} = -j\omega\mu_0 H_z^0 \tag{2.34}$$

$$\frac{1}{r}\frac{\partial H_z^0}{\partial \theta} + j\beta H_\theta^0 = j\omega\varepsilon_0 n^2 E_r^0 \tag{2.35}$$

$$\frac{\partial H_z^0}{\partial r} + j\beta H_r^0 = -j\omega\varepsilon_0 n^2 E_\theta^0 \tag{2.36}$$

$$\frac{H_\theta^0}{r} + \frac{\partial H_\theta^0}{\partial r} - \frac{1}{r}\frac{\partial H_r^0}{\partial \theta} = j\omega\varepsilon_0 n^2 E_z^0 \tag{2.37}$$

From equations 2.32–2.37, we find that all the transverse components, E_r^0, E_θ^0, H_r^0, and H_θ^0, are expressed in terms of the longitudinal components E_z^0 and H_z^0:

$$E_r^0 = -\frac{j}{\beta_t^2}\left(\beta\frac{\partial E_z^0}{\partial r} + \omega\mu_0\frac{1}{r}\frac{\partial H_z^0}{\partial \theta}\right) \tag{2.38}$$

$$E_\theta^0 = -\frac{j}{\beta_t^2}\left(\beta\frac{1}{r}\frac{\partial E_z^0}{\partial \theta} - \omega\mu_0\frac{\partial H_z^0}{\partial r}\right) \tag{2.39}$$

$$H_r^0 = -\frac{j}{\beta_t^2}\left(\beta\frac{\partial H_z^0}{\partial r} - \omega\varepsilon_0 n^2\frac{1}{r}\frac{\partial E_z^0}{\partial \theta}\right) \tag{2.40}$$

$$H_\theta^0 = -\frac{j}{\beta_t^2}\left(\beta\frac{1}{r}\frac{\partial H_z^0}{\partial \theta} + \omega\varepsilon_0 n^2\frac{\partial E_z^0}{\partial r}\right) \tag{2.41}$$

Here $\beta_t^2 = n^2 k^2 - \beta^2$ and $k^2 = \omega^2/c^2 = \omega^2\varepsilon_0\mu_0 = (2\pi/\lambda)^2$. Note that λ is the wavelength of the electromagnetic field in the vacuum. β_t is called the transverse phase constant or transverse wave number. Using equations 2.40, 2.41, and 2.37, we can derive a differential equation for E_z^0; similarly another differential equation for H_z^0 is obtained from equations 2.38, 2.39, and 2.34. The differential equations for the electric and magnetic field show an identical form, Bessel equation, given as follows:

$$\frac{\partial^2 E_z^0}{\partial r^2} + \frac{1}{r}\frac{\partial E_z^0}{\partial r} + \frac{1}{r^2}\frac{\partial^2 E_z^0}{\partial \theta^2} + \beta_t^2 E_z^0 = 0 \tag{2.42}$$

$$\frac{\partial^2 H_z^0}{\partial r^2} + \frac{1}{r}\frac{\partial H_z^0}{\partial r} + \frac{1}{r^2}\frac{\partial^2 H_z^0}{\partial \theta^2} + \beta_t^2 H_z^0 = 0 \tag{2.43}$$

Therefore, the solution of vectorial wave equations 2.26 and 2.27 is reduced to solution of Bessel equations 2.42 and 2.43 for longitudinal components assuming harmonic time dependence and propagation along z direction of the waves.

Now we will focus on the solution of Bessel function using separation of variables:

$$\begin{pmatrix} E_z^0 \\ H_z^0 \end{pmatrix} = R(r)\Theta(\theta) \tag{2.44}$$

Then the differential equations for those variables are

$$\frac{d^2\Theta}{d\theta^2} = -m^2\Theta \tag{2.45}$$

$$\frac{d^2R}{dr^2} + \frac{1}{r}\frac{dR}{dr} + \left(\beta_t^2 - \frac{m^2}{r^2}\right)R = 0 \tag{2.46}$$

where m^2 is the constant of variable separation.

The solutions of equation 2.45 are trigonometric functions and those of equation 2.46 are Bessel functions. For the given cylindrical structure (Figure 2.1), it can be shown that the longitudinal fields are obtained as follows [1,2].

In the core, we have

$$\boxed{E_{1z}^0 = AJ_m(ur)\cos(m\theta + \theta_0)} \tag{2.47}$$

$$\boxed{H_{1z}^0 = BJ_m(ur)\sin(m\theta + \theta_0)} \tag{2.48}$$

where

$$u^2 = n_1^2 k^2 - \beta^2 \tag{2.49}$$

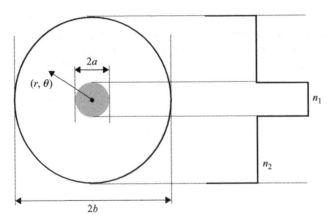

FIGURE 2.1 Waveguide structure of cylindrically symmetric optical fiber.

Note that the propagation parameter u needs to be real in order to make $J_m(ur)$ a stationary solution. Therefore, the propagation constant β has an upper boundary:

$$\beta^2 < n_1^2 k^2 \tag{2.50}$$

In the cladding, we have

$$\boxed{\begin{aligned} E_{2z}^0 &= CK_m(wr)\cos(m\theta + \theta_0) \\ H_{2z}^0 &= DK_m(wr)\sin(m\theta + \theta_0) \end{aligned}}$$

$$(2.51)$$
$$(2.52)$$

where

$$w^2 = \beta^2 - n_2^2 k^2 \tag{2.53}$$

Note that the propagation parameter w needs to be real in order to make $K_m(wr)$ an evanescent solution. Therefore, the propagation constant β has a lower boundary:

$$\beta^2 > n_2^2 k^2 \tag{2.54}$$

The functional behaviors of $J_m(x)$ and $K_m(x)$ are shown for the first few m in Figure 2.2.

Now at this stage, we need to find out four constants A, B, C, and D for electromagnetic fields and the propagation constant β to define modes.

In order to find out the coefficients, boundary conditions should be applied. In the absence of surface charge and current density at the core–cladding interface, the fields at the core–cladding dielectric interface should satisfy the continuity boundary conditions. Note that the direction normal to the interface shown as in Figure 2.1 is parallel to the unit radial vector, $\hat{r}$, and tangential directions are parallel to either $\hat{z}$ or $\hat{\theta}$

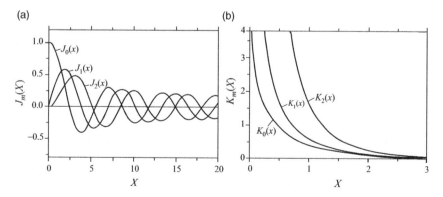

FIGURE 2.2 The plots of (a) $J_m(x)$ and (b) $K_m(x)$.

We have four boundary conditions relative to the normal vector, $\hat{r}$, and tangential vectors, $\hat{z}$ and $\hat{\theta}$, as follows:

(1) Continuity of normal component of the electric flux density, $\vec{D}$

$$\hat{r} \cdot (\vec{D}_2 - \vec{D}_1) = 0 \tag{2.55}$$

(2) Continuity of tangential component of the electric field, $\vec{E}$

$$\hat{z} \cdot (\vec{E}_2 - \vec{E}_1) = 0, \qquad \hat{\theta} \cdot (\vec{E}_2 - \vec{E}_1) = 0 \tag{2.56}$$

(3) Continuity of normal component of the magnetic flux density, $\vec{B}$

$$\hat{r} \cdot (\vec{B}_2 - \vec{B}_1) = 0 \tag{2.57}$$

(4) Continuity of tangential component of the magnetic field, $\vec{H}$

$$\hat{z} \cdot (\vec{H}_2 - \vec{H}_1) = 0, \qquad \hat{\theta} \cdot (\vec{H}_2 - \vec{H}_1) = 0 \tag{2.58}$$

The above equations can be expressed in terms of longitudinal and transverse components of the electromagnetic fields:

$$\varepsilon_0 n_1^2 E_{1r}^0 = \varepsilon_0 n_2^2 E_{2r}^0 \tag{2.59}$$

$$E_{1z}^0 = E_{2z}^0 \tag{2.60}$$

$$E_{1\theta}^0 = E_{2\theta}^0 \tag{2.61}$$

$$\mu_0 H_{1r}^0 = \mu_0 H_{2r}^0 \tag{2.62}$$

$$H_{1z}^0 = H_{2z}^0 \tag{2.63}$$

$$H_{1\theta}^0 = H_{2\theta}^0 \tag{2.64}$$

Here the subscripts 1 and 2 denote the core and the cladding, respectively. Note that all the components are evaluated at the core–cladding interface, $r = a$. It is known that all of the six boundary conditions are not independent and only four equations are independent [1]. We are going to use equations 2.60, 2.61, 2.63, and 2.64 in the following analysis.

It is right step to review what we need to pursue to find electromagnetic fields and the propagation vector β. We have four unknowns A, B, C, and D for longitudinal components of electric and magnetic fields, along with four boundary conditions so that we should be able to relate them into a linear algebraic equation. It is noteworthy to find that boundary conditions in equations 2.60 and 2.63 directly relate the longitudinal components, but equations 2.61 and 2.64 relate transverse θ components. Remember that we can express all the transverse components of electromagnetic fields in terms of the longitudinal components and we can apply the formulae in equations 2.39 and 2.41 to derive the following equations.

The azimuthal component of electric field in the core:

$$E^0_{1\theta} = \frac{\text{j}}{u^2}\left[Am\frac{\beta}{r}J_m(ur) + B\omega\mu_0\frac{\partial J_m(ur)}{\partial r}\right]\sin(m\theta + \theta_0) \qquad (2.65)$$

In the cladding

$$E^0_{2\theta} = \frac{-\text{j}}{w^2}\left[Cm\frac{\beta}{r}K_m(wr) + D\omega\mu_0\frac{\partial K_m(ur)}{\partial r}\right]\sin(m\theta + \theta_0) \qquad (2.66)$$

And similarly we get magnetic fields.
In the core

$$H^0_{1\theta} = \frac{-\text{j}}{u^2}\left[A\omega\varepsilon_0 n_1^2\frac{\partial J_m(ur)}{\partial r} + Bm\frac{\beta}{r}J_m(ur)\right]\cos(m\theta + \theta_0) \qquad (2.67)$$

In the cladding

$$H^0_{2\theta} = \frac{\text{j}}{w^2}\left[C\omega\varepsilon_0 n_2^2\frac{\partial K_m(wr)}{\partial r} + Dm\frac{\beta}{r}K_m(wr)\right]\cos(m\theta + \theta_0) \qquad (2.68)$$

Note that β_t^2 in equations 2.38–2.41 will change to u^2 and $-w^2$ in the core and cladding, respectively.

Now we have four linear equations for A, B, C, and D by the boundary conditions 2.60, 2.61, 2.63, and 2.64. By plugging equations 2.47, 2.48, 2.51, and 2.52 into equations 2.65–2.68, we have the following.
From equation 2.60, $E^0_{1z}(r = a) = E^0_{2z}(r = a)$, we get

$$AJ_m(ua) = CK_m(wa) \qquad (2.69)$$

From equation 2.61, $E^0_{1\theta}(r = a) = E^0_{2\theta}(r = a)$, we get

$$A\left(\frac{m\beta}{u^2a^2}J_m(ua)\right) + B\left(\frac{\omega\mu_0}{ua}J'_m(ua)\right) = -C\left(\frac{m\beta}{w^2a^2}K_m(wa)\right) - D\left(\frac{\omega\mu_0}{wa}K'_m(wa)\right)$$

(2.70)

From equation 2.63, $H^0_{1z}(r = a) = H^0_{2z}(r = a)$, we get

$$BJ_m(ua) = DK_m(wa)$$

(2.71)

From equation 2.64, $H^0_{1\theta}(r = a) = H^0_{2\theta}(r = a)$, we get

$$A\left(\frac{\omega\varepsilon_0 n_1^2}{ua}J'_m(ua)\right) + B\left(\frac{m\beta}{u^2a^2}J_m(ua)\right) = -C\left(\frac{\omega\varepsilon_0 n_2^2}{wa}K_m(wa)\right) - D\left(\frac{m\beta}{w^2a^2}K'_m(wa)\right)$$

(2.72)

Here

$$J'_m(x) = \frac{dJ_m(x)}{dx}, \quad K'_m(x) = \frac{dK_m(x)}{dx}$$

The above four equations can be arranged into a homogeneous linear matrix equation:

$$\overleftrightarrow{\Lambda} \cdot \vec{X} = 0$$

(2.73)

where

$$\overleftrightarrow{\Lambda} = \begin{bmatrix} J_m(ua) & 0 & -K_m(wa) & 0 \\ 0 & J_m(ua) & 0 & -K_m(wa) \\ \dfrac{m\beta}{u^2a^2}J_m(ua) & \dfrac{\omega\mu_0}{ua}J'_m(ua) & \dfrac{m\beta}{w^2a^2}K_m(wa) & \dfrac{\omega\mu_0}{wa}K'_m(wa) \\ \dfrac{\omega\varepsilon_0 n_1^2}{ua}J'_m(ua) & \dfrac{m\beta}{u^2a^2}J_m(ua) & \dfrac{\omega\varepsilon_0 n_2^2}{wa}K_m(wa) & \dfrac{m\beta}{w^2a^2}K'_m(wa) \end{bmatrix}$$

and

$$\vec{X} = \begin{bmatrix} A \\ B \\ C \\ D \end{bmatrix}$$

In order to have a nontrivial solution, the determinant of the 4×4 matrix should be equal to zero such that $\det(\overleftrightarrow{\Lambda}) = 0$. This condition leads us to the characteristic equation

$$\left(\frac{J'_m(ua)}{uaJ_m(ua)} + \frac{K'_m(wa)}{waK_m(wa)} \right) \left(\frac{n_1^2}{n_2^2} \frac{J'_m(ua)}{uaJ_m(ua)} + \frac{K'_m(wa)}{waK_m(wa)} \right)$$

$$= m^2 \left(\frac{1}{u^2 a^2} + \frac{1}{w^2 a^2} \right) \left(\frac{n_1^2}{n_2^2} \frac{1}{u^2 a^2} + \frac{1}{w^2 a^2} \right) \qquad (2.74)$$

The above equation can be further reduced to a simple form using recurrence relations in the Bessel functions [3]:

$$J_{m-1}(x) = \frac{m}{x} J_m(x) + J'_m(x) \qquad (2.75)$$

$$K_{m-1}(x) = -\frac{m}{x} K_m(x) - K'_m(x) \qquad (2.76)$$

The reduced characteristic equations are given as follows:

$$\boxed{\begin{aligned} \frac{K_{m+1}(W)}{WK_m(W)} &\left(\frac{(n_r^2+1)J_{m-1}(U)}{UJ_m(U)} - \frac{2K_{m-1}(W)}{WK_m(W)} \right) = \frac{-J_{m+1}(U)}{UJ_m(U)} \\ &\times \left(\frac{2n_r^2 J_{m-1}(U)}{UJ_m(U)} - \frac{(n_r^2+1)K_{m-1}(W)}{WK_m(W)} \right) \end{aligned}} \qquad (2.77)$$

where $n_r = n_1/n_2$ and we introduced dimensionless propagation parameters, U and W, which are derived from the definitions of u and w in 2.49 and 2.53, respectively.

$$\begin{aligned} U &= au = a\sqrt{n_1^2 k^2 - \beta^2} \\ W &= aw = a\sqrt{\beta^2 - n_2^2 k^2} \end{aligned} \qquad (2.78)$$

By the definitions of u and w given in equations 2.49 and 2.53, we note that

$$V^2 \equiv k^2 a n_1^2 2\Delta = U^2 + W^2 \qquad (2.79)$$

$$\Delta \equiv \frac{n_1^2 - n_2^2}{2n_1^2} \approx \frac{n_1 - n_2}{n_1} \qquad (2.80)$$

As defined in equation 2.78, the normalized frequency, V, contains waveguide parameters, a, n_0, and Δ and the frequency of light, $\omega = ck_0$. Here Δ is the relative refractive index between the core and the cladding. In most of optical fibers the core refractive index is raised by doping GeO_2, P_2O_5, or Al_2O_3 in silica glass and the relative index difference, Δ, is usually less than 10%. Therefore, the term $n_r = n_1/n_2$ is very close to unity in conventional optical fibers. This specific case is called "weakly guiding approximation" [4,5]. Conventional optical fibers described in this book can be well described in the weakly guiding approximation, except the fibers in silica–air guiding structures discussed in Chapter 8.

Now in the weakly guiding approximations, $n_r = n_1/n_2 \approx 1$, the characteristic equation 2.74 or 2.77 can be further reduced and classified into three cases.

For $m \neq 0$,

$$\frac{J_{m-1}(U)}{UJ_m(U)} = \frac{K_{m-1}(W)}{WK_m(W)} \quad \text{(HE mode)} \tag{2.81}$$

$$\frac{J_{m+1}(U)}{UJ_m(U)} = -\frac{K_{m+1}(W)}{WK_m(W)} \quad \text{(EH mode)} \tag{2.82}$$

Above two cases, equations 2.81 and 2.82, correspond to the hybrid HE mode and EH mode, respectively.

For $m = 0$, these two equations are identical since $J_{-1}(x) = -J_1(x)$ and $K_{-1}(x) = K_1(x)$ to result in characteristics equation for the degenerate TE and TM modes:

$$\frac{J_1(U)}{UJ_0(U)} = -\frac{K_1(W)}{WK_0(W)} \quad \text{(TE/TM mode)} \tag{2.83}$$

Using the recurrence relations among the Bessel functions,

$$\frac{m}{U}J_m(U) = \frac{1}{2}\{J_{m-1}(U) + J_{m+1}(U)\} \tag{2.84}$$

$$\frac{m}{W}K_m(W) = -\frac{1}{2}\{K_{m-1}(W) - K_{m+1}(W)\} \tag{2.85}$$

equation 2.81 becomes

$$\frac{J_{m-1}(U)}{UJ_{m-2}(U)} = -\frac{K_{m-1}(W)}{WK_{m-2}(W)} \tag{2.86}$$

We can further combine three characteristic equations 2.82, 2.83, and 2.86 into one equation by introducing an azimuthal index l. The unified characteristic equation is given in terms of l as follows:

$$\frac{UJ_{l-1}(U)}{J_l(U)} = -\frac{WK_{l-1}(W)}{K_l(W)}$$

$$l = \begin{cases} 1, & \text{TE, TM} \\ m+1, & \text{EH} \\ m-1, & \text{HE} \end{cases} \tag{2.87}$$

Note that above equation is valid for negative index for J and K functions using the mathematical relations $J_{-i}(x) = (-1)^i J_i(x)$ and $K_{-i}(x) = K_i(x)$.

Solving simultaneously the unified characteristic equation 2.87 and the normalized frequency definition in equation 2.79 in (U, W) plane for a given V, we can obtain the propagation constant, β, given by

$$\beta^2 = n_1^2 k^2 - u^2 = n_1^2 k^2 - \left(\frac{U}{a}\right)^2 \quad \text{(from equation 2.49)}$$

$$\beta^2 = n_2^2 k^2 + w^2 = n_2^2 k^2 + \left(\frac{W}{a}\right)^2 \quad \text{(from equation 2.53)}$$

In finding β for a given azimuthal index l, we notice that there are multiple solutions and we need to introduce another index, the radial index p, to designate the individual modes. The solutions in the weakly guiding approximation are called LP modes, where LP stands for "linear polarized." LP modes are represented by two indices, the azimuthal index, l, and the radial index, p, such as LP_{lp}, where $l = 0, 1, 2, \ldots$, and $p = 1, 2, 3, \ldots$, respectively.

The radial index p represents the radial variation of the electromagnetic field and is closely related to the cutoff conditions of the modes. If β varies as a function of V in the range $n_2^2 k_0^2 < \beta^2 < n_1^2 k_0^2$ as in equations 2.50 and 2.54, the arguments in the Bessel functions are real and the electromagnetic fields are confined to the core and decay exponentially in the cladding. When V varies to a certain value so that the propagation constant becomes equal to that of a plane wave propagating in the bulk cladding material:

$$\beta = n_2 k \quad \text{(cutoff condition)} \tag{2.88}$$

the electromagnetic wave is no longer guided by the core. Equation 2.88 is the cutoff condition, and the normalized frequency at this condition is called the cutoff frequency, V_c. In the cutoff condition, we notice that

$$W = 0 \tag{2.89}$$

$$U = V = V_c \tag{2.90}$$

Using the asymptotic properties of $K_l(W)$ at $W \to 0$ and equation 2.88, the unified characteristic equation at the cutoff condition leads to the condition for cutoff frequency, V_c.

$$J_{l-1}(V_c) = 0 \tag{2.91}$$

In other words, the cutoff frequencies are the roots of $J_l(x)$. Note that there are multiple roots of $J_l(x)$ for a given azimuthal index, l, due to oscillatory behavior of the Bessel functions. Those roots are denoted by the radial index, p, such that the cutoff frequency of the LP_{lp} mode is given by

$$V_c = x_{l-1, p} \tag{2.92}$$

where $x_{l-1,p}$ is the pth root of $J_{l-1}(x)$ (see Figure 2.2 to locate the roots). Note that the azimuthal indices are integers that start from $l = 0$, but the radial index starts from $p = 1$.

Utilizing the unified characteristic equation 2.87, and newly introduced indices l and, p, we can express the traditional modes in a consistent way in terms of LP_{lp} modes as follows:

$$\left. \begin{array}{ll} HE_{l+1,p}, & l \geq 0 \\ EH_{l-1,p}, & l > 1 \\ TE_{0,p}, TE_{0,p}, & l = 1 \end{array} \right\} = LP_{l,p} \qquad (2.93)$$

The LP_{lp} mode designations are summarized in Table 2.2. Note that the fundamental mode LP_{01} has the cutoff condition, $V_c = 0$, which is another solution to the unified characteristics in the limit of $W \rightarrow 0$ so as to include $J_1(0) = 0$ as the first root [2,4]. Its physical meaning is that the mode propagates and guided irrespective of the light frequency.

It is also noteworthy that few of higher order modes share the common cutoff frequencies. LP_{21} and LP_{02} have the same cutoff frequency $V_c = 3.8317$, and LP_{22} and LP_{03} share $V_c = 7.0156$.

When the normalized frequency takes the value $V < 2.405$, we will have only the fundamental mode, LP_{01}, which is called the single-mode condition. For other case when $V \geq 2.405$, we will have more than one modes guided along the optical fiber. Therefore, the propagation constants, β, of the LP modes will have different value as a function of the normalized frequency, V, which is the origin of modal dispersion. Mode designations as a function of V and V_c are schematically shown in Figure 2.3.

Before we discuss the behavior of β for different modes, it is noteworthy to confirm our derivations under the weakly guiding approximation result in the "linearly polarized" mode. We need to go back to the linear algebraic equation 2.73 for coefficients A, B, C, and D. The propagation constant β is found from the characteristic equation 2.87, as a function of V, satisfying $\det(\overset{\leftrightarrow}{\Lambda}) = 0$. For this β, three of the coefficients can be expressed in terms of the rest, for example, B, C, and D can be

TABLE 2.2   LP Mode Designations in Weakly Guiding Approximation

Modes	Indices	Cutoff Condition and Cutoff Frequency, V_c	Traditional Modes and Degeneracy
LP_{01}	$l = 0, p = 1$	First root of $J_1(V_c) = 0$, $V_c = 0.0000$	$HE_{11} \times 2$
LP_{11}	$l = 1, p = 1$	First root of $J_0(V_c) = 0$, $V_c = 2.4048$	$TE_{01}, TM_{01}, HE_{21} \times 2$
LP_{21}	$l = 2, p = 1$	Second root of $J_1(V_c) = 0$, $V_c = 3.8317$	$EH_{11} \times 2, HE_{31} \times 2$
LP_{02}	$l = 0, p = 2$	Second root of $J_1(V_c) = 0$, $V_c = 3.8317$	$HE_{12} \times 2$
LP_{31}	$l = 3, p = 1$	First root of $J_2(V_c) = 0$, $V_c = 5.1356$	$EH_{21} \times 2, HE_{41} \times 2$
LP_{12}	$l = 1, p = 2$	Second root of $J_0(V_c) = 0$, $V_c = 5.5201$	$TE_{02}, TM_{02}, HE_{22} \times 2$
LP_{41}	$l = 4, p = 1$	First root of $J_3(V_c) = 0$, $V_c = 6.3802$	$EH_{31} \times 2, HE_{51} \times 2$
LP_{22}	$l = 2, p = 2$	Third root of $J_1(V_c) = 0$, $V_c = 7.0156$	$EH_{12} \times 2, HE_{32} \times 2$
LP_{03}	$l = 0, p = 3$	Third root of $J_1(V_c) = 0$, $V_c = 7.0156$	$HE_{13} \times 2$

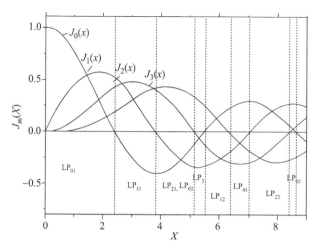

FIGURE 2.3 Bessel functions $J_l(x)$ and LP mode designations.

expressed in A. The field components of the x-polarized LP_{01} mode are summarized [6] in Table 2.3. Note that the longitudinal components are in the order of $\sqrt{\Delta}$ smaller relative to the transverse components E_x^0 and H_y^0. The longitudinal fields, therefore, can be neglected in the weakly guiding approximation, where Δ is small such that $n_1 \approx n_2$. Then the LP_{01} mode is practically "linearly polarized" in the direction of either x or y. Note that these two polarization modes are degenerate such that they have the identical propagation constant.

Furthermore, it is subsequently found that three vectors, the electric field, the propagation direction $\hat{z}$, and the magnetic field, form right angles in space and are related to one another by

$$\vec{H}_0 = \frac{1}{\eta}\hat{z} \times \vec{E}_0, \quad \text{where } \eta = \sqrt{\frac{\mu_0}{\varepsilon_i}} (i = 1, 2) \qquad (2.94)$$

These properties are identical to those of TEM plane wave solution to the wave equations in free space. Similar to the fundamental LP_{01} mode, we will have linearly

TABLE 2.3   Electromagnetic Fields of the Fundamental LP_{01} Mode

Core $(r \leq a)$	Cladding $(r \geq a)$
$E_x^0 = E_0 \dfrac{J_0(ur)}{J_0(ua)}$	$E_0 \dfrac{K_0(wr)}{K_0(wa)}$
$H_y^0 = E_0 \dfrac{n_1}{\eta_0}\dfrac{J_0(ur)}{J_0(ua)}$	$E_0 \dfrac{n_2}{\eta_0}\dfrac{K_0(wr)}{K_0(wa)}$
$E_z^0 = -jE_0 \dfrac{\sqrt{2\Delta}}{V} ua \dfrac{J_0(ur)}{J_0(ua)}\cos\theta$	$-jE_0 \dfrac{\sqrt{2\Delta}}{V} wa \dfrac{K_0(wr)}{K_0(wa)}\cos\theta$
$H_z^0 = -jE_0 \dfrac{n_1}{\eta_0}\dfrac{\sqrt{2\Delta}}{V} ua \dfrac{J_0(ur)}{J_0(ua)}\sin\theta$	$-jE_0 \dfrac{n_2}{\eta_0}\dfrac{\sqrt{2\Delta}}{V} wa \dfrac{K_0(wr)}{K_0(wa)}\sin\theta$

polarized transverse electromagnetic fields in LP_{lp} modes in the weakly guiding approximation, and the longitudinal fields are negligible, similar to TEM wave.

In Table 2.3, it is noted that all the field components are scaled with the coefficient E_0 since the nature of homogeneous algebraic equation 2.73 leaves one coefficient undetermined. In order to complete the description of the LP_{01} mode, we need to determine the coefficient E_0 using the Poynting theorem, which relates the fields to the power flow.

Poynting vector is defined as follows

$$\vec{S} = \vec{E} \times \vec{H} \tag{2.95}$$

The time average of Poynting vector is related to the direction and magnitude of energy flow density, or, equivalently, intensity I:

$$\langle \vec{S} \rangle = \frac{1}{2}\mathrm{Re}(\vec{E} \times \vec{H}*) \quad [\mathrm{W/m^2}] \tag{2.96}$$

and it is reduced to an expression in terms of the coefficient E_0 in the LP_{01} mode:

$$\langle |\vec{S}| \rangle = \langle S\hat{z} \rangle = I = \frac{1}{2\eta}|E_x^0|^2 = \begin{cases} \dfrac{n_1}{2\eta_0}\dfrac{J_0^2(ur)}{J_0^2(ua)}|E_0|^2 & (r \le a) \\[4mm] \dfrac{n_2}{2\eta_0}\dfrac{K_0^2(ur)}{K_0^2(ua)}|E_0|^2 & (r \ge a) \end{cases} \tag{2.97}$$

The power is described by the integral of the Poynting vector over the cross-section of the core and the cladding. In the LP_{01} mode, the results are given as follows [4,6]:

$$P_{\mathrm{core}} = |E_0|^2 \pi a^2 \frac{n_1}{\eta_0}\left(1 + \frac{W^2}{U^2}\frac{K_1^2(W)}{K_0^2(W)}\right) \tag{2.98}$$

$$P_{\mathrm{clad}} = |E_0|^2 \pi a^2 \frac{n_2}{\eta_0}\left(\frac{K_1^2(W)}{K_0^2(W)} - 1\right) \tag{2.99}$$

Since we have assumed that $n_1 \approx n_2$, we can calculate the total power in the LP_{01} mode:

$$P_{\mathrm{total}} = |E_0|^2 \pi a^2 \frac{n_1}{\eta_0}\frac{K_1^2(W)}{K_0^2(W)}\frac{V^2}{U^2} \tag{2.100}$$

From equation 2.96, we finally get the coefficient E_0 in the LP_{01} mode in terms of the total power contained in the mode.

In Figure 2.4, the intensity distributions of several LP modes are shown. It is noted that the fundamental mode LP_{01} mode is axially symmetric, which is same as higher order LP_{0m} modes.

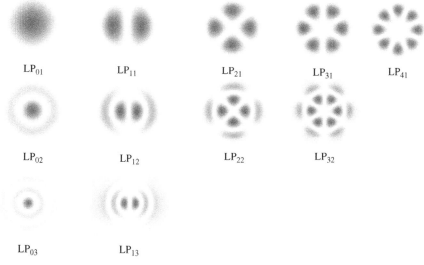

LP$_{01}$ LP$_{11}$ LP$_{21}$ LP$_{31}$ LP$_{41}$

LP$_{02}$ LP$_{12}$ LP$_{22}$ LP$_{32}$

LP$_{03}$ LP$_{13}$

FIGURE 2.4 Mode intensity distribution of selected LP modes in cylindrical optical fiber.

Thus far, we have calculated the electromagnetic fields and the propagation constant, β, for a given normalized frequency, V. From now on we will discuss the functional behavior of β as a function of V in order to discuss dispersion in the transmission of signals carried by the individual modes along the optical fiber. The term "dispersion" has been conventionally used to describe a variation of a physical quantity as a function of frequency. It is appropriate, at this point, to introduce the "physical quantities" that are to be analyzed as a function of frequency or wavelength.

In the following section, we will discuss the dispersion in the propagation constant, group delay, and chromatic dispersion.

Once the propagation constant, β, of one of LP modes is known as a function of V, $\beta(V)$, we can define the effective index of the mode as

$$n_{\text{eff}} = \frac{\beta}{k} = \frac{c\beta}{\omega} = \frac{\lambda\beta}{2\pi} \qquad (2.101)$$

where c is the speed of light in the vacuum.

Note that the effective index of a guided mode is in the range of $n_2 < n_{\text{eff}} < n_1$ and is related to the phase velocity:

$$v_p = \frac{c}{n_{\text{eff}}} = \frac{ck}{\beta} = \frac{\omega}{\beta} \qquad (2.102)$$

The group velocity and the group index are defined as follows:

$$v_g = \frac{c}{n_g} = c\left(\frac{d\beta}{dk}\right)^{-1} = \frac{d\omega}{d\beta} \qquad (2.103)$$

$$n_g = n_{\text{eff}} + k\frac{dn_{\text{eff}}}{dk} = n_{\text{eff}} - \lambda\frac{dn_{\text{eff}}}{d\lambda} \qquad (2.104)$$

Direct detection of intensity-modulated light signals recognizes dispersion effects only in the envelope of light signal, which is influenced by the group delay, τ_g.

$$\tau_g = \frac{L}{v_g} = \frac{Ln_g}{c} = \frac{L\,d\beta}{c\,dk} = \frac{L}{c}\left[n_{\text{eff}} - \lambda\frac{dn_{\text{eff}}}{d\lambda}\right] \tag{2.105}$$

where L is the fiber length.

Now let us start the discussion on dispersion from the propagation constant, β. By simultaneously solving equations 2.78 and 2.87, we get $U = U(V)$, and the propagation constant can be expressed as

$$\beta^2 = n_1^2 k_0^2 - \frac{U(V)^2}{a^2} \tag{2.106}$$

We introduce a normalized propagation constant, $b(V)$, in order to make the discussion independent of particular fiber structures.

$$b(V) = \frac{(\beta/k)^2 - n_2^2}{n_1^2 - n_2^2} = \frac{n_{\text{eff}}^2(V) - n_2^2}{n_1^2 - n_2^2} = 1 - \left(\frac{U(V)}{V}\right)^2 \tag{2.107}$$

$b(V)$ reduces to a linear equation in the weakly guiding approximation

$$b(V) \approx \frac{(\beta/k) - n_2}{n_1 - n_2} = \frac{n_{\text{eff}} - n_2}{n_1 - n_2} \tag{2.108}$$

The propagation constant is, then, expressed in the first order of Δ

$$\beta = \frac{2\pi n_{\text{eff}}}{\lambda} \approx kn_2(1 + b\Delta) = kn_2\left\{1 + \Delta\left(1 - \left(\frac{U(V)}{V}\right)^2\right)\right\} \tag{2.109}$$

Using the expression in equations 2.105 and 2.109, we will try to find the dispersion in the group delay, τ_g. When we differentiate β in terms of k in equation 2.105, we need to consider the k dependence on n_2 and Δ as well as b shown in equation 2.109. In conventional silica fibers, where the core is doped with GeO_2, it is found that Δ is independent of k. In all of high silica glass material used in optical fibers, it is also known that $k(dn/dk) \ll n$ [4,5,7]. Based on these observations, the group delay can be expressed in the first order of Δ:

$$\tau_g = \frac{L}{c}\left(\frac{d(n_2 k)}{dk} + n_2\Delta\frac{d(Vb)}{dV}\right) \tag{2.110}$$

Note that the first term on the right-hand side of equation 2.110 is the contribution from material dispersion, which will be affecting all the modes in the same magnitude.

The second term does depend on a waveguide structure (a, Δ, and n), and thus contributes the waveguide dispersion.

Differentiating both sides of the unified characteristic equation 2.87 with respect to V, we derive the equation for (dU/dV).

$$\frac{dU}{dV} = \frac{U}{V}(1 - \kappa_l(W)) \tag{2.111}$$

$$\kappa_l(W) = \frac{K_l^2(W)}{K_{l-1}(W)K_{l+1}(W)} \tag{2.112}$$

Using equations 2.107 and 2.111, we have the expression for $(d(Vb)/dV)$

$$\frac{d(Vb)}{dV} = 1 - \left(\frac{U}{V}\right)^2 (1 - 2\kappa_l) \tag{2.113}$$

Gloge [4] has derived an expression for κ_l in terms of V

$$\kappa_l \approx 1 - \frac{1}{\sqrt{V^2 - V_c^2 + l^2 + 1}} \tag{2.114}$$

where V_c is the cutoff frequency of the LP_{lp} mode, the pth root of $J_{l-1}(x)$.

And finally we have the expression for the group delay

$$\tau_g = \frac{L}{c}\left(\frac{d(n_2 k)}{dk} + n_2 \Delta\left\{1 - \left(\frac{U}{V}\right)^2(1 - 2\kappa_l)\right\}\right) \tag{2.115}$$

This clearly shows the group delay is composed of two parts: material and waveguide dispersion.

In optical fiber, the chromatic dispersion, D, is expressed in the unit of ps/(nm km) and is related to the group delay, τ_g, by the following equation:

$$D = \frac{1}{L}\frac{d\tau_g}{d\lambda} = -\frac{\lambda}{c}\frac{d^2 n_{\text{eff}}}{d\lambda^2} \tag{2.116}$$

In Figure 2.5, the propagation constant, β, is plotted as a function of V number for various optical modes. It is noted that for the case of a high Δ, we can assign HE, EH, TE, and TM modes as in Figure 2.5a. In contrast, for the case of a small Δ where the weakly guiding approximation is valid, the modes are assigned by LP modes. Major steps in the modal analysis are summarized in Table 2.4.

2.2 MATERIAL DISPERSION IN OPTICAL FIBERS

In most of commercial optical fibers, the major constituent material is vitreous silica glass, which is composed of SiO_2 molecules in a random network. In the perspective of the electromagnetic wave propagation, silica behaves as a typical dielectric material whose response is highly frequency dependent. This frequency-dependent dielectric response is characterized as material dispersion in optical fiber, whose unit is given as ps/(nm km) [8]. Material dispersion, in other words, originates

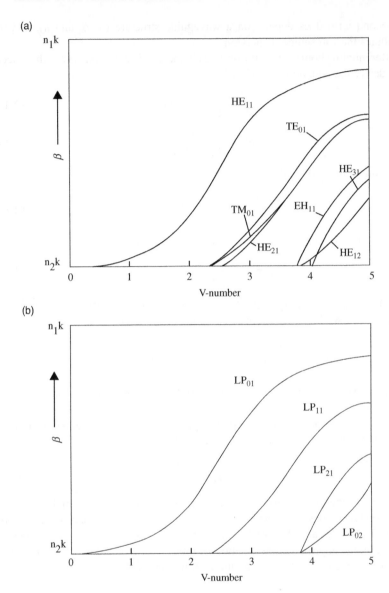

FIGURE 2.5 Propagation constant as a function of V number for selected modes: (a) exact solutions and (b) weakly guiding approximations.

from frequency-dependent electric permittivity, $\varepsilon(\omega)$, or equivalently wavelength-dependent refractive index, $n(\lambda)$.

An insulating dielectric could be modeled as a set of electron–cation pairs, where an electron is bound to a cation with a specific binding force as represented in Figure 2.6. Note that there are no free electrons that will contribute to the conductivity of the material because we assume a dielectric material.

TABLE 2.4 Major Steps in the Optical Fiber Model Analysis

Major Steps	Equations
1. Setting up differential equations for E_z and H_z	2.42, 2.43
2. Functional forms of fields in the cylindrical symmetry	
Core E_{1z}, H_{1z}	2.47, 2.48
$E_{1\theta}, H_{1\theta}$	2.65, 2.67
Cladding E_{2z}, H_{2z}	2.51, 2.52
$E_{2\theta}, H_{2\theta}$	2.66, 2.68
3. Boundary conditions at the core–cladding interface	2.59–2.64
4. Evaluating boundary conditions at $r = a$ to have a homogeneous matrix equation, $\overleftrightarrow{\Lambda} \cdot \vec{X} = 0$	2.69–2.73
5. Characteristic equation $\det(\Lambda) = 0$	2.74
6. Definition of U, W, V	2.78, 2.79
7. Characteristic equation in weakly guiding approximation	2.87

8. Numerically solve two equations for a given V
  - Characteristic equation 2.87
  - $U^2 + W^2 = V^2$	

9. *Find β from either U or W found in step 8*
  - $\beta^2 = n_1^2 k^2 - u^2 = n_1^2 k^2 - \left(\frac{U}{a}\right)^2$
  - $\beta^2 = n_2^2 k^2 + w^2 = n_2^2 k^2 + \left(\frac{W}{a}\right)^2$
	2.49, 2.53

10. *Find fields* of the LP_{01} mode	Table 2.3
11. Optical power carried by the LP_{01} mode	2.100
12. Find propagation properties	
- Effective index, $n_{\text{eff}} = \beta \frac{\lambda}{2\pi}$	2.101
- Phase velocity, $v_p = \frac{\omega}{\beta}$	2.102
- Group velocity, $v_g = \frac{d\omega}{d\beta}$	2.103
- Group delay, $\tau_g = \frac{L}{v_g}$	2.105
- Chromatic dispersion, $D = \frac{1}{L}\frac{d\tau_g}{d\lambda}$	2.115

Here we will focus on a single molecule for simplicity.

Let us assume that the binding force between the electron and the cation is approximately represented as a harmonic oscillator with a resonant frequency of ω_{0i} such that the displacement of electron from the equilibrium position, $\vec{x}_i$, is described as a harmonic oscillator. As the electron oscillates by the electric field, it will follow the harmonic oscillator without damping and the equation of motion for the electron is, therefore, given by

$$m\left(\frac{d^2\vec{x}_i}{dt^2} + \omega_{0i}^2\vec{x}_i\right) = -e\vec{E} \tag{2.117}$$

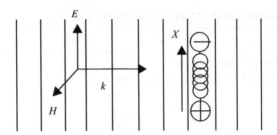

FIGURE 2.6 Modeling of a molecule under external electromagnetic wave.

Here m and e are the mass and the electric charge of the electron, respectively. Note that we have assumed a lossless dielectric with zero conductivity such that the damping term is absent. For a monochromatic plane electromagnetic wave at the frequency of ω, the electric field is given by $\vec{E}(t) = \vec{E}^0 \, e^{j\omega t}$. Assuming the similar harmonic solution for $\vec{x}_i(t) = \vec{x}_i^0 \, e^{j\omega t}$, we have the electric dipole momentum, $\vec{p}_i$, given as follows:

$$\vec{p}_i = -e\vec{x}_i = \frac{e^2}{m}(\omega_{0i}^2 - \omega^2)^{-1}\vec{E} \qquad (2.118)$$

Let us assume that the dielectric is composed of N molecules per unit volume and Z electrons per molecule with the oscillator strength of f_i, where $\sum_i f_i = Z$.

The polarization of the medium, dipole momentum per unit volume, is given by

$$\vec{P} = \sum_i f_i N \vec{p}_i = \frac{Ne^2}{m} \sum_i f_i(\omega_{0i}^2 - \omega^2)^{-1}\vec{E} \qquad (2.119)$$

By the definition of electric permittivity, ε, and susceptibility, χ_e

$$\vec{P} = \varepsilon_0 \chi_e \vec{E} = (\varepsilon - \varepsilon_0)\vec{E} \qquad (2.120)$$

we have the frequency-dependent $\varepsilon(\omega)$ of the dielectric as follows:

$$\varepsilon(\omega) = \varepsilon_0(1 + \chi_e) = \varepsilon_0 + \frac{Ne^2}{m}\sum_i f_i(\omega_{0i}^2 - \omega^2)^{-1} \qquad (2.121)$$

At wavelengths far from the resonance, the expression can be further changed in terms of λ

$$\varepsilon(\lambda) \simeq \varepsilon_0 + \frac{Ne^2}{4mc^2\pi^2}\sum_i f_i \lambda_i^2 \frac{\lambda^2}{\lambda^2 - \lambda_i^2} \qquad (2.122)$$

The refractive index of the medium is given by $n = \sqrt{\varepsilon\mu}$ and in most of dielectric the magnetic permeability, μ, is approximated as 1. Therefore, the refractive index of the medium can be expressed as Sellmeier equation [9]:

$$n^2 \simeq 1 + \sum_i \frac{A_i \lambda^2}{\lambda^2 - \lambda_i^2} \qquad (2.123)$$

The Sellmeier equation can describe the wavelength-dependent refractive index by a set of fitting parameters, A_i and λ_i, which are usually tabulated by experimental observations. A_i represents the oscillator strength and λ_i represents the resonance location in wavelength.

In silica, there are two major resonance absorptions: one is related to the electronic transition between the valence band and the conduction band of the silica and the other is the molecular vibrational absorption in the tetrahedron structure of SiO_4 in the glass networks. The electronic transition occurs in the UV region with a strong absorption intensity, whose tail extends to the visible range [10]. The silica is based on SiO_4 tetrahedron building block and its bonds are characterized by various types of vibrational modes whose resonance wavelengths and their harmonics are distributed in the mid-IR to IR region [11].

In principle, when the precise absorption, $\alpha(\lambda)$, of a given dielectric is known for the whole wavelength range, then its refractive index $n(\lambda)$ can be calculated using Kramers–Kronig relation. Kramers–Kronig equation relates the refractive index of a medium to its absorption coefficient as [12]

$$n(\lambda) = 1 + \frac{\lambda^2}{2\pi^2} P \int_0^\infty \frac{\alpha(\lambda')}{\lambda'^2 - \lambda^2} d\lambda' \qquad (2.124)$$

where P is the principal value of Cauchy integral and $\alpha(\lambda)$ the absorption coefficient of the medium.

In Figure 2.7, the origin of material dispersion of silica and doped silica is schematically represented. By Kramers–Kronig relation, the refractive index is severely modulated near the absorption peaks, where the UV and IR transitions occur,

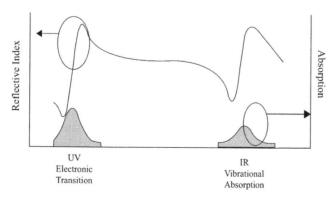

FIGURE 2.7   Schematic diagram for the origin of material dispersion in dilectric.

and its modulation will determine the dispersion at the wavelength far from the resonance spectral positions.

Conventionally, three-term Sellmeier equations have been widely used to express the refractive index of silica glass in an analytic form, especially for fiber optic applications in the range of 1.2–1.6 μm. Usually two terms correspond to absorption in deep UV and UV, while the rest is related to IR absorption:

$$n^2 \simeq 1 + \sum_{i=1}^{3} \frac{A_i \lambda^2}{\lambda^2 - \lambda_i^2} \tag{2.125}$$

Six Sellmeier coefficients, A_i and λ_i for $i = 1-3$, will characterize the refractive index, $n(\lambda)$, and subsequently the material dispersion of glass. The Sellmeier coefficients for the pure silica and doped silica are summarized in Table 2.5 [13–18].

Note that the Sellmeier equation 2.124 and the coefficients in Table 2.5 assume the wavelength, λ, is given in micrometers. The refractive indices calculated using Sellmeier equations for the glasses are plotted in Figure 2.8.

It is found that the refractive indices in all glasses decrease as the wavelength increases, and it is also noteworthy that GeO_2 and P_2O_5 increase the refractive index while fluorine and B_2O_3 decrease relative to pure SiO_2. Using this capability to change the refractive index relative to pure SiO_2, we can fabricate optical waveguides with a desired refractive index profile.

Once the refractive indices are known in the wavelength range, we can calculate group velocity that is defined as

$$V_g = \frac{c}{N_g} = \frac{c}{n - \lambda(dn/d\lambda)} \tag{2.126}$$

The results are plotted in Figure 2.9. As indicated by the refractive index, addition of GeO_2 and P_2O_5 decreases the group velocity while F and B_2O_3 increase, relative to pure SiO_2. The ability to control the group velocity enables waveguide design based on material choice for optimal bandwidth in multimode optical transmission fibers.

The first derivative of the refractive index with respect to the wavelength, $(dn/d\lambda)$, is related to the group velocity while the second derivative, $(d^2n/d\lambda^2)$, is related to the material dispersion by

$$D_m = -\frac{\lambda}{c}\frac{d^2n}{d\lambda^2} \quad [\text{ps/(nm km)}] \tag{2.127}$$

The material dispersion, along with the waveguide dispersion described in the preceding section, will determine the total chromatic dispersion of a waveguide. The material dispersion of the glasses is shown in Figures 2.10 and 2.11. It is noted that addition of GeO_2 in silica changes significantly the material dispersion as shown in Figure 2.10. Addition of GeO_2 will translate the material dispersion of SiO_2 toward the longer wavelength, and the zero-dispersion wavelength, λ_0 as indicated in the figure, increases accordingly as GeO_2 concentration increases. Silica has the

TABLE 2.5 Sellmeier Coefficients for Glasses Used in Optical Fibers [13-18]

Mole Percent			Sellmeier Coefficients					
SiO_2	GeO_2	F	A_1	λ_1	A_2	λ_2	A_3	λ_3
100	0	0	0.6961663	0.0684043	0.4079426	0.1162414	0.8974994	9.896161
95.9	4.1	0	0.68671749	0.072675189	0.43481505	0.11514351	0.89656582	10.002398
93.0	7.0	0	0.68698290	0.078087582	0.44479505	0.11551840	0.79073512	10.436628
86.5	13.5	0	0.73454395	0.086976930	0.42710828	0.11195191	0.82103399	10.846540
0	100	0	0.80686642	0.68972606	0.71815848	0.15396605	0.85416831	11.841931
99	0	1	0.69325	0.06724	0.39720	0.11714	0.86008	9.7761
98	0	2	0.67744	0.06135	0.40101	0.12030	0.87193	9.8563
SiO_2	B_2O_3	P_2O_5	A_1	λ_1	A_2	λ_2	A_3	λ_3
86.7	13.3	0	0.690618	0.061900	0.401996	0.123662	0.898817	9.098960
90.9	0	9.1	0.695790	0.061568	0.452497	0.119921	0.712513	8.656641

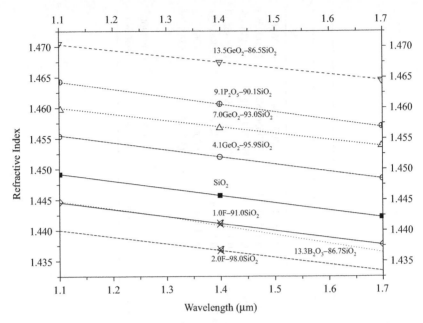

FIGURE 2.8   Refractive index dispersion, $n(\lambda)$, for the pure and doped silica glasses.

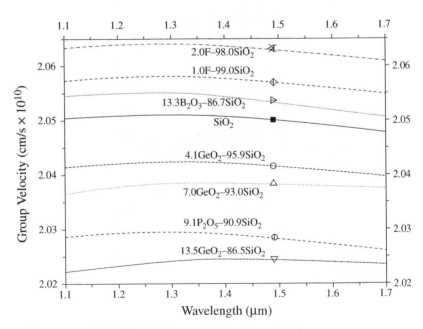

FIGURE 2.9   Group velocities in optical glasses.

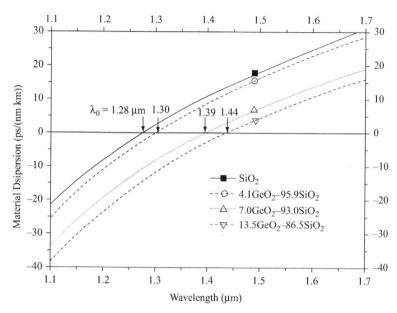

FIGURE 2.10   Material dispersion of pure silica and GeO_2-doped glass.

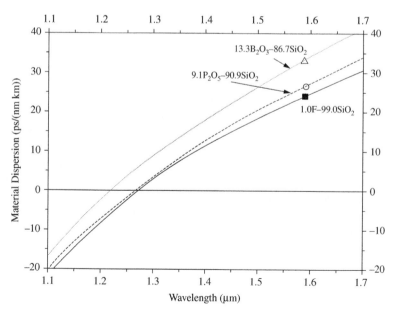

FIGURE 2.11   Material dispersion of F-, P_2O_5-, and B_2O_3-doped silica glass.

zero-dispersion wavelength near 1.28 μm while doping 13.5 mol% of GeO_2 increases λ_0 to 1.44 μm. Therefore, dispersion control using germanosilicate glasses should contain thorough consideration on both material and waveguide dispersions.

In comparison to GeO_2, the effect of doping fluorine was relatively small in terms of dispersion value and its slope. The material dispersion curves of fluorine-doped glasses are overlapping with the pure silica case as shown in Figure 2.11. The effect of addition of P_2O_5 shifts the zero-dispersion wavelength slightly to the shorter wavelength such that $\lambda_0 = 1.27$ μm and results in a higher dispersion in the longer wavelength region. It has been known that addition of B_2O_3 to a silicate glass system moves the point where material dispersion goes to zero to a shorter wavelength [18]. In Figure 2.11, the zero dispersion of the silica doped with 13.3 mol% of B_2O_3 is near 1.22 μm, as expected.

Let us now focus on GeO_2, the most widely used additive for index raise in silica optical fibers, and we will discuss how the refractive index control would be designed. In prior discussions, we only dealt with discrete data points in terms of GeO_2 concentration. Namely, we have considered only three cases of GeO_2-doped silica for tabulating Sellmeier coefficients, as shown in Table 2.5. In the table, note that Sellmeier coefficients for pure germanate glass, GeO_2, are also given [14]. The Sellmeier parameters for pure GeO_2 glass and those of pure SiO_2 glass will provide a basis for modeling the refractive index dispersion of all binary GeO_2–SiO_2 compositions. These parameters represent characteristics of the intrinsic absorption bands, and it has been shown that a three-term Sellmeier equation with two oscillations in the UV and one in the IR provides a very precise representation of refractive index dispersion. In the binary $xGeO_2$–$(1-x)SiO_2$ glass system, therefore, as the composition of a glass is changed, the Sellmeier coefficients are going to shift in a continuous fashion, especially in the case of low mole fraction. The general refractive index dispersion in $xGeO_2$–$(1-x)SiO_2$ glass system is well represented by a modified Sellmeier equation, which is dependent only on the composition parameter, x, and the equation is given by

$$n^2(\lambda) = 1 + \sum_{i=1}^{3} \frac{[SA_i + x(GA_i - SA_i)]\lambda^2}{\lambda^2 - [S\lambda_i + x(G\lambda_i - S\lambda_i)]^2} \qquad (2.128)$$

Here SA_i, $S\lambda_i$, GA_i, and $G\lambda_i$ are the Sellmeier coefficients of pure SiO_2 and GeO_2 glasses, respectively [14]. And x is the mole fraction of GeO_2 in the binary glass system. Using the above equation, the refractive index of the glass at different wavelengths is shown as a function of the mole fraction in Figure 2.12.

The relative index difference of the binary germanosilicate glass system with respect to SiO_2,

$$\Delta = \left[(n^2_{xGeO_2(1-x)SiO_2} - n^2_{SiO_2})/2n^2_{xGeO_2(1-x)SiO_2} \right]$$
$$\simeq \left[(n_{xGeO_2(1-x)SiO_2} - n_{GeO_2})/n_{xGeO_2(1-x)SiO_2} \right],$$

is shown in Figure 2.13 at 1.55 μm.

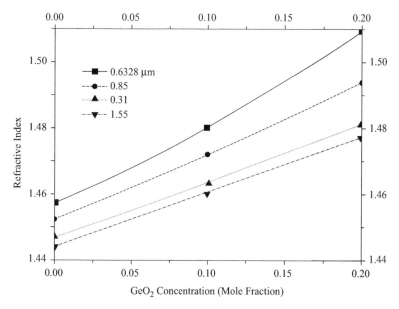

FIGURE 2.12   Refractive index of $x\text{GeO}_2\text{–}(1-x)\text{SiO}_2$ binary glass system.

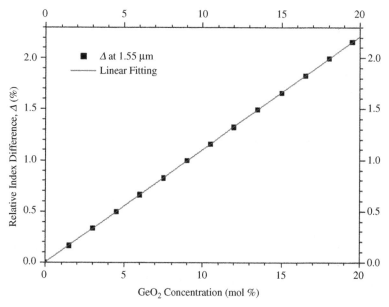

FIGURE 2.13   Relative index difference of $x\text{GeO}_2\text{–}(1-x)\text{SiO}_2$ binary glass system.

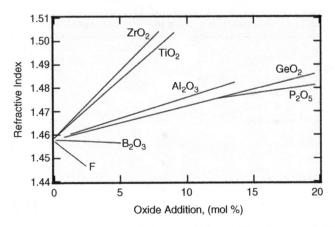

FIGURE 2.14   Relative index difference of binary silica glass system [19].

It is noted that very high linear correlation exists between Δ and the GeO_2 mol%, X, such as $\Delta \simeq 0.11 \times X$.

Refractive index is plotted as a function of cation additive mole percent in Figure 2.14 [19]. It is also noted that in these binary silica glass systems, the refractive index is almost a linear function of the cation concentration. The most effective index-raising cation is found to be Zr_2O_3 while fluorine is the most effective index-lowering element.

2.3   OPTICAL ATTRIBUTES FOR SINGLE-MODE FIBER CHARACTERIZATION AND CLASSIFICATION

As we have discussed in the preceding sections, optical fibers can be designed and fabricated with a unique combination of waveguide structure and its material composition. Their propagation characteristics will be predicted with the analysis on waveguide and material dispersion. The optical properties and characteristics of optical fibers, however, are in general described in special terms that can classify each type of optical fiber for specific applications. In this section we will briefly review those "optical attributes" that are being used in international standards such as International Telecommunication Union, Telecommunication Standardization Sector (ITU-T) and International Electrotechnical Commission (IEC). Detailed discussion of the types of telecommunication transmission optical fibers will, then, be followed in terms of those main fiber attributes.

There have been intensive efforts to characterize and standardize the attributes of optical fiber ever since its invention and subsequent massive deployment worldwide. With rapidly changing optical communication technologies in both devices and systems, the ranges of existing attributes are continuously being updated and furthermore new ones are added to cope with renewing platforms.

Both ITU-T and IEC have upgraded the main bodies of standards that relate transmission optical fibers based on the following key optical parameters [20,21]:

- mode field diameter, $2w$,
- effective area, A_{eff},
- cutoff wavelength, λ_c,
- attenuation coefficient, $a(\lambda)$,
- chromatic dispersion coefficients, $D(\lambda)$
  - minimum zero-dispersion wavelength, λ_{0min},
  - maximum zero-dispersion wavelength, λ_{0max},
  - maximum zero-dispersion slope, S_{0max},
- polarization mode dispersion coefficient

  - upper bound probability, Q,
  - maximum PMD_Q,
  - sampling cable number, M,
  - maximum differential group delay (DGD_{max})

Besides those parameters there are mechanical parameters and geometrical factors, which are beyond the scope of this book and readers are recommended to consult the references.

2.3.1   General Definitions of Fiber Parameters

2.3.1.1   Mode Field Diameter   The mode field diameter (MFD) $2w$ represents a measure of the transverse extent of the electromagnetic field intensity of the mode in a fiber cross-section and it could be defined from either the far-field intensity distribution $F^2(\theta)$, θ being the far-field angle, or the near-field intensity distribution $f^2(r)$, r being the radial coordinate.

The far-field definition of MFD is given by the following equation [22]:

$$2w = \frac{\lambda}{\pi} \left[\frac{2\displaystyle\int_0^{\pi/2} F^2(\theta)\sin\theta\,\cos\theta\,d\theta}{\displaystyle\int_0^{\pi/2} F^2(\theta)\sin^3\theta\,\cos\theta\,d\theta} \right]^{1/2} \tag{2.129}$$

A schematic diagram of the test apparatus for far-field scan is shown in Figure 2.15.

Here the stable light source is coupled to the input end of fiber and the light signal is guided in the fundamental LP_{01} mode by the cladding mode stripper. At the well-cleaved exit end of the fiber, the light will propagate to the air and its intensity is recorded by a photodetector whose locus is defined as a half circle centered at the exit

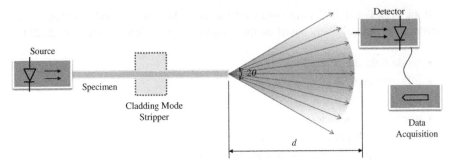

FIGURE 2.15 The far-field scan setup recommended in ITU-T G.650.1 [23].

end. The detector should keep the distance, d, from the fiber end at least 10 mm to fully observe far-field pattern.

The MFD is also determined from the near-field intensity distribution, $f^2(r)$ [22]:

$$2w = 2\left[2\frac{\int_0^\infty rf^2(r)dr}{\int_0^\infty r[(df(r))/dr]^2 dr}\right]^{1/2} \qquad (2.130)$$

The mathematical equivalence of equations 2.129 and 2.130 is valid in the approximation of small angles θ. Under this approximation, the near-field $f(r)$ and the far-field $F(\theta)$ form a Hankel pair. By means of the Hankel transform, it is possible to transform equation 2.129 to equation 2.130 and vice versa.

A schematic diagram of the test apparatus for near-field imaging is shown in Figure 2.16.

In the near-field imaging setup, the field pattern of the LP_{01} mode at the output end of the sample fiber is directly imaged to CCD camera or equivalents by a magnifying optics and its pattern is normalized for calculation of MFD.

Other than these two typical methods, there are alternatives recommended by ITU-T G.650.1, namely, the variable aperture technique and bidirectional backscatter difference method [22].

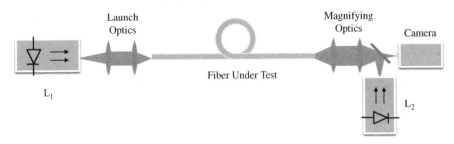

FIGURE 2.16 The near-field imaging setup recommended by ITU-T G.650.1 [23]; L_1 is the launching light source and L_2 is the illumination light source.

In the variable aperture technique, the mode field diameter is determined from the complementary aperture transmission function $a(x)$, where $x = D \cdot \tan \theta$ is the aperture radius and D the distance between the aperture and the fiber:

$$2w = \frac{\lambda}{\pi D} \left[\int_0^{\infty} a(x) \frac{x}{(x^2 + D^2)^2} dx \right]^{-1/2} \tag{2.131}$$

The mathematical equivalence among equations 2.129–2.131 is valid in the approximation of small angles θ.

In bidirectional backscatter difference method, the mode field diameter is determined from the difference in bidirectional backscatter across a splice with a dead-zone fiber with a known mode field diameter:

$$w_s = w_d 10^{(g(L_d - L_s) + f)/20} \tag{2.132}$$

where w is the mode field diameter and L is the change in backscatter (dB) across the splice. Subscripts "s" and "d" represent the dead-zone fiber and specimen fiber, respectively. g and f are wavelength- and fiber design–dependent adjustment factors.

2.3.1.2 Effective Area   Effective area is a parameter that is closely related to not only the splice loss but also optical fiber nonlinearities that will affect the transmission quality.

Effective area A_{eff} is defined as follows

$$A_{\text{eff}} = \frac{2\pi \left[\int_0^{\infty} I(r) r \, dr \right]^2}{\int_0^{\infty} I(r)^2 r \, dr} \tag{2.133}$$

where $I(r)$ is the field intensity distribution of the fundamental mode of the fiber at radius r. The integration is carried out over the entire cross-sectional area of the fiber. If we assume a Gaussian approximation such that

$$I(r) = \exp\left(\frac{-2r^2}{w^2} \right) \tag{2.134}$$

where $2w$ is the MFD, then equation 2.133 can be analytically integrated and gives

$$A_{\text{eff}} = \pi w^2 \tag{2.135}$$

The Gaussian approximation is known to be fairly accurate for conventional step-index fiber such as Corning SM-28 only near the LP_{11} cutoff. At longer wavelength far

TABLE 2.6 Range of Correction Factor *k* at Different Wavelengths

λ (nm)	1310	1550
k Factor	0.970–0.980	0.960–0.970

apart from the cutoff, the approximation is no longer valid and nonstep-index fibers cannot be described by the simple Gaussian approximation.

A more general but empirical relationship between A_{eff} and w is given as [23]

$$A_{\text{eff}} = k\pi w^2 \tag{2.136}$$

Here k is a correction factor that depends on varieties of parameters such as the wavelength, the refractive index profiles, MFD, and zero-dispersion wavelength. In conventional step-index single-mode fiber whose zero dispersion is near 1310 nm, the k factor shows the distribution as given in Table 2.6.

2.3.1.3 Cutoff Wavelength Specification of the LP_{11} mode cutoff wavelength is to assure effective single-mode operation above a specified wavelength where multimodal interference effects are absent. Widely used transmitted power technique as recommended by ITU-T G.650.1 [22] measures the variation with wavelength of the transmitted power of a short length of the fiber under test, compared to a reference transmitted power. There are two possible ways to obtain this reference power

(a) the same test fiber with a loop of smaller radius about 30 mm; or

(b) a short (1–2 m) length of multimode fiber.

The measurement is performed on a 2 m length of test fiber. The fiber is bent to form a loosely constrained loop with one full turn of a circle of 140 mm radius. The output power $P_1(\lambda)$ shall be recorded versus wavelength and similarly the reference power $P_2(\lambda)$ will be also measured depending on the option either (a) or (b). The spectral attenuation of the test specimen, relative to the reference power, is

$$a(\lambda) = 10 \log \frac{P_1(\lambda)}{P_2(\lambda)} \tag{2.137}$$

Assuming a straight-line representation of the longer wavelength region, the deviation of higher order modes from the fundamental mode is given by

$$\Delta a(\lambda) = a(\lambda) - (A_u + B_u \lambda) \tag{2.138}$$

In the transition region, higher order mode power is reduced with increasing wavelength. Fiber cutoff wavelength, λ_c, is defined as the wavelength at which the higher order mode power relative to the fundamental mode power, $\Delta a(\lambda)$, has been

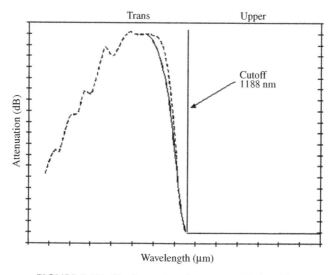

FIGURE 2.17   Single-mode reference cutoff plot [22].

reduced to 0.1 dB. Typical measurement result is shown in Figure 2.17 where dotted line and solid line represent $P_1(\lambda)$ and $P_2(\lambda)$, respectively. As an alternative test method for λ_c, the split mandrel technique is also recommended in ITU-T G.650.1 [22] and the setup is schematically shown in Figure 2.18. The measurement length of the fiber is 2 m and the fiber is bent to form a loosely constrained full-turn loop consisting of two arcs (180° each) of 140 mm radius. The output power $P_1(\lambda)$ shall be recorded versus wavelength and then, as shown in Figure 2.18, the lower semicircular mandrel moves to a certain distance to give measurement of $P_2(\lambda)$.

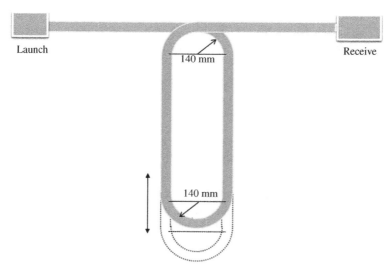

FIGURE 2.18   Cutoff wavelength measurement by the split mandrel technique [22].

The rest of calculation procedures are identical to the transmitted power technique described earlier.

2.3.1.4 Attenuation

The attenuation coefficient, in dB/km, will give the principal estimate for the power budget in optical links and therefore is regarded as a primary fiber parameter. Continuous efforts have been made to reduce the fiber attenuation by developing synthetic silica fabrication technologies [24] and chlorine dehydration process [25] as well as high purification of raw material. The origin of the intrinsic attenuation in optical fibers has been understood to come from three distinctive sources: UV absorption edge, Rayleigh scattering, and IR vibrational absorption.

The electronic transition between the valence band and conduction band of amorphous glass induces a significant absorption in the visible and near IR spectral range as well as in the UV, due to its absorption tail, commonly referred as Urbach edge [26]. The UV loss in GeO_2–SiO_2 binary glass is expressed in terms of the mole fraction of GeO_2, X, and wavelength, λ, in micrometers [27]:

$$\alpha_{UV}(\lambda) = 10^{-2} \left[\frac{1.542X}{44.6X + 60} \right] \exp\left(\frac{4.63}{\lambda} \right) \quad [dB/km] \tag{2.139}$$

Rayleigh scattering originates from the fluctuations in the dielectric constant, which could be related to density, concentration, and the composition. The Rayleigh scattering is the main contribution to the optical fiber attenuation in the communication bands from 1.2 to 1.6 μm and is highly dependent on the glass material as well as the processing conditions. In general, the Rayleigh loss is expressed as

$$\alpha_R(\lambda) = R\lambda^{-4} \quad [dB/km] \tag{2.140}$$

where R is the scattering coefficient whose value ranges from 0.7 to 1.0 dB/(km μm^4) for single-mode fibers and 1 to 1.6 dB/(km μm^4) for multimode fibers [24].

Si and Ge in GeO_2–SiO_2 binary glass form a random network based on the tetrahedron constituents and it is known that the Si–O and Ge–O bonds generate fundamental vibrational absorption near 9.0 and 11.0 μm, respectively [28]. The IR loss for all silica glass is expressed as [29]

$$\alpha_{IR} = 7.81 \times 10^{11} \exp\left(\frac{-44.48}{\lambda} \right) \tag{2.141}$$

Besides these intrinsic loss mechanisms, the most important source of extrinsic loss is related to OH vibration. Water incorporated in high silica glass as OH has its fundamental vibration at 2.73 μm [30,31] and the first overtone of OH vibration occurs near 1.38 μm. The combination of harmonics also induces optical loss at 1.24 μm. It is estimated at 1 ppm of Si–OH in weight will cause an excess loss of 48 and 2.5 dB/km at 1.38 and 1.24 μm, respectively. In commercial optical fibers OH level is maintained below ppb in weight and as an indicative measure of the dehydration process the loss at 1.38 μm is specified in international standards. Recently OH-free fibers have been developed and are being fabricated in mass production, and their specifications are explained in recent international standards such as ITU-T recommendation G.652.D.

In ITU-T G.650.1, the cutback technique is introduced as a primary method to measure the attenuation of optical fiber [22]. The technique is a direct application of the definition in which the power levels P_1 and P_2 are measured at two points of the fiber without change of input conditions. P_2 is the power emerging from the far end of the fiber and P_1 is the power emerging from a point near the input after cutting the fiber. Diagrams of suitable test equipments to obtain the loss at a single wavelength or the loss spectrum measurements are shown as examples in Figure 2.19a and b, respectively.

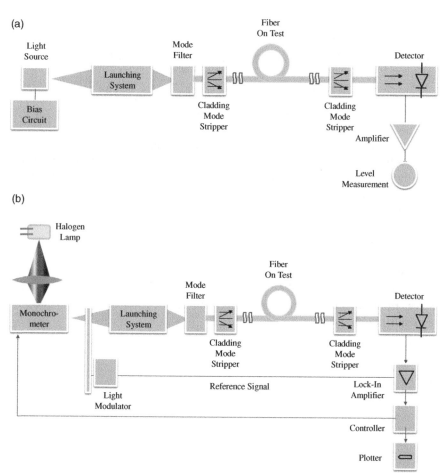

FIGURE 2.19 Optical attenuation measurement by the cutback technique [22]: (a) attenuation measurement at a single wavelength and (b) attenuation spectrum measurements.

The output power P_2 is initially recorded. Keeping the launching conditions fixed, the fiber is, then, cut to the desired cutback length, for example, 2 m from the launching point. With proper refitting of the cladding mode stripper, the output power P_1 from the cutback length is recorded. The attenuation of the fiber, with known parameters P_1 and P_2, and the length between the measurements, L, can be calculated as

$$a(\lambda) = \frac{10 \log(P_1(\lambda)/P_2(\lambda))}{L} \quad [\text{dB/km}] \qquad (2.142)$$

As an alternative to the cutback measurement, ITU-T G.650.1 [22] recommends the backscattering methods using optical time domain reflectometry and the insertion loss technique where the test fiber throughput is compared to the known reference fiber.

2.3.1.5 Chromatic Dispersion
In previous sections, we have learned that both waveguide and material dispersion contribute to the chromatic dispersion and the relation is well presented in terms of group delay as in equation 2.110 and 2.113:

$$\tau_g = \frac{L}{c}\left(\frac{d(n_2 k)}{dk} + n_2 \Delta\left\{1 - \left(\frac{U}{V}\right)^2 (1 - 2\kappa_l)\right\}\right) \qquad (2.115)$$

The first and second terms represent material and waveguide dispersion, respectively.

As a reference test method for chromatic dispersion in the unit of ps/(nm km), the phase-shift technique is recommended in ITU-TG.650.1 [22]. The fiber chromatic dispersion coefficient is derived from the measurement of the relative group delay experienced by the various wavelengths during propagation through a known length of fiber. The group delay is measured in the frequency domain, by detecting the phase shift of a sinusoidal modulating signal. A schematic diagram of the test apparatus is shown in Figure 2.20.

The phase shift between the reference signal and the channel signal at the operating wavelength is to be measured by the delay detector. The time group delay will be deduced from the corresponding phase shift ϕ through the relation $\tau = \phi/(2\pi f)$, f being the modulation frequency.

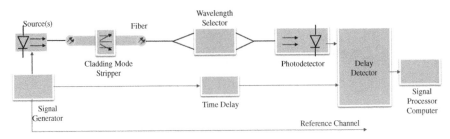

FIGURE 2.20 Experimental setup for chromatic dispersion measurements [22].

Fitting of the group delay as a function of wavelength does depend on the type of optical fibers. For the case of conventional step-index single-mode fiber (ITU-T G.652) [20], the measured group delay per unit fiber length versus wavelength is fitted by the three-term Sellmeier expression:

$$\tau(\lambda) = \tau_0 + \frac{S_0}{8}\left(\lambda - \frac{\lambda_0^2}{\lambda}\right)^2 \tag{2.143}$$

Here τ_0 is the relative delay minimum at the zero-dispersion wavelength λ_0 and S_0 is the zero-dispersion slope, that is, the value of the dispersion slope $S(\lambda) = dD/d\lambda$ at λ_0. The chromatic dispersion coefficient $D(\lambda) = d\tau/d\lambda$ can be determined from the differentiated Sellmeier expression:

$$D(\lambda) = \frac{S_0}{4}\left(\lambda - \frac{\lambda_0^4}{\lambda^3}\right) \tag{2.144}$$

Note that these equations for $\tau(\lambda)$ and $D(\lambda)$ are sufficiently accurate over the 1270–1340 nm range, but are less accurate in the 1550 nm region.

For the case of dispersion-shifted single-mode fiber (ITU-T G.653) [32], the fitting formula is optimized for zero dispersion near 1.5 μm. The measured group delay per unit fiber length versus wavelength is fitted by the quadratic expression:

$$\tau(\lambda) = \tau_0 + \frac{S_0}{2}(\lambda - \lambda_0)^2 \tag{2.145}$$

Therefore, the chromatic dispersion coefficient $D(\lambda) = d\tau/d\lambda$ can be determined from the differentiated quadratic expression:

$$D(\lambda) = (\lambda - \lambda_0)S_0 \tag{2.146}$$

These equations for $\tau(\lambda)$ and $D(\lambda)$ are sufficiently accurate over the 1500–1600 nm range and yet they are not meant to be used in the 1310 nm region.

For the case of cutoff shifted fiber (ITU-T G.654) [33], the measured group delay per unit fiber length versus wavelength shall be fitted by the quadratic expression:

$$\tau(\lambda) = \tau_{1550} + \left(\frac{S_{1550}}{2}\right)(\lambda - 1550)^2 + D_{1550}(\lambda - 1550) \tag{2.147}$$

Here τ_{1550} is the relative group delay at the wavelength $\lambda = 1550$ nm. The chromatic dispersion coefficient $D(\lambda) = d\tau/d\lambda$ can be determined from the differentiated quadratic expression:

$$D(\lambda) = S_{1550}(\lambda - 1550) + D_{1550} \tag{2.148}$$

Here S_{1550} is the (uniform) dispersion slope at 1550 nm wavelength, that is, the value of the dispersion slope $S_{1550}(\lambda) = dD/d\lambda$ at $\lambda = 1550$ nm. Also, D_{1550} denotes the dispersion values at $\lambda = 1550$ nm. These equations for $\tau(\lambda)$ and $D(\lambda)$ are sufficiently accurate over the 1500–1600 nm range. They are not meant to be used in the 1310 nm region.

For the case of nonzero dispersion-shifted fiber (NZ-DSF) (ITU-T Recommendation G.655) [34], the chromatic dispersion can be calculated from the chromatic dispersion coefficient, D_{1550}, and chromatic dispersion slope coefficient, S_{1550}, at 1550 nm assuming a linear dependence on length, and with due regard for the signs of the coefficients.

These values, together with link length, L_{Link}, can be used to calculate the typical dispersion for use in optical link design:

$$D_{Link}(\lambda) = L_{Link}[D_{1550} + S_{1550}(\lambda - 1550)] \quad [ps/nm] \qquad (2.149)$$

As alternative chromatic dispersion measurement methods, the interferometric technique and the pulse delay technique could be adopted for chromatic dispersion measurements [22]. According to the interferometric measuring principle, the wavelength-dependent time delay between the test sample and the reference path is measured by a Mach–Zehnder interferometer. The reference path can be an air path or a single-mode fiber with known spectral group delay. The interferometric test method allows the chromatic dispersion to be measured, using a short piece of fiber of several meters. This offers the possibility of measuring the longitudinal chromatic dispersion homogeneity of optical fibers. Moreover, it is possible to test the effect of overall or local influences, such as temperature changes and macrobending losses, on the chromatic dispersion.

In the pulse delay technique, chromatic dispersion coefficient is directly derived from the measurement of the relative group delay in time domain experienced by the various wavelengths during propagation through a known length of fiber.

2.3.1.6 Polarization Mode Dispersion

The fundamental LP_{01} mode in single-mode optical fiber is, in fact, a doubly degenerate mode such that both x and y polarizations of the HE_{11} mode have the same propagation constant for the perfect circular symmetry in waveguide structures. However, real fibers cannot maintain perfectly circular symmetry and are under the continuous influence of local stresses; consequently, the optical signal will propagate in local polarization mode (PM) traveling at different velocities and furthermore those PMs will no longer stay still but vary statistically by mode coupling effects. Therefore, conventional "single mode" fiber will, in fact, show multimode fiber characteristics especially in terms of PM delay and PM coupling effects. Note that PM delay in single-mode fiber is in order of picoseconds while the modal delay in conventional multimode fibers, as defined in ITU-T G.651 [35], is in the range of nanoseconds for 1 km of fiber. Therefore, it is only when the bit period of the data is comparable to that of PM delay as in high data rate over 10 Gbps, the effects of polarization mode dispersion (PMD) become the main concern.

Even though PMD has inherent statistical nature, there always exist two polarization states, called principal states of polarization (PSP) [36], for a given arbitrarily deployed fiber at a given time and optical frequency, such that the pulse spreading due to PMD vanishes, if only one PSP is excited. On the contrary, the maximum pulse spread due to PMD occurs when both PSPs are equally excited, and are related to the difference in the group delays associated with the two PSPs. Note that local birefringence in conventional single-mode fibers changes along the fiber, and the PSP depends on the fiber length, contrary to hi-birefringence polarization maintaining fibers where two PSPs are maintained without mode coupling.

DGD, denoted by $\tau(\omega)$, is the time difference in the group delays of the PSPs such that it has the unit of picosecond. In other words, DGD is PM delay between PSPs. Under normal conditions, DGDs are random functions of optical frequency and time, and vary at random from one fiber to the other. The statistical nature of DGD is known to follow Maxwell distribution [37,38] and its functional form is given as

$$P(\tau, \Delta\tau) = \frac{\tau^2}{\sqrt{\pi/2}(\Delta\tau)^3} \exp\left(-\frac{\tau^2}{2(\Delta\tau)^2}\right) \qquad (2.150)$$

Here $P(\tau, \Delta\tau)$ is the probability to have the DGD of τ when the mean PM delay is $\Delta\tau$. Definitions of the mean PMD will be discussed in the following section.

The variation of Maxwell distributions is shown in Figure 2.21. (Note that dimensionless numbers are shown for both x and y axes.) As shown in the figure, the distribution is highly dependent on the mean PM delay, $\Delta\tau$, especially the width of distribution is sensitive to $\Delta\tau$. In general, a higher $\Delta\tau$ will result in a broader distribution such that the probability for the case τ exceeds a certain value will

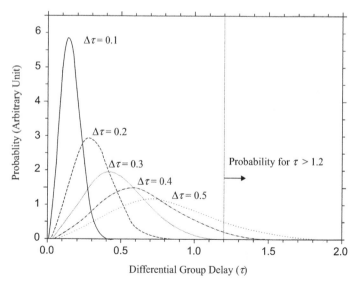

FIGURE 2.21 Maxwell probability distribution function for various parameters $\Delta\tau$.

increase. See the dotted vertical line in Figure 2.21. The probability that τ exceeds 1.2, for instance, will significantly increase as $\Delta\tau$ increases from 0.1 to 0.5. Abrupt increase in the probability to have a high time delay in the Maxwell distribution is, in fact, imposing a transmission limit for high speed data transmission.

It is noteworthy to pay attention to system performance impacts of PMD. The instantaneously high values of DGD limit the transmission capacity of digital systems very similar to the chromatic dispersion–related penalties such as inter-symbol interferences (ISI), while in analogue systems, the derivative of the differential group delay with respect to the wavelength limits the signal-to-noise ratio. Therefore, the statistical distribution of the differential group delays in the time and frequency domains plays an important role in predicting real system performance.

Having found the significance of $\Delta\tau$, the mean PM delay, we will discuss it following the most accepted ITU-T G.650.2, "Definitions and test methods for statistical and non-linear attributes of single-mode fiber" [39].

Essentially there are two different types of definition for the mean PM delay depending on which of the coherence time of light signal or the PM delay is shorter.

The first definition assumes that the coherence time of the signal is negligibly shorter than the PM delay and it follows closely the definition of modal dispersion for multimode fibers [35]. Remember that a "single" mode fiber actually supports two orthogonal polarization modes. In the sense of multimode fibers, the natural definition for PM delay is the mean square deviation of the time of flight of the various modes:

$$\text{PM delay}_1 = 2\left(\langle t^2 \rangle - \langle t \rangle^2\right)^{1/2} = 2\left(\frac{\int I(t)t^2\,dt}{\int I(t)dt} - \left(\frac{\int I(t)t\,dt}{\int I(t)dt}\right)^2\right)^{1/2} \quad (2.151)$$

where $I(t)$ is the intensity of the light pulse after propagating through the fiber, assuming an initially very short pulse similar to a delta function. It is shown that for arbitrary random PM coupling, the mean PM delay, $\Delta\tau$, is expressed in terms of the fiber length L, the mean PM coupling length h, and the average modal birefringence B [40]:

$$(\Delta\tau)^2 = \frac{B^2 h^2}{2}\left[\frac{2L}{h} - 1 + \exp\left(-\frac{2L}{h}\right)\right] \quad (2.152)$$

The second definition is consistent to the case where the laser source coherence time is larger than the PM delay as in most of optical communication applications. The outcoming polarization modes will thus interfere and the interference pattern produces two distinctive pulses that correspond to PSPs of the fiber [36]. It is therefore natural to consider, in addition to the output distribution $I(t)$, the DGD, $\tau(\omega)$, between two PSPs. As $\tau(\omega)$ will depend on the signal frequency, the second definition for the PM delay will be the mean of $\tau(\omega)$ over an appropriately broad range of optical frequency:

$$\text{PM delay}_2 = \langle \tau(\omega) \rangle = \frac{\displaystyle\int_{\omega_1}^{\omega_2} \tau(\omega)d\omega}{\omega_2 - \omega_1} \tag{2.153}$$

In the cases of main practical interest, such as low and high PM coupling, these two definitions are related as

For low PM coupling: PM delay$_2$ = PM delay$_1$.

For high PM coupling: PM delay$_2$ = PM delay$_1$/1.085.

The "constant" 1.085 ($\approx \sqrt{3\pi/8}$) comes from the Maxwellian statistics of the delay between the principal states [41].

In order to eliminate the discrepancy between the above two definitions for the high and low PM coupling, another definition, root mean square of DGD over a frequency range, is provided:

$$\text{PM delay}_3 = \sqrt{\langle \tau(\omega)^2 \rangle} = \left(\frac{\displaystyle\int_{\omega_1}^{\omega_2} \tau(\omega)^2 d\omega}{\omega_2 - \omega_1} \right)^{1/2} \tag{2.154}$$

It is shown in both theories and experiments that the third definition of PM delay is identical to the first one for any degree of PM coupling. This relation is fundamental for the understanding of PMD, since it relates measurement results with high and low coherent sources.

Once the mean PM delay is defined, the PMD coefficient, PMD_c, is subsequently defined, once again depending on PM coupling strength

- Weak PM coupling (short conventional single-mode fibers $L \ll h$)

$$\text{PMD}_c \, [\text{ps/km}] = \frac{\text{PM delay}}{L} \tag{2.155}$$

- Strong PM coupling (long conventional single-mode fibers $L \gg h$)

$$\text{PMD}_c \left[\text{ps}/\sqrt{\text{km}} \right] = \frac{\text{PM delay}}{\sqrt{L}} \tag{2.156}$$

In contrast, high birefringent fibers or equivalently polarization maintaining fiber (PMF) do not show a statistical distribution of DGD because there is almost no, or at

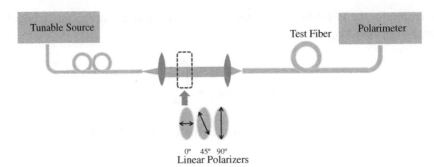

FIGURE 2.22 Schematic diagram of a typical polarization measurement setup [39].

best very weak, coupling between the fast axis and slow axis modes and DGD remains constant.

There are various methods to measure PMD and ITU-T G.650.2 [39] recommends the Stokes parameter evaluation technique as the reference test method. The change in the output state of polarization (SOP) with wavelength is determined and characterized through Jones matrix eigenanalysis (JME), or the rotation of the SOP vector on the Poincare sphere (PS). Schematic diagram of a typical measurement system is shown in Figure 2.22.

Here a narrow linewidth tunable source is used and the light emerging from the test fiber remains polarized under all conditions of the measurement. A polarization adjuster follows the laser and is set to provide roughly circularly polarized light to the polarizers. Three linear polarizers, at relative angles of approximately 45°, are arranged to be inserted into the light beam in turn. With a proper input optics the signal is, then, sent to polarimeter to measure the three output states of polarization corresponding to insertion of each of the three polarizers.

Typical measurement of DGD as a function of wavelength is shown in Figure 2.23. Data may also be displayed in a histogram as shown in Figure 2.24.

Note that a Maxwell curve is superimposed on each histogram and the curves of measured DGD versus wavelength are also shown for a reference. In these measurements, PM delays are measured in terms of the root mean square of DGD over a frequency range, PM delay$_3$, as defined in equation 2.154.

In contrast to deterministic parameters such as attenuation and chromatic dispersion, PMD is a statistical attribute such that it requires appropriate specifications representing the inherent natures.

Following the definition in equation 2.154, PMD is the average of measured DGD values across a range of wavelengths. Since the DGD values are random across time and wavelength, there is a theoretical lower limit to achievable reproducibility of the reported PMD value of approximately ±15%. This feature implies that it is not appropriate to select individual fibers or cables to a specification that is tighter than the capability of measurement process. Therefore, a specification on the overall process distribution is more reasonable and acceptable. In the most recent ITU-T G.650.2 [39], the upper confidence limit is named PMD$_Q$ and it is a PMD metric to

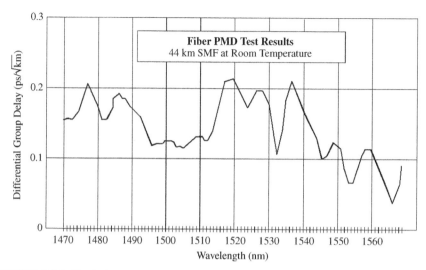

FIGURE 2.23 Typical measurement results of DGD for 44 km single-mode fiber at room temperature [39].

be compared with measured PMD that is associated with the 99.99% level of accumulated probability, or equivalently upper bound probability vale of 0.01%. The upper boundary probability of 0.01% is usually designated as Q associated with PMD_Q. The optical fibers or cables that show the PMD at the upper bound probability value of 0.01%, less than the specified PMD_Q, are known to "pass Method 1" [39].

Even though PMD_Q will serve as an indicative parameter for PMD, most of system impairments are directly related to instantaneously high DGD in the upper distribution tail. See Figure 2.21, for instance. It is therefore highly informative to specify what would be the maximum DGD for a given probability limit that is acceptable to overall system performance. In ITU-T G.650.2, the probability limit for maximum allowed DGD, DGD_{max}, is set at 6.5×10^{-8} ps based on various system considerations including the presence of other PMD generating components that may be in the links [39]. The DGD_{max} value is established for a broad range of distribution shapes and is specified as another PMD metric. If the probability that DGD in a fiber or cable is greater than specified DGD_{max} but is less than the specified maximum probability of 6.5×10^{-8}, then the fiber cable is known to "pass Method 2."

Method 1 is a metric that is based on what is measured and is therefore somewhat more straightforward for use in trade and commerce as a normative requirement. Method 2 is a means of extrapolating the implications for system design and is therefore included as information for system design.

These PMD metrics, PMD_Q and DGD_{max} in transmission optical fibers will be discussed in Section 2.4 along with current international standards.

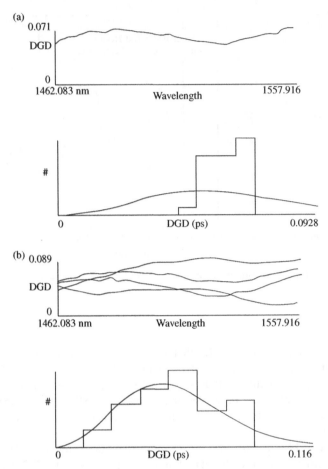

FIGURE 2.24 Examples of DGD data in histogram format [39]. (a) Single 24-interval measurement of a spooled fiber; (b) four 24-interval measurements of the same spool conducted at different oven temperatures.

2.4   INTERNATIONAL STANDARDS FOR SINGLE-MODE FIBERS

Two major standard bodies, ITU-T and IEC, are actively involved in issuing recommendations and specifications over optical fibers, which are being commonly referenced in commercial markets as well as academia. In order to design and fabricate specialty fibers with certain novelties, it is important to be fully aware of and understand the standards of conventional single-mode fibers, simply because those specialty fibers or devices based on them will have to be compatible with standard ones in terms of optical and mechanical properties.

In Section 2.3, we have reviewed major optical attributes that are being used to characterize and specify the types of single-mode optical fibers. In this section, we

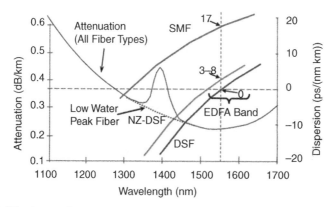

FIGURE 2.25   Attenuation and dispersion spectra of conventional single-mode fibers [43]; SMF, single-mode fiber; DSF, dispersion-shifted fiber; NZ-DSF, nonzero dispersion-shifted fiber; EDFA, erbium-doped fiber amplifier. The numbers are the chromatic dispersion values at $\lambda = 1550$ nm [42].

will review the current standards for transmission single-mode fibers recommended by both ITU-T and IEC. In a recent review paper from Corning [42], the attenuation and chromatic dispersion spectra of transmission single-mode fibers were compared as in Figure 2.25. In terms of attenuation, all the fibers have similar low-loss spectra and only the low water peak fiber has the reduced attenuation near 1380 nm by adopting additional dehydration process. The lowest loss of $\sim$0.2 dB/km is obtained near $\lambda = 1550$ nm.

It is the chromatic dispersion that clearly distinguishes the listed fiber types schematically shown in Figure 2.25, and in international standards. Conventional single-mode fibers (SMFs) have a typical chromatic dispersion $\sim$17 ps/(nm km) at $\lambda = 1550$ nm and zero dispersion near $\lambda = 1310$ nm. Dispersion-shifted fibers (DSFs) show zero dispersion at $\lambda = 1550$ nm, which can take advantage of low attenuation and low dispersion in high speed telecommunications [43–45]. NZ-DSFs show a low chromatic dispersion, typically 3–8 ps/(nm km) at $\lambda = 1550$ nm, which can take advantage of optical gain band of erbium-doped fiber amplifier (EDFA) [46,47] and high immunity to four-wave mixing for wavelength division multiplexing systems [48–50]. These fibers have their own specific optical transmission system applications that take full advantage of their chromatic dispersion characteristics.

In ITU-T, there are six different kinds of categories for transmission single-mode fibers, which are mainly characterized by the operating spectral window, chromatic dispersion, and attenuation requirements, as well as macrobending performances:

(1) ITU-T Recommendation G.652.A, B, C, and D [20]
Characteristics of a single-mode optical fiber cable (SMF).

(2) ITU-T Recommendation G.653.A and B [32]
Characteristics of a dispersion-shifted single-mode optical fiber cable (DSF).

TABLE 2.7   Correspondence Between ITU-T and IEC Standard Fibers

IEC B1.1	↔	ITU-T G.652.A, B, C, D	↔	Conventional SMF
IEC B1.2	↔	ITU-T G.654.A, B, C	↔	Cutoff shifted SMF
IEC B1.3	↔	ITU-T G.652.C, D	↔	Low water peak SMF
IEC B2	↔	ITU-T G.653.A, B	↔	DSF
IEC B4	↔	ITU-T G.655.A, B, C, D, E	↔	NZ-DSF
IEC B5	↔	ITU-T G.656	↔	Wide band NZ-DSF
IEC B6	↔	ITU-T G.657	↔	Bend-insensitive SMF

(3) ITU-T Recommendation G.654.A, B, and C [33]

Characteristics of a cut-off shifted single-mode optical fiber and cable (cut-off shifted SMF).

(4) ITU-T Recommendation G.655.A, B, C, D, and E [34]

Characteristics of a non-zero-dispersion shifted single-mode optical fiber cable (NZ-DSF).

(5) ITU-T Recommendation G.656 [51]

Characteristics of a fiber and cable with non-zero-dispersion for wideband optical transport (wide band NZ-DSF).

(6) ITU-T Recommendation G.657.A and B [52]

Characteristics of a bending loss insensitive single mode optical fibre and cable for the access network (bend insensitive SMF).

In the IEC 60793-2-50, specification for class B single-mode fiber types [53] can be grouped with that corresponding to ITU-T standards as given in Table 2.7.

As the optical fiber fabrication technologies mature, the OH radical vibrational absorption near 1380 nm, commonly known as water peak, has been significantly reduced to result in a broad band of optical communication window ranging from 1260 to 1675 nm. ITU-T has defined six communication bands as a spectral standard as summarized in Table 2.8.

Most of single-mode fiber types developed until 1990s were designed to operate in O band and conventional band (C band). After advent of EDFA [46,47] the fiber design rapidly expanded the spectral window of operation to long band (L band), and then S band where optical gain was provided by modification of C-band EDFAs [54–57].

TABLE 2.8   ITU-T Spectral Bands in Optical Communications

Descriptor	O Band (Original)	E Band (Extended)	S Band (Short)	C Band (Conventional)	L Band (Long)	U Band (Ultralong)
λ_{min} (nm)	1260	1360	1460	1530	1565	1625
λ_{min} (nm)	1360	1460	1530	1565	1625	1675
Bandwidth (nm)	100	100	70	35	60	50

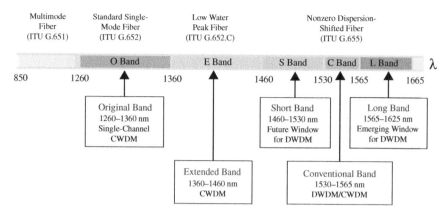

FIGURE 2.26   Optical transmission bands and corresponding fiber standards.

Recent mass production of low water peak fibers has opened the E band [58,59]. U band is also proposed to be utilized for system monitoring signals and actual data channels [60,61]. In Figure 2.26 these optical transmission windows and corresponding fiber standards are summarized.

There are various specifications in optical fiber standards such as geometrical dimension, mechanical reliability, and environmental requirements as well as optical attributes. Those specifications are important and essential in practical deployment and commercial markets, yet they are beyond the scope of this book. Only the optical attributes will be discussed to keep the content of this book self-contained. The readers are requested to consult the relevant journal articles and books for these subjects [62–64].

Major optical attributes of conventional SMF, DSF, and NZ-DSF in the ITU-T standards are summarized and compared in Tables 2.9–2.11. Optical fibers defined in ITU-T G.652.A represent the most commonly used conventional SMFs such as Corning SM-28, where the chromatic dispersion is zero near 1310 nm while the loss shows the minimum near 1550 nm. G.652.B is an upgrade of G.652.A in terms of PMD and attenuation. As shown in Table 2.9, G.652.B has a lower PMD_Q of 0.2 ps/$\sqrt{km}$ compared with 0.5 ps/$\sqrt{km}$ in G.652.A. Attenuation coefficients in G.652.B are also reduced to 0.4 and 0.35 dB/km at 1310 and 1550 nm, respectively. Furthermore, G.652.B defines the attenuation coefficient at 1625 nm to be the equal level as in 1310 nm in order to facilitate the usage of L band, 1565–1625 nm. These enhancements in PMD and attenuation attributes will make the SMF in G.652.B type more suitable for higher data rate and longer link distance communications. The detailed relations between the PMD_Q, DGD_{max}, and the corresponding link length and channel bit rate are summarized in Table 2.12 and will be discussed later.

Compared with standard G.652.A and B, an additional communication window in E band is included in G.652.C and D by eliminating or significantly reducing OH absorption peak near 1383 nm, which is commonly referred as water peak. Accordingly the attenuation coefficient specification at 1383 ± 3 nm has been added in the standards as in Table 2.9 to specify the attenuation level in E band, whose value is

TABLE 2.9 Optical Attributes in ITU-T Standards for Conventional Single-Mode Fibers

Attributes	Details	Standard SMF G.652.A	Low PMD SMF G.652.B	Low OH SMF G.652.C	Low OH, PMD SMF G.652.D	Cutoff Shifted SMF G.654.A	Cutoff Shifted SMF G.654.B	Bend-Insensitive SMF G.657.A	Bend-Insensitive SMF G.657.B
Fiber									
Mode field diameter	Wavelength (nm)	1310	1310	1310	1310	1550	1550	1310	1310
	Nominal values (µm)	8.6–9.5	8.6–9.5	8.6–9.5	8.6–9.5	9.5–10.5	9.5–13.0	8.6–9.5	6.3–9.5
	Tolerance (µm)	±0.6	±0.6	±0.6	±0.6	±0.7	±0.7	±0.4	±0.4
Chromatic dispersion coefficient	λ_{0min} (nm)	1300	1300	1300	1300	$D_{1550max}=20$	$D_{1550max}=22$	1300	1300
	λ_{0max} (nm)	1324	1324	1324	1324	–	–	1324	1420
	S_{0max} (ps/(nm²km))	0.092	0.092	0.092	0.092	$S_{1550max}=0.07$	$S_{1550max}=0.07$	0.092	0.1
Macrobend loss	Radius (mm)	30	30	30	30	30	30	15 10	10 7.5
	Number of turns	100	100	100	100	100	100	10 1	10 1
	Maximum at 1550 nm (dB)	0.1	0.1	0.1	0.1	0.5	0.5	0.25 0.75	0.03 0.5
	Maximum at 1625 nm (dB)	–	–	–	–	–	–	1.0 1.5	0.1 1.0

TABLE 2.9 Optical Attributes in ITU-T Standards for Conventional Single-Mode Fibers

Attributes	Details	Standard SMF G.652.A	Low PMD SMF G.652.B	Low OH SMF G.652.C	Low OH, PMD SMF G.652.D	Cutoff Shifted SMF G.654.A	Cutoff Shifted SMF G.654.B	Bend-Insensitive SMF G.657.A	Bend-Insensitive SMF G.657.B
Cable									
Cutoff wavelength	Maximum (nm)	1260	1260	1260	1260	1530	1530	1260	1260
Attenuation coefficient (dB/km)	Maximum at 1310 nm	0.5	0.4	0.4	0.4	–	–	0.4	0.5
	Maximum at 1383 ± 3 nm[a]	–	–	0.4	0.4	–	–	0.4	–
	Maximum at 1550 nm	0.4	0.35	0.3	0.3	0.22	0.22	0.3	0.3
	Maximum at 1625 nm	–	0.4	0.4	0.4	–	–	0.4	0.4
PMD coefficient	M (number of cables)	20	20	20	20	20	20	20	–
	Q (%)	0.01	0.01	0.01	0.01	0.01	0.01	0.01	–
	Maximum PMD_Q (ps/$\sqrt{km}$)	0.5	0.2	0.5	0.2	0.5	0.2	0.2	–

Note: $D_{1550max}$ and $S_{1550max}$ are the maximum chromatic dispersion and dispersion slope at $l = 1550$ nm and given in the unit of ps/(nm km) and ps/(nm² km), respectively.
[a] The sampled attenuation average at this wavelength shall be less than or equal to the maximum value specified for the range 1310–1625 nm after hydrogen aging according to [b-IEC 60793-2-50] regarding the B1.3 fiber category.

equivalent to those of O band at 1310 nm and the L band at 1625 nm. Note that this OH-free SMF will provide the widest bandwidths of optical communication windows, O, E, S, C, and L bands in silica fibers. Requirement in a lower PMD_Q of 0.2 ps/$\sqrt{km}$ is added in G.652.D to enhance the high bit rate capacity as is in the case for G.652.B.

Optical fibers in ITU-T G.653, G655, and G656 are dispersion-shifted fibers, whose zero-dispersion wavelength is shifted from 1310 nm to the spectral region near 1550 nm, where the optical loss is minimum and EDFA gain is available. Therefore, their MFDs are referenced to $\lambda = 1550$ nm while the conventional SMFs in G.652 type are at 1310 nm.

DSF was initially developed to make zero-dispersion wavelength coincide with the spectral position where the minimum attenuation is obtained, in order to facilitate low loss and low dispersion simultaneously for high capacity transmission. ITU-T G.653. A fibers in Table 2.10 are a typical example for early DSFs and they have zero-dispersion wavelength, λ_0, in the range of 1500–1600 nm along with the positive dispersion slope at λ_0 given by $S_0 \leq 0.085$ ps/(nm^2 km). The LP_{11} mode cutoff is set to 1270 nm, which is longer than that of G.652 fibers by 10 nm. Macrobend loss limit is increased to 0.5 dB for 30 m bending radius and 100 turns, in comparison to 0.1 dB of G.652 fibers. The maximum attenuation coefficient of 0.5 dB/km at 1550 nm is also higher than 0.3–0.4 dB/km in G.652 fibers. Typical range of chromatic dispersion of the IUT-T G.653 fibers over S + C + L bands is shown in Figure 2.27. In the spectral range of 1525 nm $< \lambda <$ 1575 nm, the maximum magnitude of chromatic dispersion is 3.5 ps/(nm km) and the zero-dispersion wavelength should lie between 1500 and 1600 nm.

As wavelength division multiplexing (WDM) techniques were rapidly and widely adopted in optical communications, zero dispersion at 1550 nm has been pinpointed as a most detrimental penalty, rather than ultimate merit, due to nonlinear optical interactions among WDM channels such as four-wave mixing (FWM) and cross-phase modulation (XPM). These nonlinear interactions dramatically increase in the wavelength conversion efficiency when the zero-dispersion wavelength coincides with WDM channels [44,48]. In order to balance the trade-off between the linear ISI penalty at a high chromatic dispersion value and the nonlinear penalties at zero dispersion, a new type of dispersion-controlled fiber, NZ-DSF, has been invented especially for long-haul WDM applications by allowing optimally small chromatic dispersion in the operating windows [65–67].

Specifications of NZ-DSFs defined in ITU-T G.655 [34] and G.656 [51] are listed in Tables 2.10 and 2.11. There are six different NZ-DSF types: G.655.A, B, C, D, and E, and G.656. All of these NZ-DSFs have the LP_{11} mode cutoff at 1450 nm so that single-mode optical transmission is not allowed in either O or E band.

ITU-T G.655.A is the earliest version of NZ-DSF dedicated to C band of operation, where the chromatic dispersion lies in the range of 0.1–6.0 ps/(nm·km) for the spectral range of 1530–1565 nm. The attenuation coefficient is defined only for C band and is lower than 0.35 dB/km at 1550 nm. PMD_Q and DGD_{max} are the same as G.652.A, the conventional SMF.

Operation windows extend to cover both C and L bands in G.655.B and C so that the attenuation coefficients are defined to be lower than 0.4 dB/km at 1625 nm as well

TABLE 2.10 Optical Attributes in ITU-T Standards for Dispersion-Shifted and Nonzero Dispersion-Shifted Single-Mode Fibers—1

Attribute	Detail	C-Band DSF G.653.A	C-Band NZ-DSF G.655.A	C + L–Band NZ-DSF G.655.B[a]	Low PMD C + L–Band NZ-DSF G.655.C[a]
Fiber					
Mode field diameter	Wavelength (nm)	1550	1550	1550	1550
	Nominal values (μm)	7.8–8.5	8–11	8–11	8–11
	Tolerance (μm)	±0.8	±0.7	±0.7	±0.7
Chromatic dispersion coefficient	λ_{min} (nm)	1525	1530	1530	1530
	λ_{max} (nm)	1575	1565	1565	1565
	D_{min} (ps/(nm km))	–	0.1	1.0	1.0
	D_{max} (ps/(nm km))	3.5	6.0	10.0	10.0
	$D_{max} - D_{min}$ (ps/(nm km))	–	–	≤5.0	≤5.0
	λ_{0min} (nm)	1500	–	–	–
	λ_{0max} (nm)	1600	–	–	–
	S_{0max} (ps/(nm^2 km))	0.085	–	–	–
Macrobend loss	Radius (mm)	30	30	30	30
	Number of turns	100	100	100	100
	Maximum at 1550 nm (dB)	0.5	0.5	0.5	0.5
Cable					
Cutoff wavelength	Maximum (nm)	1270	1450	1450	1450
Attenuation coefficient (dB/km)	Maximum at 1550 nm	0.35	0.35	0.35	0.35
	Maximum at 1625 nm	–	–	0.4	0.4
PMD coefficient	M (number of cables)	20	20	20	20
	Q (%)	0.01	0.01	0.01	0.01
	Maximum PMD_Q (ps/$\sqrt{km}$)	0.5	0.5	0.5	0.2

[a]The dispersion coefficients for 1565–1625 nm are being discussed.

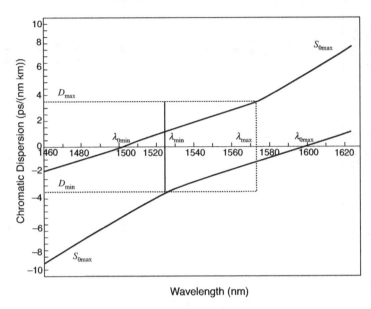

FIGURE 2.27 Chromatic dispersion boundaries for ITU-T G.653 fibers [33].

as 0.35 dB/km at 1550 nm. The permitting chromatic dispersion value is also increased to adopt even faster data rate and narrower channel spacing so that it lies in the range of 1–10 ps/(nm km). It is noted, however, that the difference between D_{max} and D_{min}, $D_{max} - D_{min}$, is set to be lower than 5 ps/(nm km), which indicates the newer versions require a shallow chromatic dispersion slope in the operating band. Typical dispersion values of G.655.B and C are listed in Table 2.13. PMD_Q in G.655.C is reduced to 0.2 ps/$\sqrt{km}$, for further transmission capacity increase.

Dispersion slope and maximum dispersion values have been further adjusted in G.655.D so that both C and L bands could be utilized [34]. The window has been further expanded to $S + C + L$ bands in G.655.E [34] and G.656 [51] to provide a positive nonzero dispersion from 1460 to 1625 nm. The dispersion boundaries of these NZ-DSFs are summarized in Figure 2.28. It is noted that by expanding operation windows to $S + C + L$ bands, the dispersion at the long wavelength boundary increased to ~14 ps/(nm km) in G.655.E and G.656, which will significantly reduce the FWM among WDM channels.

Cutoff shifted fibers, ITU-T G.654.A and B [33], primarily distinguish themselves from G.652 conventional SMFs in the LP_{11} mode cutoff wavelength, whose maximum is given at 1530 nm. This long cutoff wavelength will ensure a lower bending loss in the cabling process and deployment of fiber cables in C bands. In order to dedicate to the C-band communications, the attenuation coefficient of G.654 shows the tightest standard less than 0.22 dB/km at $\lambda = 1550$ nm among all fiber standards. Requirements for chromatic dispersion and PMD are expressed in terms of maximum dispersion (ps/(nm km)) and slope (ps/(nm^2 km)) at $\lambda = 1550$ nm, represented as $D_{1550max}$ and $S_{1550max}$, respectively, in Table 2.9, which are close to values in G.652.A. G.654.B has a smaller PMD_Q of 0.2 ps/$\sqrt{km}$ to ensure a higher capacity.

TABLE 2.11 Optical Attributes in ITU-T Standards for Dispersion-Shifted and Nonzero Dispersion-Shifted Single-Mode Fibers—2

Attribute	Detail	Low PMD C-Band DSF G.653B	Low PMD C+L-Band NZ-DSF G.655.D	Low PMD Low Slope, C+L-Band NZ-DSF G.655.E	Low PMD S+C+L-B and NZ-DSF G.656
Fiber					
Mode field diameter	Wavelength (nm)	1550	1550	1550	1550
	Nominal values (μm)	7.8–8.5	8–11	8–11	8–11
	Tolerance (μm)	±0.6	±0.6	±0.6	±0.7
Chromatic dispersion coefficient (ps/(nm km))	$D_{min}(\lambda)$: 1460–1525 nm	$0.085 \times (\lambda - 1525) - 3.5$	–	–	–
	$D_{min}(\lambda)$: 1525–1625 nm	$\frac{3.5}{75} \times (\lambda - 1600)$			
	$D_{min}(\lambda)$: 1460–1550 nm	–	$\frac{7}{90} \times (\lambda - 1460) - 4.2$	$\frac{5.42}{90} \times (\lambda - 1460) + 0.64$	$\frac{2.6}{90} \times (\lambda - 1460) + 1.0$
	$D_{min}(\lambda)$: 1550–1625 nm		$\frac{2.97}{75} \times (\lambda - 1550) + 2.8$	$\frac{3.3}{75} \times (\lambda - 1550) + 6.06$	$\frac{0.98}{75} \times (\lambda - 1550) + 3.6$
	$D_{max}(\lambda)$: 1460–1575 nm	$\frac{3.5}{75} \times (\lambda - 1550)$	–	–	–
	$D_{max}(\lambda)$: 1575–1625 nm	$0.085 \times (\lambda - 1575) + 3.5$			
	$D_{max}(\lambda)$: 1460–1550 nm	–	$\frac{2.91}{90} \times (\lambda - 1460) + 3.29$	$\frac{4.65}{90} \times (\lambda - 1460) + 4.66$	$\frac{4.68}{90} \times (\lambda - 1460) + 4.6$
	$D_{max}(\lambda)$: 1550–1625 nm		$\frac{5.06}{75} \times (\lambda - 1550) + 6.2$	$\frac{4.12}{75} \times (\lambda - 1550) + 9.31$	$\frac{4.72}{75} \times (\lambda - 1550) + 9.28$
Macrobend loss	Radius (mm)	30	30	30	30
	Number of turns	100	100	100	100
	Maximum at 1550 nm (dB)	0.1	0.1	0.1	0.5

(continued)

TABLE 2.11 (*Continued*)

Attribute	Detail	Low PMD C-Band DSF G.653B	Low PMD C + L–Band NZ-DSF G.655.D	Low PMD Low Slope, C + L–Band NZ-DSF G.655.E	Low PMD S + C + L–B and NZ-DSF G.656
Cable					
Cutoff wavelength	Maximum (nm)	1270	1450	1450	1450
Attenuation coefficient (dB/km)	Maximum at 1460 nm	–	–	–	0.4
	Maximum at 1550 nm	0.35	0.35	0.35	0.35
	Maximum at 1625 nm	–	0.4	0.4	0.4
PMD coefficient	M (number of cables)	20	20	20	20
	Q (%)	0.01	0.01	0.01	0.01
	Maximum PMD_Q (ps/$\sqrt{km}$)	0.2	0.2	0.2	0.2

TABLE 2.12 Relation among PMD$_Q$ and Link Length versus Allowed DGD$_{max}$ and Channel Bit Rate

Maximum PMD$_Q$ (ps/√km)	Link Length (km)	Implied Fiber Induced Maximum DGD (ps)	Channel Bit Rates (Gbit/s)	SDH/SONET	Bit Period (ps)	Tolerance (ps)
No specification	–	–	Up to 2.5	Up to STM-16/OC-48	401.9	40
0.5	400	25.0	10	STM-64/OC-192	100.5	10
	40	19.0	10	STM-64/OC-192	100.5	10
	2	7.5	40	STM-256/OC-768	25.12	6.25
0.20	3000	19.0	10	STM-64/OC-192	100.5	10
	80	7.0	40	STM-256/OC-768	25.12	6.25
0.10	>4000	12.0	10	STM-64/OC-192	100.5	10
	400	5.0	40	STM-256/OC-768	25.12	6.25

Note: Cable section length is 10 km except for the 0.10 ps/√km, >4000 km link, where it is set to 25 km; the probability level is 6.5×10^{-8}.

TABLE 2.13   Commercially Available Fibers Corresponding to ITU-T G.655. B and C [34]

Sample ID	D_{min} (ps/(nm km))	D_{max} (ps/(nm km))	Sign	Typical Dispersion Coefficient at 1550 nm (ps/(nm km))	Typical Dispersion Slope at 1550 nm (ps/(nm² km))
A	1.3	5.8	+	3.7	0.070
B	2.0	6.0	+	4.2	0.085
C	2.6	6.0	+	4.4	0.045
D	5.0	10.0	+	8.0	0.058
E	1.0	6.0	−	−2.3	0.065

Bend-insensitive fiber, ITU-T G.657 [52], is the latest addition in single-mode optical fiber standards to provide nominal specification for fiber to the home (FTTH) markets. In the FTTH environment, relatively short length of drop cables is used to connect to customer premises and this inevitably introduces macrobending of optical fiber with small radius, which results in large bending loss in conventional SMFs. Bend-insensitive single-mode fibers are attractive for FTTH to lower the installation costs and improve system performance [67]. Recently various techniques have been proposed to reduce the bending loss such as reducing the MFD [20], depressed cladding [68], a lower refractive index concentric trench [69,70], hole-assisted fiber designs [71,72], and adding a nanostructured ring layer in the cladding [73]. Figure 2.29 shows schematics of these fiber cross-sections. Among them, hole-assisted fibers offer superior bending performance but they are not fully compliant with recommendations in ITU-T G.652. Other than hole-assisted fibers, fibers with a nanostructured ring layer provide the lowest bending loss as indicated in Figure 2.30. This type of fiber also fully complies with recommendations in G.652 bending.

In the specification of G.657.A and B listed in Table 2.9, it is noted that bending radii are decreased to 15, 10, and 7.5 mm and the corresponding macrobend loss is significantly reduced to represent FTTH installation environment. Especially in the case of G.657.B the mode field diameter is reduced to 6.3–9.5 mm, and for a tight bend loss of 7.5 mm radius the maximum allowed loss is 0.5 and 1.0 dB at 1550 and 1625 nm, respectively.

Among the optical attributes in Tables 2.9–2.11, the PMD coefficients will provide information on maximum transmission data rate. The allowed WDM channel bit rate for PMD_Q values are tabulated in Table 2.12 for various link distances. For a given PMD_Q and a link length L, the maximum induced DGD_{max} can be calculated by

$$DGD_{max}\,[ps] = PMD_Q\left[ps/\sqrt{km}\right] \times \sqrt{L\,[km]} \qquad (2.157)$$

and this DGD should be less than the bit period. For example, a PMD_Q of 0.5 ps/$\sqrt{km}$ as in G.652.A in Table 2.9 induces DGD_{max} of 25 and 19.0 ps for fiber link length of 400 and 40 km, respectively, which will permit 10 Gbit/s channel bit rates. But for a

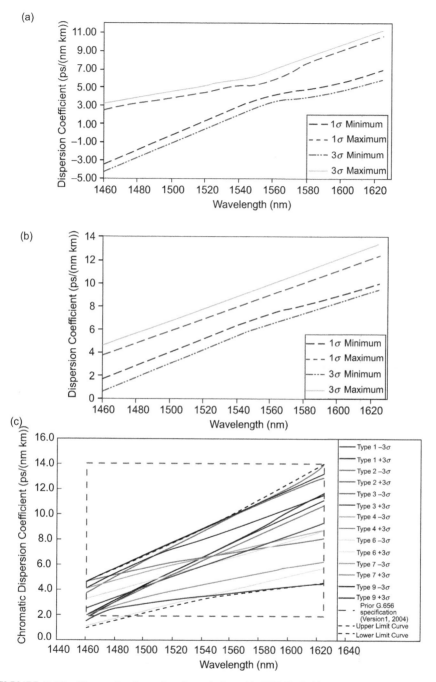

FIGURE 2.28 Chromatic dispersion boundaries: (a) ITU-T G.655.D, (b) G.655.E [34], and (c) G.656 fibers [51].

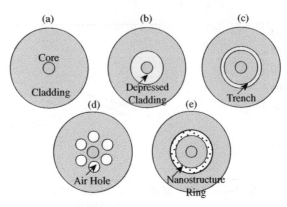

FIGURE 2.29   Fiber designs for reducing bending loss: (a) reduced MFD design, (b) depressed cladding design, (c) trench design, (d) hole-assisted design, and (e) nanostructure ring design [73].

higher data rate 40 Gbit/s, the allowed link distance for this PMD_Q of 0.5 ps/$\sqrt{km}$ is only 2 km. In order to provide sufficiently long link for 40 Gbit/s data rate, the PMD_Q should be reduced. At the data rate of 40 Gbit/s, fibers with PMD_Q of 0.2 ps/$\sqrt{km}$, as in G.652.B, for instance, can allow 80 km link distance. Therefore, in Tables 2.9–2.11, fiber with a lower PMD_Q will support higher data rate.

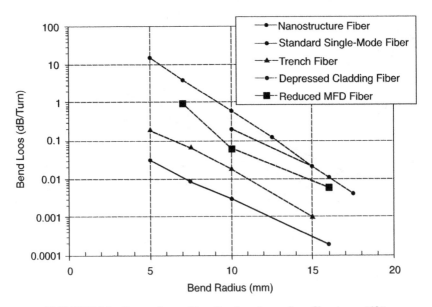

FIGURE 2.30   Comparison of bending loss for various fiber types [42].

REFERENCES

[1] T. Okoshi, "Optical fibers," Academic Press, 1982.

[2] P.-A. Bélanger, "Optical fiber theory – a supplement to applied electromagnetism," World Scientific Publishing, 1993.

[3] G. B. Arfken and H.-J. Weber, "Mathematical methods for physicists," Harcourt/ Academic Press, 2001.

[4] D. Gloge, "Weakly guiding fibers," Applied Optics, vol. 10, no. 10, pp. 2252–2258, 1971.

[5] D. Marcuse, "Theory of dielectric optical waveguides," Academic Press, 1974.

[6] A. W. Snyder and J. D. Love, "Optical waveguide theory," Kluwer Academic Publishers, 1983.

[7] L. B. Jeunhomme, "Single mode fiber optics," Marcel Dekker, Inc., 1983.

[8] J. A. Buck, "Fundamentals of optical fibers," Wiley Series in Pure and Applied Optics, John Wiley and Sons, 2004.

[9] J. Fleming, "Material dispersion in lightguide glasses," Electronics Letters, vol. 14, no. 11, pp. 326–328, 1978.

[10] P. C. Schultz, "Ultraviolet absorption of titanium and germanium in fused silica," Proceedings of 11th International Congress on Glass, vol. 3, pp. 155–158, 1977.

[11] T. Izawa, N. Shibata, S. Sudo, and F. Hanawa, "Optical attenuation in pure and doped fused silica in the IR wavelength region," Applied Physics Letter, vol. 31, no. 1, pp. 33–35, 1977.

[12] H. Bach and N. Neuroth, "The optical properties of optical glass," Springer, 1995.

[13] I. H. Malitson, "Interspecimen comparison of the refractive index of fused silica," Journal of the Optical Society of America, vol. 55, no. 10, pp. 1205–1209, 1965.

[14] J. W. Fleming, "Dispersion in GeO_2–SiO_2 glasses," Applied Optics, vol. 23, no. 24, pp. 4486–4493, 1984.

[15] S. Kobayashi, S. Shibata, N. Shibata, and T. Izawa, "Refractive-index dispersion of doped fused silica," International Conference on Integrated Optics and Optical Fibre Communication, Technical Digest, pp. 309–310, 1977, paper B8-3.

[16] C. R. Hammond and S. R. Norman, "Silica-based binary glass systems: refractive index and composition in optical fibers," Optical and Quantum Electronics, vol. 9, no. 5, pp. 399–409, 1977.

[17] J. W. Fleming and D. L. Wood, "Refractive index dispersion and related properties in fluorine doped silica," Applied Optics, vol. 22, no. 19, pp. 3102–3104, 1983.

[18] J. W. Fleming, "Material and mode dispersion in GeO_2.B_2O_3–SiO_2 glasses," Journal of American Ceramic Society, vol. 59, pp. 503–507, 1976.

[19] B. G. Bagley, C. R. Krkjian, J. W. Mitchell, G. E. Peterson, and A. R. Tynes, "Material, properties, and choices," Chapter 7, Optical Fiber Telecommunications, S. E. Miller and A. G. Chynoweth (eds), Academic Press, 1979.

[20] ITU-T G.652, "Characteristics of a single-mode optical fiber cable," 2009.

[21] IEC 60793-1-1, "Optical fibres – part 1-1: generic specification – section 1: general and guidance," 2008.

[22] ITU-T G.650.1, "Definition and test methods for linear, deterministic attributes of single-mode fibre and cable," 2004.

[23] Y. Namihira, "Relationship between non-linear effective area and mode field diameter for dispersion shifted fibers," Electronics Letters, vol. 30, no. 3, pp. 262–263, 1994.

[24] S. Nagel, "Fiber materials and fabrication method," Chapter 4, Optical Fiber Telecommunication II, S. E. Miller and I. P. Kaminow (eds), Academic Press, 1988.

[25] T. Moriyama, O. Fukuda, K. Sanada, K. Inada, T. Edahiro, and K. Chida, "Ultimately low OH content VAD optical fibres," Electronics Letters, vol. 16, no. 18, pp. 698–699, 1980.

[26] K. Saito and A. J. Ikushima, "Absorption edge in silica glass," Physical Review B, vol. 62, no. 13, pp. 8584–8587, 2000.

[27] T. Miya, Y. Terunuma, T. Hosaka, and T. Miyashita, "Ultimate low-loss single-mode fibre at 1.55 mm," Electronics Letters, vol. 15, no. 4, pp. 106–109, 1979.

[28] H. Osanai, T. Shioda, T. Moriyama, S. Araki, M. Horiguchi, T. Izawa, and H. Takata, "Effect of dopants on transmission loss of low-OH content optical fibers," Electronics Letters, vol. 12, pp. 549–540, 1976.

[29] T. Miya, Y. Terunuma, T. Hosaka, and T. Miyashita, "Ultra low-loss single-mode fibre at 1.55 mm," Review of Electrical Communication Laboratory, vol. 27, no. 7–8 pp. 497–505, 1979.

[30] D. B. Keck, R. D. Maurer, and P. C. Schultz, "On the ultimate lower limit of attenuation on glass optical waveguides," Applied Physics Letters, vol. 22, no. 7, pp. 307–309, 1973.

[31] P. Kaiser, "Spectral losses of unclad fibers made from high-grade vitreous silica," Applied Physics Letters, vol. 23, no. 1, pp. 45–47, 1973.

[32] ITU-T G.653, "Characteristics of a dispersion-shifted single-mode optical fibre and cable," 2010.

[33] ITU-T G.654, "Characteristics of a cut-off shifted single-mode optical fibre and cable," 2010.

[34] ITU-T G.655, "Characteristics of a non-zero dispersion-shifted single-mode optical fibre and cable," 2009.

[35] ITU-T G.651.1, "Characteristics of a 50/125 μm multimode graded index optical fibre cable for the optical access network," 2007.

[36] C. D. Poole and R. E. Wagner, "Phenomenological approach to polarization dispersion in long single-mode fibres," Electronics Letters, vol. 22, no. 19, pp. 1029–1030, 1986.

[37] B. L. Heffner, "Automated measurement of polarization mode dispersion using Jones matrix eigenanalysis," Photonics Technology Letters, vol. 4, no. 9, pp. 1066–1068, 1992.

[38] F. Curti, B. Diano, G. de Marchis, and F. Matera, "Statistical treatment of the evolution of the principal states of polarization on a low-birefringence terrestrial fiber cable," Journal of Lightwave Technology, vol. 8, no. 8, pp. 1162–1166, 1990.

[39] ITU-T G650.2, "Definitions and test methods for statistical and non-linear related attributes of single mode fibre and cable," 2007.

[40] N. Gisin, "Solutions of the dynamical equation for polarization dispersion," Optics Communications, vol. 86, no. 5, pp. 371–373, 1991.

[41] G. J. Foschini and C. D. Poole, "Statistical theory of polarization dispersion in single mode fibers," Journal of Lightwave Technology, vol. 9, no. 11, pp. 1439–1456, 1991.

[42] M. J. Lee and D. A. Nolan, "Optical transmission fiber design evolution," Journal of Lightwave Technology, vol. 26, no. 9, pp. 1079–1092, 2008.

[43] T. Croft, J. Ritter, and V. A. Bhagavatula, "Low-loss dispersion-shifted single mode fiber manufactured by the OVD process," Journal of Lightwave Technology, vol. LT-3, no. 5, pp. 931–934, 1985.

[44] K. Inoue, "Four-wave mixing in an optical fiber in the zero-dispersion wavelength region," Journal of Lightwave Technology, vol. 10, no. 10, pp. 1553–1561, 1992.

[45] M. J. Holmes, D. L. Williams, and R. J. Manning, "Highly nonlinear optical fiber for all optical processing applications," Photonics Technology Letters, vol. 7, no. 9, pp. 1045–1047, 1995.

[46] E. Desurvier, "Erbium-doped fiber amplifiers: principles and applications," Wiley-Interscience, 2002.

[47] P. C. Becker, N. A. Olsson, and J. R. Simpson, "Erbium doped fiber amplifiers: fundamentals and technology," Academic Press, 1999.

[48] A. R. Chraplyvy, "Limitations on lightwave communications imposed by optical-fiber nonlinearities," Journal of Lightwave Technologies, vol. 8, no. 10, pp. 1548–1557, 1990.

[49] M. Eiselt, L. D. Garrett, and R. W. Tkch, "Experimental comparison of WDM system capacity in conventional and nonzero dispersion shifted fiber," Photonics Technology Letters, vol. 11, no. 2, pp. 281–283, 1999.

[50] B. Mukherjee, "Optical WDM networks," Springer, 2006.

[51] ITU-T G.656, "Characteristics of a fibre and cable with non-zero dispersion for wideband optical transport," 2010.

[52] ITU-T G.657, "Characteristics of a bending loss insensitive single mode optical fibre and cable for the access network," 2006.

[53] IEC 60793-2-50, "Optical fibres – part 2-50: product specifications – sectional specification for class B single mode fibres." 2008.

[54] J. F. Massicott, J. R. Armitage, R. Wyatt, B. J. Ainslie, and S. P. Craig-Ryan, "High gain, broadband, 1.6 μm Er^{3+} doped silica fibre amplifier," Electronics Letters, vol. 26, no. 20, pp. 1645–1646, 1990.

[55] H. Ono, M. Yamada, T. Kanamori, S. Sudo, and Y. Ohishi, "1.58-μm band gain-flattened erbium-doped fiber amplifiers for WDM transmission systems," Journal of Lightwave Technology, vol. 17, no. 3, pp. 490–496, 1999.

[56] M. A. Arbore, Y. Zhou, G. Keaton, and T. Kane, "36 dB gain in S-band EDFA with distributed ASE suppression," Optical Amplifiers and their Applications, J. Nagel, S. Namiki, and L. Spiekman (eds), OSA Trends in Optics and Photonics Series, vol. 77, Optical Society of America, 2002, paper PD4.

[57] S. W. Harun, K. Dimyati, K. K. Jayaplan, and H. Ahmad, "An overview on S-band erbium doped fiber amplifiers," Laser Physics Letters, vol. 4, no. 4, pp. 10–15, 2007.

[58] A. Gordon, B. I. Shraiman, P. F. Glodis, and M. J. Stephen, "Towards the clarity limit in optical fibre," Nature, vol. 404, pp. 262–264, 2007.

[59] F. Hanawa, S. Sudo, M. Kawachi, and M. Nakahara, "Fabrication of completely OH-free VAD fibre," Electronics Letters, vol. 16, no. 18, pp. 699–700, 1980.

[60] N. Nakao, H. Izumita, T. Inoue, Y. Enomoto, N. Araki, and N. Tomita, "Maintenance method using 1650-nm wavelength band for optical fiber cable networks," Journal of Lightwave Technology, vol. 19, no. 10, pp. 1513–1520, 2001.

[61] N. Honda, H. Izumita, and M. Nakamura, "Spectral filtering criteria for U-band test light for in-service line monitoring in optical fiber networks," Journal of Lightwave Technology, vol. 24, no. 6, pp. 2328–2335, 2006.

[62] C. R. Kurkjian, J. T. Krause, and M. J. Matthewson, "Strength and fatigue of silica optical fibers," Journal of Lightwave Technology, vol. 7, no. 9, pp. 1360–1370, 1989.

[63] D. K. Paual,ed., "Fiber optics reliability and testing," Critical Reviews of Optical Science and Technology, vol. CR50, SPIE Optical Engineering Press, 1994.

[64] M. Gdonna and T. Volotinen (eds), "Reliability of optical fibres and components: final report of COST 246," Springer Verlag, 1999.

[65] OFS, TrueWave® REACH, RS fiber, http://www.ofsoptics.com/fiber/category.php? txtCategoryID=1022260265447.

[66] Corning, LEAF® fiber, http://www.corning.com/opticalfiber/products/LEAF_fiber.aspx.

[67] R. E. Wagner, J. R. Igel, R. Whitman, M. D. Vaughn, A. B. Ruffin, and S. Bickham, "Fiber-based broadband-access deployment in the United States," Journal of Lightwave Technology, vol. 24, no. 12, pp. 4526–4540, 2006.

[68] F. Wu, et al., "A new G.652D, zero water peak fiber optimized for low bend sensitivity in access networks," IWCS, Providence, RI, November 12-15, 2006.

[69] S. Matsuo, M. Ikeda, and K. Himeno, "Bend-insensitive and low-splice-loss optical fiber for indoor wiring in FTTH," Optical Fiber Communication Conference, Technical Digest (CD), Optical Society of America, 2004, paper ThI3.

[70] S. Matsuo, T. Nunome, T. Yoshida, T. Hamada, and K. Himeno, "Design optimization of trench index profile for the same dispersion characteristics with SMF," Optical Fiber Communication Conference and Exposition and the National Fiber Optic Engineers Conference, OSA Technical Digest Series (CD), Optical Society of America, 2007, paper JWA2.

[71] H. Shinohara, "Broadband access in Japan: rapidly growing FTTH market," IEEE Communication Magazine, vol. 43, no. 9, pp. 72–78, 2005.

[72] Y. Bing, K. Ohsono, Y. Kurosawa, T. Kumagai, and M. Tachikura, "Low-loss holey fiber," Hitachi Cable Review, no. 24, pp. 1–5, 2005.

[73] M.-J. Li, P. Tandon, D. C. Bookbinder, S. R. Bickham, M. A. McDermott, R. B. Desorcie, D. A. Nolan, J. J. Johnson, K. A. Lewis, and J. J. Englebert, "Ultra-low bending loss single-mode fiber for FTTH," Journal of Lightwave Technology, vol. 27, no. 3, pp. 376–382, 2009.

CHAPTER THREE

Preform Fabrication and Optical Fiber Drawing Process

3.1 PREFORM FABRICATION BASED ON CHEMICAL VAPOR DEPOSITION PROCESS

In order to achieve the desired optical properties in optical fibers, extensive research and development efforts have been continuously dedicated to provide accurate and reproducible fabrication processes. Fabrication process for optical fibers can be divided in large by two separate processes:

(1) Preform fabrication and postprocessing
(2) Fiber drawing

"Preform" is the terminology conventionally used among optical fiber manufacturing industries and it is a glass rod where the optical fiber is drawn from. The preform can be regarded as a large-scale optical fiber with the same refractive index and waveguide structure as the drawn optical fiber. Usually optical fiber preform has the diameter in the range of 20–300 mm and length of 1 m while the conventional optical fiber has the diameter of 125 μm. By the mass conservation law, the lengths of optical fiber that can be drawn from one preform will depend on its volume, or equivalently its diameter and length, and there have been significant achievements in producing large volume optical fiber preform. We will briefly review the fundamental aspects of preform fabrication in terms of chemical process and dopant types.

Silica Optical Fiber Technology for Devices and Components: Design, Fabrication, and International Standards, First Edition. By Kyunghwan Oh and Un-Chul Paek.
© 2012 John Wiley & Sons, Inc. Published 2012 by John Wiley & Sons, Inc.

In the fiber drawing process, glass fiber with a certain diameter, usually 125 μm, is continuously drawn from the neck-down zone of the preform that is formed by a high temperature furnace. As the bare glass fiber cools down, protective coatings are applied over its surface to provide mechanical strength in the final optical fiber. In order to cope with mass production and high uniformity in the final optical fibers, high speed and high strength drawing techniques have been developed. Even though most of the guiding properties of optical fibers are dominantly affected by the optical preform designs, geometrical, mechanical, and some of optical properties can be significantly affected in the drawing process. We will discuss the key natures of optical fiber drawing technologies with an emphasis on mechanical aspects, and review the types of coating material along with their applications in the next section.

3.1.1 Principles of Chemical Vapor Deposition for Optical Fiber Preform Fabrication

The fundamental principle of chemical vapor deposition (CVD) lies on the fact that highly selective vaporization of a certain precursor to form a high purity material can be realized if there exists significant difference in the vapor pressure between the desired precursor and the rest [1]. In optical fiber preform fabrication development, it was a natural route to adopt CVD to ensure the high purity in silica glass and subsequently long transmission distance for optical communication applications. Recipes to make high purity silica using various precursors have been known long before Kao and Hockham's seminal proposal of silica optical fiber [2]. In the early days, one of crucial issues in preform fabrication process was the impact of transition metals over the transmission quality. In Table 3.1 [3], the attenuation induced by transition metal impurities is summarized, where very low level, part per billion (ppb), of them results in significantly high loss in wide spectral ranges. Therefore, it was imperative to reduce the trace of transition metals in glass-forming process. Among known precursors of silica glass or vitreous SiO_2, it was soon found out that chlorides would benefit CVD process for impurity control in atmospheric

TABLE 3.1 Optical Loss Induced by Transition Metal Impurities in Silica Optical Fibers

Impurity	Peak Absorption Wavelength (nm)	Optical Loss (dB/km) for One Part Per 10^9
Cr^{3+}	625	1.6
C^{2+}	685	0.1
Cu^{2+}	850	1.1
Fe^{2+}	1100	0.68
Fe^{3+}	400	0.15
Ni^{2+}	650	0.1
Mn^{3+}	460	0.2
V^{4+}	725	2.7

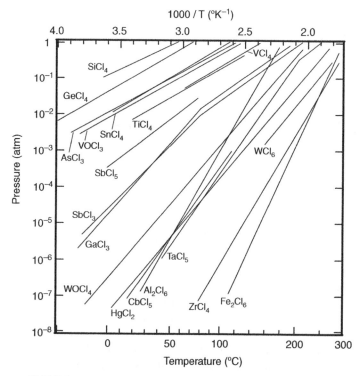

FIGURE 3.1   Vapor pressure curves for chloride precursors [4].

pressure environment. Due to strong demand in large volume of high purity glass fabrication for long length of optical fiber, atmospheric pressure environment in the fabrication process was one of critical demands.

The vapor pressure curves of a few chlorides are shown in Figure 3.1 [4]. It is noted that $SiCl_4$ and $GeCl_4$, the precursors of silica and germanosilicate glass, show a vapor pressure higher than other chlorides by several orders of magnitudes. For instance, in the case of $SiCl_4$, vapor pressure reaches 1 atm in the range of 50–60°C, while that of transition metal chlorides remains far less than a fraction of 1 atm. When the vapor pressure reaches 1 atm, $SiCl_4$ can be readily vaporized from the liquid phase in the atmospheric pressure. This significant difference in vapor pressure between $SiCl_4$ and transition metal chlorides enables selective vaporization of $SiCl_4$ in a certain temperature range at the atmospheric pressure while keeping other transition metals in solid or liquid phase. It is therefore possible to completely separate the transition metal from the vapor stream of $SiCl_4$ to result in high purity silica glass formation and subsequently low attenuation in optical fibers.

In order to implement the vapor delivery of high vapor pressure chlorides such as $SiCl_4$, $GeCl_4$, and $POCl_3$, two types of methods have been developed: (1) bubbler and (2) flash evaporator. These methods are schematically shown in Figure 3.2.

In the bubbler method, a carrier gas is fed through the liquid precursor to generate bubbles. The flow of the carrier is controlled by mass flow controller (MFC). As the

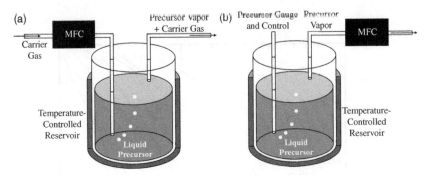

FIGURE 3.2   Precursor vapor delivery methods: (a) bubbler and (b) flash evaporator. MFC, mass flow controller.

bubbles break up at the liquid surface, they generate the precursor vapor, which will be further delivered over the carrier gas flow. The mass transfer rate of the precursor vapor will depend on: (1) bubbler temperature, (2) bubbler pressure, and (3) carrier gas flow rate.

In the flash evaporator method, no carrier gas is used but the pressure inside the precursor container is built to a high level. In order to maintain the vapor pressure above the process pressure, precise control of the container temperature is required. The amount of precursor vapor is directly controlled by a high temperature MFC at the output of the container. The delivery tube should be kept in a temperature high enough to avoid the condensation of the vapor precursor. The mass transfer rate of the precursor vapor will depend on: (1) evaporator's thermal capacity and (2) vapor MFC flow rate.

In most of the current optical fiber fabrication processes, the bubbler technologies are being widely used due to less stringent thermal requirements compared with the evaporator technologies. Dry oxygen, O_2, is being widely used as a carrier gas and as an oxidizing agent to form oxide glass layers.

With the same convective delivery of high purity precursor vapor, four different routes of preform fabrication processes have developed that differ in oxide-forming reaction mechanism, deposition mechanism, and consolidation route. These processes are modified chemical vapor deposition (MCVD), plasma chemical vapor deposition (PCVD), outside vapor deposition (OVD), and vapor axial deposition (VAD). Detailed comparative study for these processes has been reported by Nagel et al. [5]. A brief comparison table for the preform fabrication processes is shown in Table 3.2.

The schematics of the individual processes are shown in Figure 3.3. Both MCVD and PCVD are using substrate tubes and the silica layers are deposited over the inner surface of the tube. In the case of OVD and VAD, silica bodies are formed over the mandrel and bait rod, respectively.

In this book, we mainly discuss MCVD due to its versatile and flexible process control that makes MCVD widely adapted for specialty optical fiber fabrications. Furthermore, detailed process parameters are not confined only to MCVD but can be shared and applicable to other processes with suitable modifications.

TABLE 3.2 Comparison of Optical Fiber Preform Fabrication Processes Based on CVD Using Chloride Precursors [5]

Feature	MCVD	PCVD	OVD	VAD
Basic approach	Deposition of core layers inside substrate tube, which becomes outer cladding	Deposition of core layers inside substrate tube, which becomes outer cladding	Deposition of core and then cladding on a removable mandrel	End-on deposition of core on bait rod, deposition of clad from side
Reaction mechanism	High temperature gas-phase oxidation of chlorides	Microwave plasma–initiated oxidation of chlorides	Flame hydrolysis of chlorides using methane fuel	Flame hydrolysis of chlorides using H_2/O_2 flame
Deposition mechanism	Thermophoretic deposition of particles on inner tube wall	Heterogeneous nucleation at inner tube wall with tube temperature control	Thermophoretic deposition of particles on mandrel, and soot over it	Thermophoretic deposition of particles on bait rod, and soot over it
Sintering	Viscous sintering of particles simultaneous with deposition of layer	No sintering necessary, but with tube temperature control to assure vitreous layer	Separate viscous sintering step for soot body	Separate viscous sintering step for soot body

In MCVD, there are distinctive key processes in the preform fabrication and they are as follows:

- vapor delivery
- chemical reaction of vapor
- deposition of soot layer
- sintering
- collapse and sealing

In addition to these basic steps, postprocesses such as "elongation" and "rod in tube" are being frequently used to achieve the desired dimensions of waveguide structures. Especially air–silica holey fibers discussed in Chapter 8 are fabricated by modification of rod in tube method. In this section, we will review these processes to understand how to realize the waveguide designs that we discussed in previous sections into silica glass optical fiber.

3.1.1.1 Vapor Delivery   The conventional precursors used for MCVD are listed in Table 3.3 along with their products after appropriate chemical reactions.

Note that the precursors in the liquid phase at room temperature are delivered using the bubbler technology along with dry O_2 gas. The carbofluoride gases such as CF_4, C_2F_6, and C_3F_8 can be used in both F doping and etching of SiO_2 glass layer by controlling the reaction temperature [6,7]. The flow rates of the gas-phase precursors are usually controlled by high-precision mass flow controllers and the gas streams are convectively fed inside the substrate silica glass as schematically shown in Figure 3.3a. The silica glass substrate tube is rotating during the process and therefore a specially designed rotary seal is used to connect the substrate tube and the vapor delivery pipe, leak tight.

TABLE 3.3 Precursors Used in MCVD and Their Characteristics

Precursors	Phase	Delivery Method	Product
$SiCl_4$	Liquid	Bubbler with O_2 carrier	Vitreous SiO_2
$GeCl_4$	Liquid	Bubbler with O_2 carrier	GeO_2 doping in SiO_2
$POCl_3$	Liquid	Bubbler with O_2 carrier	P_2O_5 doping in SiO_2
BBr_3	Liquid	Bubbler with O_2 carrier	B_2O_3 doping in SiO_2
BCl_3	Gas	Direct delivery from high pressure gas tank	B_2O_3 doping in SiO_2
SiF_4	Gas	Direct delivery from high pressure gas tank	Heavy F doping in SiO_2
CF_4, C_2F_6, C_3F_8	Gas	Direct delivery from high pressure gas tank	Medium F doping in SiO_2, Etching of SiO_2
SF_6	Gas	Direct delivery from high pressure gas tank	Medium F doping in SiO_2, Etching of SiO_2

The phases of precursors are specified at room temperature and atmospheric pressure.

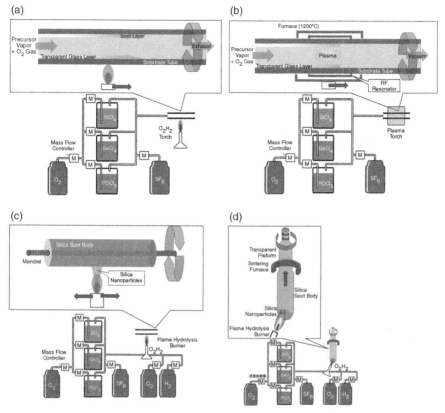

FIGURE 3.3 Schematics of preform fabrication processes: (a) MCVD, (b) PCVD, (c) OVD, and (d) VAD.

3.1.1.2 Chemical Reaction of Vapor

More than 90 mol% of glass fabricated in MCVD is silica glass, and other elements such as GeO_2, P_2O_5, B_2O_3, and F are processed as dopants. All of these oxide dopants are glass-forming elements and they are highly miscible with silica glass over 10 mol% without causing the phase separation. Main chemical reactions are described as follows:

$$\text{Oxidation of } SiCl_4 : \quad SiCl_4 \text{ (g)} + O_2 \rightarrow SiO_2 \text{ (s)} + 2Cl_2 \tag{3.1}$$

$$\text{Oxidation of } GeCl_4 : \quad GeCl_4 \text{ (g)} + O_2 \rightarrow GeO_2 \text{ (s)} + 2Cl_2 \tag{3.2}$$

$$\text{Oxidation of } POCl_3 : \quad 2POCl_3 \text{ (g)} + \frac{3}{2}O_2 \rightarrow P_2O_5 \text{ (s)} + 3Cl_2 \tag{3.3}$$

$$\text{Oxidation of } BCl_3 : \quad 2BCl_3 \text{ (g)} + \frac{3}{2}O_2 \rightarrow B_2O_3 \text{ (s)} + 3Cl_2 \tag{3.4}$$

$$\text{F doping in } SiO_2 : \quad SiF_4 \text{ (g)} + 3SiO_2 \text{ (s)} \rightarrow 4SiO_{1.5}F \text{ (s)} \tag{3.5}$$

F doping in SiO_2 : $SF_6 \ (g) + 4SiO_2 \ (s) \rightarrow 4SiO_{1.5}F \ (s) + SO_2 \ (g) + F_2 \ (g)$

$$(3.6)$$

SiO_2 etching : $CF_4 \ (g) + SiO_2 \ (s) \rightarrow SiF_4 \ (g) + CO_2 \ (g) \qquad (3.7)$

SiO_2 etching : $SF_6 \ (g) + \dfrac{3}{2} SiO_2 \ (s) \rightarrow \dfrac{3}{2} SiF_4 \ (g) + SO_2 \ (g) + \dfrac{1}{2} O_2 \ (g) \quad (3.8)$

In the above equations, (g) and (s) stand for gas and solid phases of the substances in the reaction. Reaction equations 3.1–3.4 represent thermal oxidation of chloride gas precursors to result in solid oxides. Fluorine doping of SiO_2 glass is shown in the reaction equations 3.5 and 3.6 and the etching process is represented by the reaction equations 3.7 and 3.8.

The individual oxidation process is chemically complete at a certain thermal condition such that all the precursors transform into oxide glasses. However, thermal equilibrium constants for individual reactions are different, which makes the chemical kinetics complicated for the case when the precursor gases are mixed together. Especially when the core layers are formed in MCVD, both $SiCl_4$ and $GeCl_4$ are carried over O_2 into a thermal reaction zone formed by oxyhydrogen torch.

Wood et al. [8] reported detailed chemical kinetics for GeO_2 doping in silica using MCVD process when three precursors, $SiCl_4$, $GeCl_4$, and $POCl_3$, were simultaneously carried over O_2. The composition of effluent gas after thermal oxidation was analyzed for various reaction temperatures, as shown in Figure 3.4. The reactions for $SiCl_4$ (reaction equation 3.1) and $POCl_3$ (reaction equation 3.3) were complete at the reaction temperatures over 1550 and 1450 K, respectively. Over these temperatures the partial pressure of $SiCl_4$ and $POCl_3$ decreased below the detection limit, which indicates that all the chloride gas precursors were turned into solid oxides. Consequently, these reactions will generate significant amount of Cl_2 according to reaction equations 3.1 and 3.3, which will affect the chemical equilibrium for oxidation of $GeCl_4$ in reaction equation 3.2. In fact, the maximum amount of oxidation was found to be only 20% for the input $GeCl_4$ and furthermore oxidation was obtained in a relatively narrow reaction temperature range of 1650–1900 K.

In the following discussion, we will focus on $SiCl_4$ and $GeCl_4$ reaction kinetics to understand the germanium doping process in the core layer deposition using MCVD. In addition to reaction equations 3.1 and 3.2, there are two more reactions that affect germanium doping process:

$$SiCl_4 \ (g) + O_2 \rightarrow SiO_2 \ (s) + 2Cl_2 \qquad (3.1)$$

$$GeCl_4 \ (g) + O_2 \rightarrow GeO_2 \ (s) + 2Cl_2 \qquad (3.2)$$

Cl_2 dissociation : $Cl_2 \ (g) \rightarrow 2Cl \ (g) \qquad (3.9)$

GeO_2 evaporation : $GeO_2 \ (s) \rightarrow GeO \ (g) + \dfrac{1}{2} O_2 \qquad (3.10)$

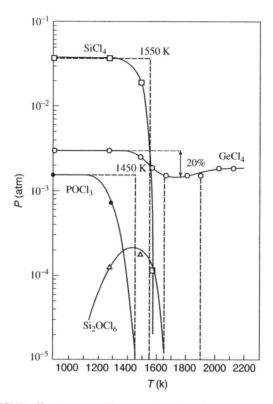

FIGURE 3.4 MCVD effluent composition as a function of reaction zone temperature [8].

The reaction 3.9 represents thermal dissociation of chlorine gas, and this might affect the partial pressure of Cl_2 in the reaction equation 3.2. Volatilization of GeO_2 is shown in reaction 3.10, where solid GeO_2 is evaporated into gaseous GeO. Calculated equilibrium constants for these reactions are listed in Table 3.4 [8].

TABLE 3.4 Equilibrium Constants for Reactions for Germanium Doping in Silica Glass [8]

Temperature (K)	K_1 (SiCl$_4$)	K_2 (GeCl$_4$)	K_3 (Cl$_2$)	K_4 (GeO$_2$)
300	2.38×10^{40}	1.40×10^{11}	2.9×10^{-37}	2.1×10^{-81}
600	7.66×10^{18}	6.41×10^{3}	4.3×10^{-16}	1.1×10^{-33}
900	5.51×10^{11}	2.49×10^{1}	5.9×10^{-9}	6.1×10^{-18}
1200	1.63×10^{8}	1.69	2.3×10^{-5}	3.6×10^{-10}
1500	1.33×10^{5}	0.471	3.5×10^{-3}	1.1×10^{-5}
1800	5.84×10^{4}	0.240	1.0×10^{-1}	8.6×10^{-2}

Here K_1 is for the reaction of SiCl$_4$ in equation 3.1, K_2 for GeCl$_4$ in equation 3.2, K_3 for Cl$_2$ in equation 3.9, and K_4 for GeO$_2$ in equation 3.10.

As indicated in Table 3.4, the reactions of chlorine dissociation and germania evaporation require higher temperature than oxidation processes of $SiCl_4$ and $GeCl_4$. In fact, the central refractive index dip in normal MCVD preforms is mainly attributed to the volatilization of GeO_2 in reaction equation 3.10. These processes, therefore, would be a concerning factor only in the tube collapse and sealing process where a very high temperature is maintained to deform the silica glass for a significantly long time. For the deposition process, the reactions in equations 3.9 and 3.10 can be neglected and only reactions in equations 3.1 and 3.2 will be considered hereafter.

We will find out the relation of the final mole fraction of GeO_2 in silica glass SiO_2, X_G, for the initial flow rates of $SiCl_4$ (Q_s^0), $GeCl_4$ (Q_G^0), Cl_2 (Q_{Cl}^0), and O_2 (Q_O^0).

The mole fraction of GeO_2 can be expressed in terms of the differences in the gas flows before and after the reactions, in equations 3.1 and 3.2:

$$X_G = \frac{\text{reacted } GeCl_4 \text{ flow}}{\text{reacted } SiCl_4 \text{ flow} + \text{reacted } GeCl_4 \text{ flow}} = \frac{Q_G^0 - Q_G}{(Q_s^0 - Q_s) + (Q_G^0 - Q_G)} \quad (3.11)$$

Here Q_l^0 and Q_l represent the initial and final flow rates of vapors, respectively.

Because the reaction for $SiCl_4$ (equation 3.1) is complete, we can have the following information:

$$Q_s = 0, \qquad Q_{Cl}^0 = 2Q_s^0 \quad (3.12)$$

$$X_G = \frac{Q_G^0 - Q_G}{Q_s^0 + Q_G^0 - Q_G} \quad (3.13)$$

or equivalently

$$Q_G = Q_G^0 + \frac{X_G}{X_G - 1} Q_s^0 \quad (3.14)$$

From the oxidation of $GeCl_4$, as in equation 3.2 along with the known equilibrium constant K_2 in Table 3.4, we have another expression for X_G:

$$X_G = \frac{K_2}{\gamma_2} \frac{Q_O \cdot Q_G}{Q_{Cl}^2} = \frac{K_2}{\gamma_2} \frac{(Q_O^0 - Q_s^0 - Q_G^0 + Q_G) \cdot Q_G}{(Q_{Cl}^0 + 2Q_G^0 - 2Q_G)^2} \quad (3.15)$$

Here γ_2 is the activity coefficient pertinent to $GeCl_4$ and it was experimentally determined that (K_2/γ_2) is equal to 0.015. In above equations, we have two unknowns, Q_G and X_G; these two unknowns can be found by solving equations 3.14 and 3.15 simultaneously:

$$X_G = \frac{K_2}{\gamma_2} \frac{[(X_G - 1)Q_O^0 + Q_s^0][(X_G - 1)Q_G^0 + X_G Q_s^0]}{(2Q_s^0)^2} \quad (3.16)$$

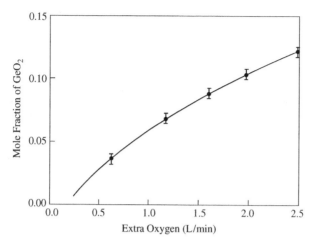

FIGURE 3.5 Mole fraction of GeO_2, in glass fiber preforms produced with various extra oxygen flow rate, keeping constant the flows of $SiCl_4$ at 1.20 g/min, $GeCl_4$ at 0.78 g/min, and $POCl_3$ at 0.021 g/min, at the total flow at 2490 cm^3/min with nitrogen and nonreactive diluents. Hot zone temperature was 1690 K [8].

Solving the above quadratic equation for X_G with given initial flows of $SiCl_4$, $GeCl_4$, and O_2, we will find the relationship between the mole fraction of GeO_2 in silica and process gas flow rates.

Comparisons of theoretical predictions and experimental measurements for the mole fraction of GeO_2 in silica are plotted as a function of flow rate for the extra oxygen and $GeCl_4$ in Figures 3.5 and 3.6, respectively.

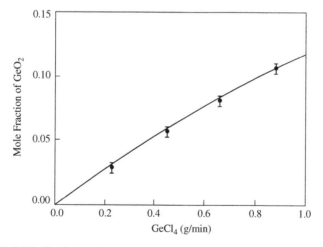

FIGURE 3.6 Mole fraction of GeO_2, in glass fiber preforms produced with various $GeCl_4$ reactant flow rate, keeping constant the flows of $SiCl_4$ at 1.27 g/min, $POCl_3$ at 0.021 g/min, and O_2 at 2025 cm^3/min. Hot zone temperature was 1690 K [8].

3.1.1.3 Deposition of Soot Layer   In the previous section, we have reviewed the chemical reactions to form oxide glasses in MCVD process. The glasses are formed in few tens of nanometer-sized particles when the appropriate thermal conditions are met, and then grow in size in the range of 0.1–0.5 µm during the Brownian coagulation in the gas–particle stream. These submicrometer oxide particles are under the influence of temperature distribution along their path and the gas flow. The gas flow rates inside the substrate tube in MCVD process are carefully controlled to maintain the laminar flow in all cross-sections along the substrate tube suppressing turbulences. Therefore, the primary impetus to drive the particle deposition in practical MCVD is the temperature distribution within and along the substrate tube.

The main mechanism for soot layer deposition in MCVD has been attributed to thermophoresis [9], which has been applied also in OVD and VAD processes to explain their deposition mechanisms. The basic mechanism of thermophoresis is that a suspended particle in a temperature gradient experiences a net force in the direction of decreasing temperature. The thermophoretic velocity, V_T, of a particle can be expressed as

$$V_T = -K\frac{\nu}{T}\nabla T \tag{3.17}$$

where ν is the kinematic viscosity, T is the temperature in absolute degrees, and K is the thermophoretic coefficient. A value of $K = 0.9$ was found to be most appropriate for the conditions encountered in the MCVD process.

Typical temperature profile inside the silica substrate tube is shown along its axial direction in Figure 3.7. As SiCl$_4$ vapor passes the reaction isotherm or any point where

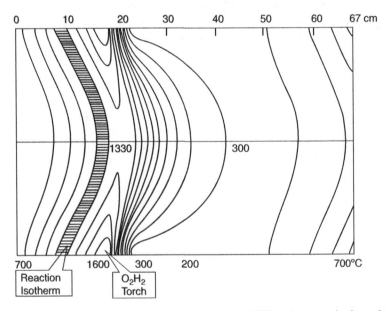

FIGURE 3.7 Typical temperature distribution inside MCVD substrate tube heated by an external O$_2$H$_2$ torch [9].

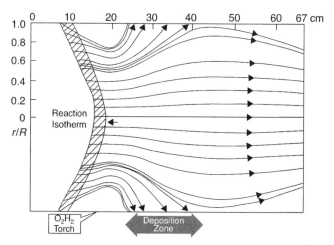

FIGURE 3.8 Silica particle trajectories resulting from the temperature field in Figure 3.7. Here r is the radial coordinate and R is the radius of substrate tube [9].

the temperature $T > T_{rxn}$, thermal oxidation process in O_2 environment results in silica nanoparticles. Here the effective reaction temperature, T_{rxn}, was found to be around 1300°C for the reaction in equation 3.1. These SiO_2 particles are then under the influence of the thermophoretic force and trajectories of particles are schematically shown in Figure 3.8. Further downstream the substrate tube cools down and the wall temperature is lower than the gas temperature. The gas and the wall equilibrate at a temperature, T_e, resulting in the particle moving toward the tube wall. Typical T_e is in the range of 200–400°C. Some trajectories near the wall result in particle deposition, while other particles near the center are swept out of the tube.

The deposition efficiency, E, is defined as the fraction of the $SiCl_4$ deposited as SiO_2. It was determined to be only a function of the reaction temperature, T_{rxn}, and the downstream tube–gas equilibrium temperature, T_e. It has the approximate relation

$$E \cong 0.8 \left[1 - \frac{T_e}{T_r} \right] \tag{3.18}$$

where the temperatures are in Kelvin [9]. Typical deposition efficiency, E, in MCVD is in the range of 0.5–0.65. The deposition efficiency is most sensitive to the equilibrium temperature, T_e, which depends strongly on the torch traverse length, torch traverse speed, the ambient temperature, and the tube wall thickness. Deposition process will determine the cross-sectional area of the glass layers, which is one of the critical information for fiber preform design.

In addition to the deposition efficiency, we need to consider another aspect of deposition process. The deposition rate is defined as the weight of SiO_2 deposited in the unit time (g/min) and it determines the efficiency of the fiber manufacturing process. The deposition rates of the current fabrication processes are listed as follows:

SiO$_2$ Glass Depositing Rate (g/min)	
OVD	15–30
MCVD	1.0–2.0
VAD	15–30
PCVD	0.5–1

The MCVD and PCVD processes have substantially lower deposition rates and therefore VAD or OVD has competitive advantages for mass-producing fiber pre-forms in large diameters or quantities. The capital investment for MCVD and PCVD is relatively lower than the other processes, however, and the production process is easier to operate and control to make them the preferred process for complicated refractive index profiles in specialty fibers.

3.1.1.4 Sintering Deposition of silica particles will form a porous soot layer on the inner wall of the substrate tube and in order to make the layer optically transparent, a sintering process should be followed with optimal process conditions. Figure 3.9 schematically shows the sintering of a silica soot layer. In most of MCVD process, thermophoretic soot deposition is immediately followed by the viscous sintering using a traversing oxyhydrogen torch. And this sequential deposition–sintering process is repeated to achieve a desired dimension. In MCVD process, substrate tube collapse will follow these repeated deposition–sintering processes and a very high temperature environment is provided, which will once again affect the final sintering quality of the glass layers.

Main concern in the sintering process is, therefore, to avoid bubbles in the glass layers and process conditions such as glass composition, sintering temperature, and gas composition should be optimized.

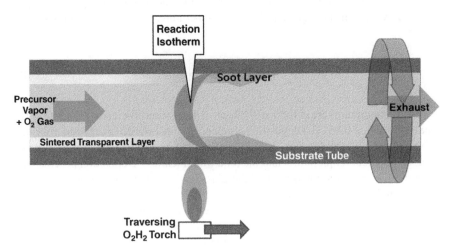

FIGURE 3.9 Schematic diagram of sintering process for the silica soot layer in MCVD.

Two problems in the sintering process are usually encountered in MCVD, which will eventually result in formation of bubbles in the glass layer. The first problem is characterized by a cloudy appearance of the tube walls after the torch has passed. This is usually attributed to too thick deposition layer or/and too low sintering temperature. If this deposition–sintering is repeated, distinct bubbles may become visible during collapse of the substrate tube. The second problem is characterized by the sudden appearance of distinct bubbles during the deposition–sintering of a layer. This is mainly attributed to too high temperature.

Walker et al. [10] have reported the sintering mechanism in MCVD process for various experimental parameters, such as glass composition (SiO_2, SiO_2–GeO_2, SiO_2–GeO_2–B_2O_3), sintering temperature of 1540–1630°C, and process gas (He, O_2). Phenomenologically they described the cause of the bubble during the sintering process by the balance between the vapor pressure, P_V, and the pressure due to surface tension, P_S. The P_V is the pressure of the most volatile component in equilibrium with the glass and it is an increasing function of the temperature and concentration of this component in the glass. A bubble, or closed pore, with an equivalent radius of curvature will have an interior pressure, P_S, equal to $2(\sigma/R_B)$, where σ is the surface tension of the glass and R_B the bubble radius (see Figure 3.10).

Bubbles arise when a net flux of gas occurs into, rather than out of, a closed pore. The growth of a closed pore into a bubble is most likely to be observed at high temperatures in a glass layer with a high dopant concentration. If $P_V < P_S$, the flux of gas will be out of the pore and the pore will shrink in size. P_S will further increase due to the decreasing radius of curvature. The pore will eventually disappear if the temperature is high enough to keep the glass viscosity low. If $P_V > P_S$, there will be a net flux of gas into the pore. The rate of bubble growth will be a function of the rate of diffusion and the viscosity of the glass. The viscosity of silica glass and glass-forming oxides is shown in Figure 3.11.

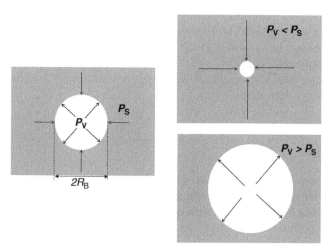

FIGURE 3.10 Phenomenological explanation of bubble formation.

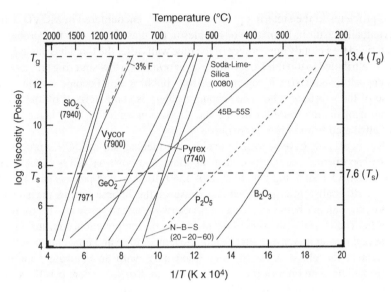

FIGURE 3.11   Viscosity of glasses as a function of temperature [11].

In Figure 3.11, it is found that doping pure silica with other glass-forming oxides, GeO_2, P_2O_5, and B_2O_3, will significantly reduce the viscosity in the MCVD process temperature range, 1500–1700°C. In most of low-loss optical fibers, the viscosity in the core glass where GeO_2 is doped will have a higher value than that of inner cladding where P_2O_5 and F are codoped. The inner cladding layers are made by doping a low concentration of P_2O_5 to reduce the process temperature and achieve "bubble-free" sintering. Fluorine should be codoped to compensate the index raise from P_2O_5 (see Figure 2.14). Therefore, the sintering temperature should be optimized separately for the core and the inner cladding layers. Among inert gases, *helium* was found to be effective to get rid of bubble in MCVD, OVD, and VAD. Improvement in consolidation by He was attributed to its high thermal conductivity and escape mobility within the viscous flows, which uniformly reduces the viscosity of glass layers without trapped bubbles [10].

Consolidation rate was found to be inversely proportional to "modified capillary number," which is defined as

$$C = \frac{\eta(T)l_0(1-\varepsilon_0)^{1/3}}{\sigma(T)t_s} \tag{3.19}$$

Here η is glass viscosity, l_0 is the size of initial bubble regions, ε_0 is the initial bubble fraction, σ is the surface tension, and t_s is the characteristic sintering time. Note that the sintering rate is heavily dependent on thermomechanical properties such as the viscosity and surface tension, which can be controlled by proper glass compositions. Lower viscosity, higher surface tension of the deposited glass layer, and longer sintering time are favored sintering conditions.

3.1.1.5 Collapse and Sealing   The substrate tube with glass layers deposited and sintered on its inner wall should be collapsed to a solid glass rod that is conventionally called as a preform. The preform collapse involves a slow viscous flow of the glass driven by surface tension and differential pressures on the inner and outer tube surfaces. The standard collapse process in MCVD is commonly executed on a horizontal glass lathe. While the tube rotates to preserve the cylindrical symmetry, an oxyhydrogen torch traverses along its axial length producing a local hot zone, where the glass softens sufficiently at a high temperature to permit viscous flow of the tube walls.

During the collapse process two issues have drawn very keen attentions: (1) noncircularity in the deposited layers and (2) GeO_2 volatilization at high temperature. These issues directly give significant impacts on the optical performances of the resulting fibers and therefore collapse process parameters should be carefully optimized along with additional process such as gaseous etching.

The dominant driving force for collapse was found to be the surface tension. Geyling et al. [12] have analyzed the collapse process both numerically and experimentally to understand how the noncircular modes of deformation may grow or decay. Depending on process parameters, the resulting core and inner cladding of the collapsed glass rod may show a finite ellipticity, which will immediately decouple the degenerate fundamental LP_{01} mode into nondegenerate HE_{11x} and HE_{11y}. This birefringence in the fundamental mode will further affect the polarization mode dispersion (PMD) for long-haul high–data rate transmissions and the polarization coupling in fiber optic sensors. In most of applications, optical fiber preforms require high circularities and it is imperative to optimize the collapse process to minimize the instabilities that could lead to magnifying the ellipticity.

In Geyling et al.'s model [12], two layers of glass with different viscosities were assumed as in Figure 3.12 to simulate the collapse of MCVD preform. As discussed in Section 3.1.1.4, it is necessary to have a certain glass composition in the deposited glass layers other than pure silica in order to achieve the "bubble-free" sintering condition. In general, the soft inner glass layer represents the doped glass layers that will form either core or inner cladding, while hard outer layer is the undoped silica substrate tube.

The noncircularity in these layers is given by

$$a = a_0 + a_2 \cos2\theta, \qquad \varepsilon_a = \frac{a_2}{a_0}$$

$$b = b_0 + b_2 \cos2\theta, \qquad \varepsilon_b = \frac{b_2}{b_0} \qquad (3.20)$$

$$c = c_0 + c_2 \cos2\theta, \qquad \varepsilon_c = \frac{c_2}{c_0}$$

As an example, the collapse stability versus interface radius is shown in Figure 3.13. The model furthermore predicted that the ellipticity of both the core, ε_c, and outer cladding, ε_b, grows with increasing pressure difference, viscosity difference, interface radius, and deposit thickness while it decays with the wall thickness of

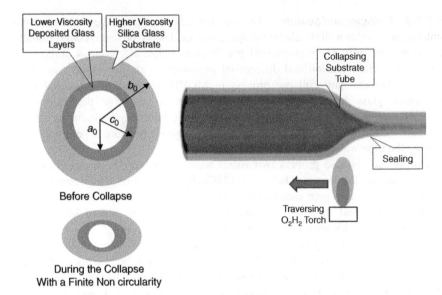

FIGURE 3.12 Schematic collapse process of MCVD tube. Here a is the inner radius, b is the outer radius, and c is the interface radius. The deposited glass layers fabricated by MCVD process over a high viscosity silica substrate tube.

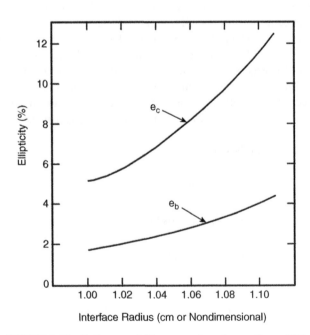

FIGURE 3.13 Collapse stability versus interface radius, c [12].

substrate tube. It was also found that the core ellipticity is usually higher than that of cladding by several folds.

During the collapse process, the chemical and thermal environments will favor the GeO_2 volatilization, which leaves behind refractive index dip at the center:

$$GeO_2 \text{ (s)} \rightarrow GeO \text{ (g)} + \frac{1}{2}O_2 \qquad (3.10)$$

The central index dip in optical fibers will deviate the final refractive index profile from the desired waveguide design affecting various optical characteristics—chromatic dispersion, mode field diameter, and cutoff, to name a few. In the case of multimode optical fiber, the central dip caused more severe problem, differential modal delay (DMD), to limit the bandwidth of transmission [13]. The central index dip can be partly reduced or completely removed by gaseous etching process using carbofluoride gases during the final collapse pass [7]:

$$CF_4 \text{ (g)} + SiO_2 \text{ (s)} \rightarrow SiF_4 \text{ (g)} + CO_2 \text{ (g)} \qquad (3.7)$$

$$SF_6 \text{ (g)} + \frac{3}{2}SiO_2 \text{ (s)} \rightarrow \frac{3}{2}SiF_4 \text{ (g)} + SO_2 \text{ (g)} + \frac{1}{2}O_2 \text{ (g)} \qquad (3.8)$$

In the above processes the fluorine-containing reactants turn solid SiO_2 into gaseous SiF_4. Therefore, the part of core layers where GeO_2 evaporates can be effectively removed by this etching process by flowing the fluorine gases during final collapse passes.

This process is schematically shown in Figure 3.14. Detailed information on etching efficiency, in terms of etched cross-sectional area, fluorine gas flow rate, temperature, and torch traverse speed should be quantified for individual preforms.

3.2   POSTPROCESSES FOR GEOMETRICAL MODIFICATION OF PREFORM

Once a preform with a certain core/cladding dimension is fabricated using one of CVD processes in Figure 3.3, we can endow further a new degree of freedom to modify the geometrical dimension of the final preform from which the optical fiber is eventually drawn. This degree of freedom is achieved by two postprocesses, namely, elongation (or stretching) process [14,15] and rod in tube (RIT) process along with an appropriate overjacketing silica tube [16–18].

In the elongation process, a segment of the initial preform is heated near the softening point and the diameter is reduced by precise control of pulling mechanism. The schematic diagram is shown in Figure 3.15 [14,15].

Elongation process is commonly used in VAD, OVD, and MCVD processes to adjust preform sizes and the mass conservation will determine the final dimension of the preform:

$$\rho\pi r_i^2 V_{torch} = \rho\pi r_f^2 (V_{pull} + V_{torch}) \qquad (3.21)$$

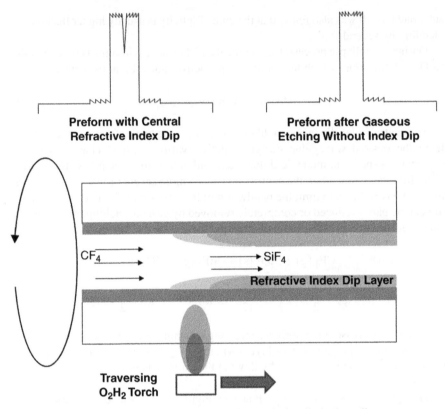

FIGURE 3.14   Gaseous etching of central refractive index dip in the collapse process.

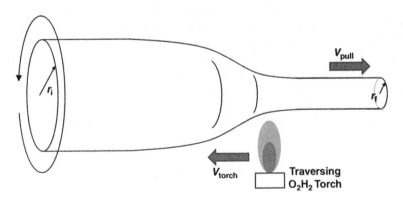

FIGURE 3.15   Schematic diagram of elongation process.

where ρ is the density of silica and r_i and r_f are the initial and final diameters of the preform, respectively. V_{torch} and V_{pull} are the torch traverse speed and the preform pull speed. Usually the preform pull is controlled by adjusting the receding speed of the rotating chuck where preform is placed. In conventional MCVD process, r_i ranges from 15 to 20 mm and r_f can be decreased below 10 mm.

The basic mechanism of elongation is very close to that of fiber drawing that will be discussed in the next sections, in a sense that softening glass with an initial radius is pulled to a uniform solid glass with a reduced dimension. The elongation process is usually accompanied by RIT process, where an overjacketing tube and the elongated preform are fused together to achieve a final preform with a desired core/cladding ratio. In recent process RIT and elongation are processed in single equipment.

Schematic diagram of combined RIT/elongation is shown in Figure 3.16 along with stepwise processes: (a) assemble overjacketing tube and handle, (b) assemble core rod, (c) weld second handle, (d) collapse and elongation, and (e) preform taper and removal.

The hot zone is usually formed by circular oxyhydrogen torch and a low-level vacuum is applied for seamless fusion between the core rod and overjacketing tube.

RIT process has been intensively developed for large preform fabrication based on MCVD process and presently 60–80 mm diameter preform is routinely produced and 120 mm preform could be also fabricated [18]. Utilizing these elongation and RIT processes, specialty fibers with complex refractive index profile and doping profile can be efficiently fabricated without modifying basic chemical vapor deposition systems.

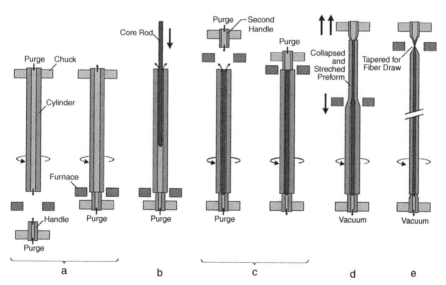

FIGURE 3.16 RIT procedure [18]: (a) assemble overjacketing tube and handle, (b) assemble core rod, (c) weld second handle, (d) collapse and elongation, and (e) preform taper and removal.

3.3   OPTICAL FIBER DRAWING

Since the low optical transmission loss (20 dB/km) in silica optical fiber was first demonstrated by Kapron et al. [19], research and development efforts for the fiber technology have led to remarkable achievements. Particularly, the manufacturing technology of silica glass optical fibers has advanced rapidly to meet the demand for very large quantities of low-cost fibers for many applications such as transoceanic cable, terrestrial cable, loop, customer premises penetration, and so on.

The high volume fiber fabrication basically requires two critical technologies: one is to produce a large size of an optical fiber preform (>80 mm in its diameter) that is partly covered in the previous section. The other is to draw a 125 μm diameter fiber from the preform and to coat it simultaneously with polymer materials at a high draw speed (>20 m/s), which will be dealt with in this section.

Until the late 1970s, the fiber draw speed was typically in the range of 1–2 m/s. Paek and Schroeder [20] first demonstrated the draw and coating speed of 5 m/s, using a UV curable resin. Prior to this work, Paek and Runk [21] had investigated the neck-down region of a silica glass preform in a furnace during fiber drawing process, using a simplified model to understand the deformation mechanics. Paek and Schroeder [22] investigated a method for high volume mass production that can provide cost-effective manufacturing of optical fibers. Yin and Jaluria [23] took a rigorous theoretical approach to describe the radiative heat transfer of a large and transparent fiber preform in a graphite furnace. This calculation was based on a three-dimensional model including convection, conduction, and radiation transport between the furnace muffle and the necked preform. The results provide useful information on a large preform drawing, namely, a 10 cm diameter preform drawn into a 125 μm diameter fiber at a draw speed of 15 m/s.

In this section, we will specifically discuss and review fiber drawing technique in terms of its principal constitutive processes.

3.3.1   Principal Processes in Fiber Drawing

An actual fiber drawing system is composed of functionally different instruments and apparatus such as a high temperature furnace, fiber diameter measurement system, coating applicator, centering device, UV curing apparatus, draw tension gauge, and take-up capstan in a mechanically stable tower that is normally taller than 8 m. Principal processes in fiber drawing can be divided into three major zones, namely: heating zone, cooling zone, and coating zone, as shown in Figure 3.17.

The heating zone that includes a high temperature furnace and a precision temperature controller softens an optical fiber preform to form a neck-down region, where bare glass fiber is continuously drawn. The softening point of silica glass ranges from 1400 to 2350°C. The viscosity of silica glass in the temperature of 1935–2322°C varies from $10^{5.86}$ to $10^{4.63}$ P [24]. In order to heat the silica preform to this temperature range, an electric high temperature furnace is used with an optimal heating zone configuration. A broad temperature distribution along the furnace yields a mild temperature gradient across the preform diameter to form a slowly varying

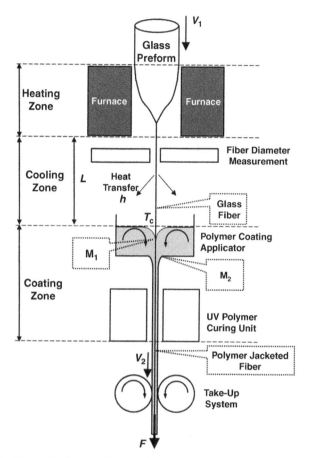

FIGURE 3.17   Schematic diagram for optical fiber drawing process along with three principal zones. Here V_1 and V_2 are the preform feed speed and fiber drawing speed, respectively; L is the cooling zone length from the bottom of the furnace to the top of polymer coating applicator; h is the heat transfer coefficient; T_c is the coating temperature; M_1 and M_2 are the meniscus at the input and output of the coating applicator; and F is the drawing tension.

neck-down region and thus provides a low fiber draw tension. On the other hand, the narrow heat zone induces an extremely high draw tension especially in case of drawing a fiber from a large preform, inducing strength degradation due to the fatigue effect and thus producing an extremely weak fiber and causing the fiber to break during drawing [25]. Temperature distribution and its consequential neck-down region formation are key parameters in the heating zone.

Pristine bare glass fiber shows a very high strength, approximately 70,000 kg/cm^2 (1 million psi) in air [26]. However, it is easily degraded due to surface damage when a bare fiber comes in contact with a foreign particle or matter. Thus, it is imperative to put a coating material on an optical fiber to protect its surface. In high speed drawing, the temperature of the fiber entering coating material in the coating applicator, T_c, plays a decisive role to determine the quality of the coating [20]. T_c must be maintained below a certain point to prevent the coating material from overheating. Otherwise the meniscus formed by the fiber entering the coating material collapses to result in improper wetting and burning the material or losing the integrity of material. Therefore, the cooling zone right after the heating zone should be followed to control T_c to ensure high speed coating.

Finally, the coating zone consists of the coating die containing liquid prepolymer and the polymer curing unit. Detailed viscous flow of liquid prepolymer in a coating die can significantly affect the final coating quality and mechanical strength. In high speed coating, the shear rate becomes so large that a fiber coated at a high speed often exhibits poor strength. It is suspected that high draw tension and viscous friction on the fiber surface by prepolymer solution in the coating die may contribute to the weak fiber strength. The common practice for reducing a high shear force in fiber manufacturing is to heat the coating applicator to reduce the polymer viscosity. The viscosity level set for coating is normally in the range of 10 P.

The three zones (preform heating, fiber cooling, and fiber coating) of a fiber draw system as shown in Figure 3.17 will be systematically described in the following sections.

3.3.1.1 Heating Zone
Silica glass is one of a few high temperature materials with noncrystalline properties. Its softening point ranges from 1400 to 2300°C. Furthermore, a particle-free environment of furnace cavity is imperative in order to produce a pristine state of fiber surface for high strength fiber. For fiber drawing applications, two types of electric furnaces are commonly used. They are graphite furnace and zirconia furnace. The zirconia induction or resistance furnace is designed for operation under atmospheric environment, while the graphite furnace must be operated under a nonoxidation atmosphere sustained by introduction of inert gases (typically argon, nitrogen, and helium) protecting the graphite muffle from burning.

The strength of convection current within the furnace is the sum of forced convection by gas blowing from the furnace bottom and natural convection by high temperature heating. On the other hand, the zirconia furnace induces only natural convection between the preform and the muffle. Therefore, the heat transfer coefficient around the preform in the zirconia furnace is substantially different from that in the graphite furnace. Within the draw furnace, cooling of an optical fiber preform is predominantly due to convection, but heating is thermal radiation emitting from the furnace wall. In consideration of those complex dynamics, designing of a reliable furnace that can not only provide a long furnace lifetime but also produce a low-loss and high strength fibers is challenging.

Draw Tension and Hot Zone Profile Inside a Furnace   To understand the relationship between draw tension and neck-down shape of the preform in the furnace, we take an elongational and Newtonian flow model. It assumes that

$$\sigma_{zz} = 3\eta\left(\frac{\partial V_z}{\partial z}\right) \tag{3.22}$$

where V_z is the velocity component in the z direction within the neck-down region in the furnace and we assume V_z is a function of z only. σ_{zz} is the axial (z) stress component acting on the surface normal to the z direction. This is a valid assumption for a long neck-down region with a high viscosity. η is the silica viscosity and strongly depends on temperature, which may be expressed in terms of activation energy ΔE by the following formula:

$$\eta = \eta_0 \exp\left(\frac{\Delta E}{R_g T}\right) \tag{3.23}$$

where η_0 is pre-exponential constant and R_g is the gas constant. When a fiber of diameter d_2 is pulled in the z direction at a speed of V_2, draw tension F can be formulated from the above σ_{zz} expression together with the mass conservation $V_2(d_2/2)^2 = V_z(z)R(z)^2$ into the following form [21]:

$$F = -\frac{3}{2}\pi\eta(T)V_2 d_2^2 \left(\frac{1}{R}\right)\frac{dR}{dz} \tag{3.24}$$

In the above equation, silica viscosity η and neck shape $R(z)$ are evaluated at the point where the maximum furnace temperature T_f occurs in the furnace (see Figure 3.18). A longer neck shape obtained with a longer heating zone gives a lower value of $(1/R)dR/dz$ and viscosity, thereby reducing the tension F. Figure 3.18 shows a qualitative description of two neck shapes formed by the two different heating elements presenting a longer neck formation under the broader heat zone. The temperature profiles along the susceptor produced by the two- and four-heating elements are also given in the figure, where the dotted curve represents the temperature profile obtained from the two heating elements and the solid line from the four elements [27]. Usually the heating elements are made of graphite and thinner wall zone forms a resistive heating element. The heat zone is determined by the measurement of temperature distribution along the furnace wall [28].

As shown in Figure 3.18, the broad temperature distribution along the furnace wall yields a mild temperature gradient across the preform diameter and thus provides a low fiber draw tension. On the other hand, the narrow heat zone as indicated by the dotted line in the figure induces an extremely high draw tension in case of drawing a fiber from a large preform. Therefore, to achieve a high speed draw, a low tension at 1 m/s is essential because tension is linearly proportional to draw speed. In case of drawing the nominal diameter fiber (125 μm), the tension at the draw speed of 1 m/s should be less than 10 g regardless of preform size. A high tension induces strength degradation due to the fatigue effect, producing an extremely weak fiber and causing the fiber to break during drawing [25]. Sakaguchi and Kimura [29] obtained a draw speed as high as 20 m/s. Later, Kobayashi et al. [30] achieved ~22 m/s (1300 m/min) that has been so far known to be the maximum draw speed in a two-layer coating on the fiber in technical literature.

Neck-Down Region in Furnace   In order to optimize heating zone design, it is imperative to understand how the neck-down region affects the geometry or the

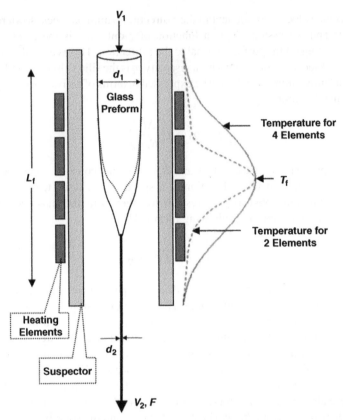

FIGURE 3.18 Furnace configuration with two heating elements (dotted line) and four elements (solid line), and schematic illustration of neck elongation by pulling.

properties of drawn fibers, namely, fiber diameter variations, fiber strength, optical loss, core index change, germanium defect generation, and so on [31–33]. Lee and Jaluria [34,35] formulated the complicated governing equations with the minimum approximations and assumptions to solve the thermal radiation problem of transparent material in the fiber draw process. Also, Kaminski [36] studied the heat transfer problem with combined radiation and convection in a fiber draw furnace, using a two-dimensional, axial-symmetric energy equation. In this case, the temperature distribution in a glass preform was calculated based on a measured neck shape. However, the main interest of this section is to present the accurate prediction of a neck shape in a draw furnace that can be achieved by solving the energy and momentum equations simultaneously. Note that the two equations are coupled by the viscosity equation of silica.

Most optical fibers are drawn from a silica preform by peripherally heating as described in Figure 3.19. Given the rod feed and fiber draw speeds, the drawdown ratio is determined by the conservation of mass. As a fused silica rod having a diameter d_1 proceeds through the heat zone of the furnace, its temperature and viscosity will vary

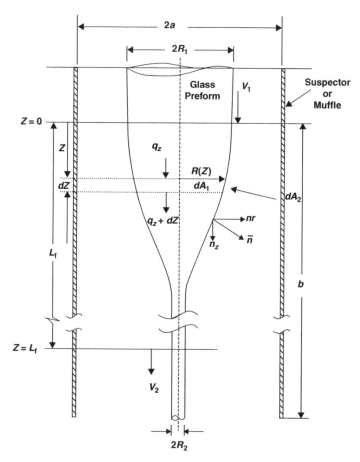

FIGURE 3.19 Physical model and its coordinate system for optical fiber drawing furnace.

along its axial direction z due to the furnace heat distribution. The extensional deformation of the preform rod by pulling yields a "neck-down" shape within the heating zone. This neck-down shape described by $R(z)$ is influenced by different drawing conditions and in turn influences the fiber diameter uniformity [37,38]. The $R(z)$ is determined by the draw parameters such as furnace temperature, draw tension, draw speed, preform diameter, fiber diameter, and so on, and is found by solving the mass, momentum, and energy equations simultaneously.

In the drawing process, the viscosity in the viscoelastic flow of silica glass is so high that the Reynolds number of the flow field is small and the inertia terms of the momentum equations can be ignored. Therefore, the momentum equation can be written in the following way [21]:

$$\rho g + \frac{\partial \sigma_{zz}}{\partial z} + \frac{1}{r}\frac{\partial}{\partial r}(r\sigma_{rz}) = 0 \tag{3.25}$$

$$\frac{\partial \sigma_{rr}}{\partial r} + \frac{\partial \sigma_{rz}}{\partial z} + \frac{\sigma_{rr} - \sigma_{\theta\theta}}{r} = 0 \qquad (3.26)$$

where ρ and g are the density of silica glass and the acceleration due to gravity, respectively. The boundary conditions of the neck-down region are

$$\left. \begin{array}{ll} V_z = V_1, & \text{at } z = 0 \\ V_z = V_2, & \text{at } z = L_f \end{array} \right\} \qquad (3.27)$$

The boundary conditions are set to be the two points where the deformation of the preform starts and ends. Therefore, the beginning of the neck-down at $z = 0$ maintains the preform size ($R_1 = d_1/2$) at which the feeding speed is V_1, while its end at $z = L_f$ maintains the fiber size ($R_2 = d_2/2$) at which the draw speed is V_2, as shown in Figure 3.19. The free boundary condition at $r = R(z)$ is given by

$$\sigma_{ij}\bar{n} + 2\alpha H_c \bar{n} = 0 \qquad (3.28)$$

where α is the surface tension, H_c is the mean curvature of the surface, and σ_{ij} is the stress tensor defined by

$$\sigma_{ij} = \begin{vmatrix} \sigma_{zz} & \sigma_{rz} & 0 \\ \sigma_{rz} & \sigma_{rr} & 0 \\ 0 & 0 & \sigma_{\theta\theta} \end{vmatrix} \qquad (3.29)$$

Note that the subscripts i and j of σ_{ij} refer to a stress component acting in the i direction on the plane perpendicular to the j direction and $\sigma_{ij} = \sigma_{ji}$. The H_c can be written as

$$H_c = \frac{1}{2}\left\{\frac{1}{R} + \frac{1}{R}\left(\frac{dR}{dz}\right)^2 - \frac{d^2R}{dz^2}\right\}\left\{1 + \left(\frac{dR}{dz}\right)^2\right\}^{-(3/2)} \qquad (3.30)$$

From equation 3.30, the free boundary condition can be divided into two equations in terms of z and r components:

$$\sigma_{zz}n_z + \sigma_{rz}n_r + 2\alpha H_c n_z = 0 \qquad (3.31)$$

$$\sigma_{rz}n_z + \sigma_{rr}n_r + 2\alpha H_c n_r = 0 \qquad (3.32)$$

Here $\bar{n} = n_z\bar{z} + n_r\bar{r}$. The z and r components of the unit vector $\bar{n}$ can be expressed as

$$n_z = -\frac{R'}{(1 + R'^2)^{1/2}} \qquad (3.33)$$

$$n_r = \frac{1}{(1 + R'^2)^{1/2}} \qquad (3.34)$$

where $R' = (dR/dz)$.

By multiplying equation 3.25 by $2\pi r\, dr$ and integrating it from 0 to $R(z)$, one obtains

$$\rho g R^2 + 2\int_0^R r\frac{\partial\sigma_{zz}}{\partial z}\,dr + 2R\sigma_{rz}(R) = 0 \tag{3.35}$$

From equation 3.31, the solution of equation 3.35 can be obtained subject to the boundary conditions at both ends of neck-down region:

$$V_z(z) = C_1\int_0^z\frac{dz}{\eta R^2} - \frac{\rho g}{3}\int_0^z\frac{1}{\eta R^2}\left(\int_0^z R^2\,dz\right)dz - \frac{4\alpha}{3}\int_0^z\frac{1}{\eta R^2}\left(\int_0^z R'RH_c\,dz\right)dz_1 + V_1 \tag{3.36}$$

where

$$C_1 = \left\{(V_2 - V_1) + \frac{\rho g}{3}\int_0^{L_f}\frac{1}{\eta R^2}\left(\int_0^z R^2\,dz\right)dz + \frac{4\alpha}{3}\int_0^{L_f}\frac{1}{\eta R^2}\left(\int_0^z R'RH_c\,dz\right)dz\right\}\left(\int_0^{L_f}\frac{dz}{\eta R^2}\right)^{-1}$$

The velocity component V_r of $\overline{V} = \{V_z, V_r\}$ can be determined from the equation of continuity, $V_r = -(1/2)rV'_z$.

Now, let us consider the formulation of energy equation that is coupled with the momentum equation through the viscosity of the material. From Figure 3.19, the heat flux leaving differential area dA_2 of the susceptor will be partially absorbed by differential area dA_1 in the neck-down region of the preform. A portion of the absorbed energy will be conducted through the neck-down region while the rest will be lost to the surroundings in the form of thermal radiation and convection. For such a diathermanous or transparent material the radiative transfer within the material is significant, especially at elevated temperatures. The conduction due to radiation contribution must be added to the nonradiative heat conduction. Therefore, the apparent thermal conductivity, K, will be the sum of the nonradiative conductivity, K_c, and radiative conductivity, K_r. The K_r is expressed as

$$K_r = \frac{16n^2\sigma T^3}{3\gamma} \tag{3.37}$$

where n is the refractive index of silica glass, σ is the Stefan–Boltzmann constant, and γ is the absorption coefficient of silica glass. For the value of $\gamma = 0.2\,\mathrm{cm}^{-1}$ the conductivity K_r is calculated to be approximately 0.25 W/(cm °C) in the temperature range of 1500–2000°C and the value of K_c is ~0.05 W/(cm °C). The density and specific heat of fused silica are $\rho = 2.2\,\mathrm{g/cm^3}$ and $C_p = 0.25\,\mathrm{cal/(g\,°C)}$ (see Table 3.5). When the heat transfer coefficient surrounding the preform in the furnace is defined by h, the Biot number ($hd_1/4K$) is calculated to be less than 0.1. Thus, the temperature

TABLE 3.5   Material Properties of Silica [39]

Thermal conductivity, K_c (cal/(s cm °C))	0.0064
Density, ρ (g/cm³)	2.2
Specific heat, C_p (cal/(g °C))	0.25
Elastic modulus, E (kg/cm²)	8.25×10^5
Linear expansion coefficient, α_e (°C^{-1})	5.5×10^{-7}
Surface tension, α (dyne/cm) at 2000°C	310

gradient across the preform rod with its diameter d_1 can be neglected. If the rod diameter is less than 1.5 cm, the condition can be easily satisfied and the energy equation can be reduced to a one-dimensional problem. However, for a large size preform, modeling requires the formulation of a three-dimensional radiative heat conduction equation. Since the analysis of the neck-down region in a three-dimensional case is extremely complex and beyond the scope of this book, the one-dimensional coupled radiative heat conduction equations will be illustrated to describe the temperature distribution along the elongated silica glass rod.

The heat flux q_z and heat conduction in the z direction with a variable cross-section $(A = \pi R^2)$ can be formulated by the energy balance in the control volume (Figure 3.19) as

$$q_z = -K_c \frac{dT}{dz} - K_r(T) \frac{dT}{dz} \tag{3.38}$$

$$\rho C_p V_z(z) \frac{dT}{dz} + \frac{d(q_z A)}{A\,dz} + \frac{2\{\varepsilon\sigma(T^4 - T_0^4) + h(T - T_0)\}}{R} = \frac{2Q_{2-1}}{R} \tag{3.39}$$

where ε is the emissivity of silica glass and $V_z = R_1^2 V_1 / R^2$. A detailed expression of Q_{2-1} is given in Ref. [40]. We need to define two proper boundary conditions within the neck-down region in order to solve the coupled first-order differential equations 3.38 and 3.39. The softening point T_s of silica glass is taken at two boundary points, $z=0$ and L_f at which the deformation of the preform is to start and end as mentioned earlier. Thus, the region between these two points will maintain a temperature higher than T_s ($\sim$1600°C):

$$\left.\begin{array}{ll} T = T_s, & \text{at } z = 0 \\ T = T_s, & \text{at } z = L_f \end{array}\right\} \tag{3.40}$$

However, it must be noted that the values of emissivity ε and heat transfer coefficient h depend strongly on the thickness of a sample. For example, the emissivity of silica glass changes significantly if the thickness is less than 0.2 cm [41]. Since the dimensional reduction in the neck-down region is very large (1.5–0.01 cm), the emissivity is taken to be 0.6 in the region where the diameter of the neck-down region is greater than 0.2 cm, while in the fiber region it is assumed to be 0.1.

In the intermediate region the two values are smoothly connected with a polynomial. Similarly, the heat transfer coefficient h in the intermediate region defined above is also obtained by smoothly connecting the value of 1.5×10^{-2} W/(cm^2 °C) in the bulk region to that of 3×10^{-2} W/(cm^2 °C) in the fiber region [21]. Thus, the following iterative computational scheme is used to calculate the temperature distribution $T(z)$ along the neck-down region, velocity $V_z(z)$, and the neck shape:

(i) An arbitrary neck shape $R(z)$ is assumed for a set of drawing conditions (V_2, d_1, d_2, T_f, F).

(ii) The temperature $T(z)$ can be obtained from equations 3.38 and 3.39 using the arbitrary neck shape $R(z)$. Then, the viscosity $\eta(T)$ is determined from the relation between the viscosity and the temperature of silica glass, namely,

$$\eta(T) = \eta_0 \exp\left(\frac{\Delta E}{R_g T}\right)$$

(iii) The viscosity and the assumed neck shape are substituted into equation 2.3.1.15 to obtain the velocity $V_z(z)$, which will yield a new neck shape $R(z)$ from the conservation of mass:

$$R(z) = R_1 \sqrt{\frac{V_1}{V_z}} \tag{3.41}$$

The iterative steps of (i)–(iii) are repeated until the neck shape is converged upon. Three or four iterations are sufficient to reach the final solutions and the predictions are compared with experimental measurement as in Figure 3.20.

Similarly, Choudhury et al. [42] also developed an analytic and numerical model to calculate the shape of the neck-down region formed during fiber drawing in the furnace. The result is shown in Figure 3.20b. Figure 3.21 shows the temperature distributions $T(z)$ for three different values of draw speeds. Particularly in this experiment, a 100 μm diameter fiber was drawn from a 1.2 cm diameter fused silica rod at three different drawing speeds ($V_2 = 100$, 200, 300 cm/s). The corresponding draw tensions are measured to be 16, 32, and 48 g, respectively.

The tension can be also calculated from equation 3.24 by using the values of $R(z)$, dR/dz, and viscosity at the T_f point. The measured and the calculated tensions are compared and given in Figure 3.22a. The results of this low speed drawing case indicated that the draw tension was in a linear proportion with the draw speed. Also, Choudhury et al. [42] modeled rigorously the radiation transport in a fiber draw furnace and predicted the tension at the draw speed of 1 m/s to be ~10 g. Rajala et al. [43] reported the experimental results of high speed drawing. The measured data for draw tension versus drawing speed in high speed drawing also show a linear relationship as given in Figure 3.22b. In these low speed draw cases, the lengths of neck-down region were $L_f = 20$, 25, and 29 cm, respectively. Vaidya and Mihalacopoulos [44] investigated the neck-down profile in high speed drawing. It was shown

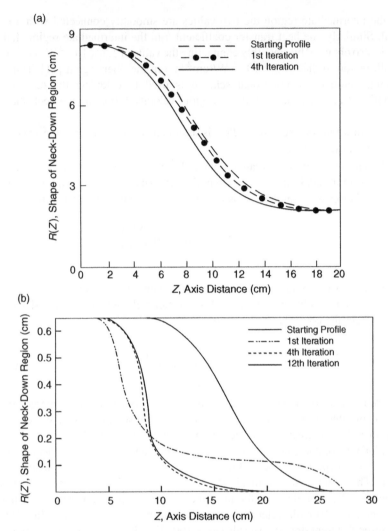

FIGURE 3.20   (a) Neck shapes: (solid line) after fourth iteration; (solid and dotted line) after first iteration; and (broken line) measured neck shape. (b) Predicted by Choudhury et al. [42].

from both low and high speed drawing cases that the length of neck-down also varies linearly with draw speed.

3.3.1.2 Cooling Zone   The fiber exiting from the bottom of the furnace by free drawing is cooled rapidly by the surrounding air or by blowing a gas (helium). Most of analyses dealing with fiber cooling were formulated on the basis of an approximated method such as Reynolds analogy or boundary layer approximations. However, it must be noted that the calculation of fiber cooling based on the boundary layer approximations may not be applicable because the boundary layer thickness is of the

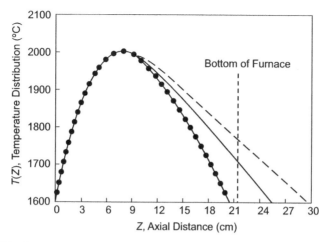

FIGURE 3.21   Temperature distributions along the neck-down region for three different draw speeds ($V_2 = 100$, 200, 300 cm/s).

same order of magnitude as the fiber diameter. Therefore, a new novel analysis on the heat transfer problem associated with flow over a needle-like body such as a fiber should be explored, since the fiber cooling time depends mainly on an accuracy of the value of heat transfer coefficient, h [45,46]. Paek and Kurkjian [39] developed a method for calculating the fiber cooling time while a fiber is being drawn. The analysis is applicable to obtaining the cooling rate regardless of convection modes (natural convection or forced convection), provided that the heat transfer coefficient is known. Conversely, using the equation of fiber cooling, one can also obtain accurately the heat transfer coefficient from a measured fiber temperature at a certain position and draw parameters. Importantly, it is worthy to remember that the cooling time is the key factor determining a fiber coating speed.

In the cooling rate calculation, the fiber temperature near the bottom of the furnace is designated as T_s, the softening point. Setting $z = 0$ at the bottom of the furnace as an origin, we can formulate the governing equation along the fiber by taking energy balance (see the cooling zone in Figure 3.17). The physical size of the fiber diameter is so small that the emissivity of the silica fiber is also negligibly small [41] and, thus, the radiative contribution to cooling can be ignored. The transverse temperature gradient can also be neglected because of small dimension of fiber diameter. Therefore, one-dimensional steady-state heat conduction equation can be formulated based on the energy balance from a local control volume in the fiber cooling zone as [22]

$$V_2 \frac{dT}{dz} = \kappa \frac{d^2T}{dz^2} - \frac{4h(T - T_0)}{\rho C_p d_2} \qquad (3.42)$$

where κ is the thermal diffusivity of silica glass. Since the fiber is to cool down to ambient temperature T_0, the boundary conditions can be set as

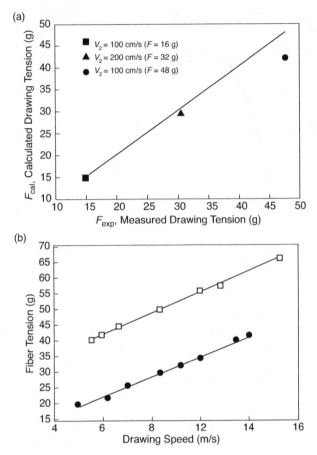

FIGURE 3.22 (a) Comparison of calculated and measured drawing tension and (b) draw tension versus drawing speed with two furnace input power, 85% (open square) and 90% (solid circle) [43].

$$T = T_{\rm s}, \quad \text{at } z = 0 \atop T = T_0, \quad \text{at } z = \infty \Bigg\} \tag{3.43}$$

The following nondimensional variables and parameters are introduced:

$$T^* = \frac{T - T_0}{T_{\rm s} - T_0}, \qquad z^* = \frac{z}{d_2}$$

$$\Lambda = \frac{V_2 d_2}{\kappa}, \qquad H = \frac{4 h d_2}{K_{\rm c}} \tag{3.44}$$

With equations 3.43 and 3.44, equation 3.42 can be rewritten in the following forms:

$$\frac{d^2T^*}{dz^{*2}} - \Lambda\frac{dT^*}{dz^*} - HT^* = 0 \tag{3.45}$$

$$\left.\begin{array}{ll} T^* = 1, & \text{at } z^* = 0 \\ T^* = 0, & \text{at } z^* = \infty \end{array}\right\} \tag{3.46}$$

Assuming that all material properties remain constant during the drawing period (see Table 3.5), one can find the solution of equation 3.45, namely,

$$T^* = \exp\left\{ -\frac{1}{2}(\sqrt{\Lambda^2 + 4H} - \Lambda)z^* \right\} \tag{3.47}$$

Since a fiber draw speed V_2 is much faster than thermal diffusion speed, equation 3.47 can be further simplified shown as follows:

$$\left.\begin{array}{c} \Lambda^2 \ll 4H \\[2mm] \text{or} \\[2mm] \left(V_2\dfrac{d_2}{\kappa}\right)^2 \gg \dfrac{16hd_2}{K_c} \end{array}\right\} \tag{3.48}$$

Hence, it can be reduced to a simpler form:

$$\frac{T - T_0}{T_s - T_0} = \exp\left(-\frac{4hz}{\rho C_p d_2 V_2} \right) \tag{3.49}$$

From drawing experiments, h is found to be 7.2×10^{-3} cal/(cm^2 s °C). T_s and T_0 are chosen to be 1600 and 25°C, respectively [22]. The draw speed V_2 can be replaced with z/t, where t is the cooling time. Then, the following expression for fiber cooling is obtained:

$$\frac{T - T_0}{T_s - T_0} = \exp\left(-\frac{4ht}{\rho C_p d_2} \right) \tag{3.50}$$

Numerous calculations and measurements have been made for the cooling rate of fibers produced at a moderate temperature ($\sim$1250°C) [33]. Glicksman [47] estimated the Nusselt number by using the boundary layer approximation to calculate the time required for cooling from 1750 to 200 and 100°C. Even though all the properties of materials involved are assumed to be independent of temperature, it is difficult to exactly calculate the heat transfer coefficient h around an extremely thin body like a fiber. For example, Arridge and Prior [46] measured the cooling rate of silica fibers of different diameters drawn at different speeds. The experimental data are given in Figure 3.23a.

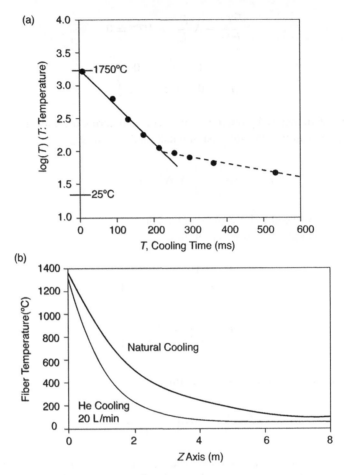

FIGURE 3.23   (a) Comparison of calculated and experimental cooling rate: (circles) experiments [46]; (solid line) theoretical calculation with $h = 7 \times 10^{-3}$; (dotted line) theoretical calculation with $h = 1 \times 10^{-3}$. (b) Computed fiber temperature as a function of the cooling condition [48].

   The heat transfer coefficient h was calculated from these measured data by using equation 3.50 and its value was obtained to be $\sim 7 \times 10^{-3}$ cal/(s cm^2 °C) for the high temperature region ($T > 100$°C) and 1×10^{-3} cal/(s cm^2 °C) for $t > 300$ ms in the region of $T < 100$°C. Subsequently, these results were also experimentally confirmed [49,50], and Vaskopoulos et al. [51] further studied active cooling of optical fibers. In this active cooling study [49,51], a cooling apparatus was designed for opposing flow to increase the heat transfer coefficient h. Subsequently, Vaskopoulos et al. [49] found experimentally that the heat transfer effect is approximately fivefold increased simply by changing nitrogen to helium gas. This is believed to be due to the increase in the thermal conductivity approximately six times. Gossiaux et al. [48]

studied numerically on fiber cooling by using a water-cooled tube in which a gas is introduced to increase heat transfer. The calculated results are shown in Figure 3.23b.

In addition to the method using equation 3.50, there are two other ways of determining the heat transfer coefficient h accurately by using equation 3.49. The methods are called the one- and two-point measurement techniques, which can be applied to both natural cooling and forced cooling. In the one-point measurement technique, the temperature measurement system is placed at the distance L between the furnace and the coating applicator to measure the fiber temperature, while the draw speed is changed (see Figure 3.18). For example, when a fiber is drawn at two different speeds of V_2^1 and V_2^2, the corresponding temperatures, T_1 and T_2, will be recorded by the temperature measurement system. Thus, the substitution of the measured data into equation 3.49 gives the following relation:

$$\frac{T_2 - T_0}{T_1 - T_0} = \exp\left\{ - \frac{4hL}{\rho C_p d_2} \left(\frac{1}{V_2^2} - \frac{1}{V_2^1} \right) \right\} \tag{3.51}$$

From the above equation, the heat transfer coefficient h can be obtained since all the variables and parameters are known except the coefficient h.

Similarly, the two-point measurement technique can also determine the heat transfer coefficient h by measuring the fiber temperatures T_1 and T_2 at two different locations, L_1 and L_2, while a fiber is being drawn at a speed of V_1. The following expression can be obtained from equation 3.49, namely,

$$\frac{T_1 - T_0}{T_2 - T_0} = \exp\left\{ - \frac{4h}{\rho C_p d_2 V_2} (L_1 - L_2) \right\} \tag{3.52}$$

Therefore, the heat transfer coefficient h can be found from the above equation. Considering the difficulties of analyzing high temperature heat transfer and flow problems associated with fiber drawing, we found that the applications of equation 3.49 have led to the convenient and simple ways of determining the heat transfer coefficient h as demonstrated by equations 3.51 and 3.52 [39].

As discussed earlier, let us define the fiber temperature at the entrance of coating applicator as T_c. From equation 3.49, the T_c values for different coating speed V_2 in terms of L are compared with the measured values and shown in Figure 3.24.

At this point it must be noted that coating speed entirely depends on the temperature T_c. In fact, the value of T_c for polymer coating materials is generally less than $100°C$. As mentioned earlier, its value is taken to be $80°C$. Putting $T = T_c$ and $z = L$ into equation 3.49, the following relation can be obtained:

$$L = \left(\frac{\rho C_p d_2 \Omega}{4h} \right) V_2 \tag{3.53}$$

where $\Omega = \ln[(T_s - T_0)/(T_c - T_0)]$.

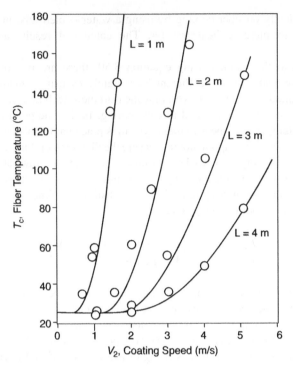

FIGURE 3.24 Fiber temperature T_c versus coating speed V_2 for different locations of coating applicator.

The equation shows that distance L is linearly proportional to draw speed. This gives us the important result that the height of a draw tower increases linearly with a draw speed. Thus, this allows us to estimate a draw tower height required for achieving a desired draw speed. Therefore, in designing a high speed coating system the following considerations must be taken: (1) to provide a sufficient distance between the furnace and coating applicator and (2) to attach a gas-blowing device around the fiber to cool it effectively [22,48,49]. Nowadays, it is common to find the draw towers higher than 22 m in fiber manufacturing plants.

3.3.1.3 Coating Zone   The application of a thin layer coating on an optical fiber is to provide a protection of its surface and to preserve the pristine state of its strength. Many candidates for coating materials can be grouped into two categories: one includes organic materials, for example, UV curable material (epoxy acrylate, urethane), thermally curable material (silicone), hot melt coating (EVA), and so on. The other includes inorganic materials such as metallic coating, silicon nitride, carbon coating, and so forth. However, all those materials must meet a variety of the stringent coating requirements for mechanical and optical properties to assure a good performance of coated fibers. To avoid possible contamination and damage on the fiber surface, the coating apparatus is in-line with the drawing system so that the fiber can be immediately coated while it is being drawn.

Since the UV curable polymer materials show much faster response in the process of photon-assisted curing compared with thermal curing, they are found to be the choice of high speed coating among different polymerization schemes; nowadays, they are exclusively used in fiber manufacturing. In implementing a high speed fiber coating technology, there are two fundamental drawbacks that limit the draw speed and affect fiber strength as well. One is the fiber cooling rate problem. The other is the high shear rate problem as draw speed increases. Therefore, this section will only focus on the polymer coating technology practiced in the fiber production.

Coating Mechanics   Figure 3.25 shows a schematic diagram of a pressurized coating applicator. In the pressurized coating die with a length d, the coating material velocity u along the z direction can be expressed as [52],

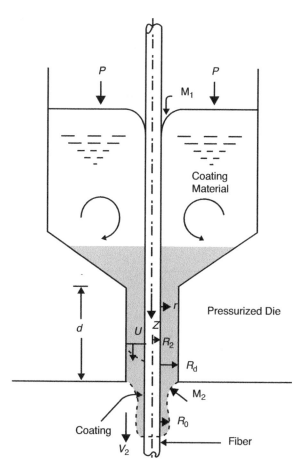

FIGURE 3.25   Schematic diagram of pressurized coating applicator [52].

$$\frac{\partial^2 u}{\partial r^2} + \frac{1}{r}\frac{\partial u}{\partial r} = \frac{1}{\mu}\frac{\partial p}{\partial z} \tag{3.54}$$

where μ is the viscosity of a coating material and p is the pressure in the die. The boundary conditions are

$$\left.\begin{array}{ll} u = V_2, & \text{at } r = R_2 \\ u = 0, & \text{at } r = R_d \end{array}\right\} \tag{3.55}$$

where $2R_2$ and $2R_d$ are the diameters of the fiber and die, respectively.

From the above equation, the pressure gradient can be obtained

$$\frac{\partial p}{\partial z} = \frac{4\mu V_2}{R_2^2 - R_d^2 - 2R_2^2\ln(R_2/R_d)} \tag{3.56}$$

When the pressure gradient is replaced by $\Delta p/d$, it can be noticed that the pressure is directly related to the viscosity and the draw speed. Therefore, for a given viscosity at a set temperature, the optimum pressure can be chosen to coat a fiber at a desired speed. However, we notice that a high shear force damages the fiber surface, resulting in a weak strength of the fiber. To avoid this situation, the coating material is normally heated to reduce its viscosity as mentioned earlier.

The solution u can be obtained as

$$u = \frac{V_2}{R_2^2 - R_d^2 - 2R_2^2\ln(R_2/R_d)}\left(r^2 - R_d^2 - 2R_2^2\ln\frac{r}{R_d}\right) \tag{3.57}$$

Therefore, the flow rate Q, which can be obtained from the expression $2\pi\int_{R_2}^{R_d} ur\,dr$, is needed to determine the coated fiber diameter $2R_0$. From the conservation of mass, the coated fiber diameter $2R_0$ is found to be

$$R_0^2 = -\frac{(R_2^2 - R_d^2)^2}{2(R_2^2 - R_d^2 - 2R_2^2\ln(R_2/R_d))} \tag{3.58}$$

This equation shows that when the pressure is optimized, the coated fiber diameter only depends on the sizes of fiber and die [52]. The coating thickness defined by $H_\infty = (R_0 - R_2)$ is nominally $50\,\mu m$, which must be maintained uniformly on the fiber surface regardless of fiber draw speed. Kobayashi et al. [30] developed high speed UV curable resins that were successfully applied for coating a fiber at a rate of $21.7\,m/s$ ($1300\,m/min$). These coated fibers demonstrated the satisfactory results that met the stringent requirements for the mechanical properties and the transmission characteristics. Koaizawa et al. [53] studied the mechanism of the coating diameter decrease due to insufficient cooling, while the pressure was not optimized. The two-layer coating model provides useful information on a concentric geometry of two different viscosities around a fiber moving at a high speed.

Shear Force in Coating Application   In coating, it is always observed that two menisci are formed. One is at the free surface where the fiber enters, and the other at the fiber exiting side of the coating die. They are designated by M_1 and M_2, as shown in Figure 3.25. When the fiber drawing speed exceeds a certain point, the meniscus (M_1) at the fiber entrance side collapses due to the T_c rise beyond the tolerable point. The meniscus collapse causes no wetting on the fiber surface, consequently inhibiting fiber coating. The second meniscus (M_2) formed at the exiting side also controls the uniformity of a coating. The stability of the lower meniscus is due largely to the shear induced within the applicator. The applicator die of length d, where the coating thickness is controlled, has a thin tube shape with a diameter $2R_d$. In this case, the flow motion in the coating die is well described by Chida et al. [54]. With the dimensionless variable ζ defined by $\zeta = (r - R_2)/(R_d - R_2)$, the profiles of normalized velocity $u^* = u/V_2$ are similar along the axis. In the above, r refers to the radial distance measured from the fiber axis and R_2 is the fiber radius. The shear rate $\dot{\gamma}$ acting on the fiber surface can be expressed as

$$\dot{\gamma} = \frac{\partial u}{\partial r}\bigg|_{R_2} = \frac{\partial u^*}{\partial \zeta}\bigg|_0 \frac{V_2}{R_d - R_2} \qquad (3.59)$$

The value of $(\partial u^*/\partial \zeta)|_0$ is constant for a given pressure p and designated as $B(p)$.

From equation 3.59, it can be found that the shear rate induced during high speed coating is also linearly proportional to the fiber draw speed, V_2. The shear stress τ_{rz} is given by

$$\tau_{rz} = \mu\dot{\gamma} \qquad (3.60)$$

Thus, the pulling tension F_c in coating can be calculated by integrating the shear stress τ_{rz} over the fiber surface ($2\pi R_2 d$) in the die. Finally, the coating tension F_c can be expressed by

$$F_c = \frac{2\pi R_2 \mu B(p) V_2 d}{R_d - R_2} \qquad (3.61)$$

The above equation explicitly shows the functional relationships among the parameters and variables, helping in understanding the effects of the parameters such as coating speed, fiber radius, viscosity of coating material, and coating die dimensions on the tension F_c [20]. As the draw tension F is a linear dependence on the draw speed, so the coating tension F_c varies linearly with the draw speed. In high speed coating, this shear must be minimized to produce a desirable coating. According to equation 3.61, there are two ways to lower the value of F_c. One is simply to shorten the die length d. The other is to heat the coating material to effectively reduce its viscosity. However, the heating temperature is normally limited to 60°C because of the gelation problem of polymer. At 50°C the viscosity becomes 9 P, while it is 60 P at 25°C. Therefore, the total pulling tension, F_T, in coating is the sum of F and F_c. Sakaguchi and Kimura [29] demonstrated successfully high speed coating using the pressurized applicator shown in Figure 3.25. They achieved a coating speed as high as 20 m/s with a pressure of $P = 7 \, \text{kg/cm}^2$ and $F_T = 120 \, \text{g}$. The coating application parameters used were $2R_2 = 125 \, \mu\text{m}$, $2R_d = 0.35 \, \text{mm}$, die length $(d) = 0.2 \, \text{mm}$, and $\mu(T) = 9 \, \text{P}$.

Photopolymerization of Coating Material   In a prepolymer solution mixed with a photoinitiator, polymerization process can be described in three steps: (1) initiation, (2) propagation, and (3) termination. The first step is that the initiator ($\overline{S}$) absorbs the photon ($h\nu$) and a sufficient absorption cleaves the chemical bonds of molecules to form the free radicals [57–59]. The excitation and fragmentation processes are described in the following way:

$$\overline{S} + h\nu \rightarrow \overline{S}*, \qquad \overline{S}* \rightarrow 2\overline{R}\bullet \tag{3.62}$$

where $\overline{R}\bullet$ refers to the free radical. In the subsequent step, these radicals react rapidly with a monomer $\overline{M}$, and progressively grow the chains to cross-link into the polymer network [59]:

$$\overline{M} + \overline{R}\bullet \rightarrow \overline{RM}\bullet, \qquad \overline{RM}\bullet + \overline{M} \rightarrow \overline{RMM}\bullet \quad \text{(propagation)} \tag{3.63}$$

This propagation continues until it is terminated by hydrogen. However, the required UV power is determined by the initial step described by equation 3.62. Let us consider a case of fiber drawing with a speed of V_2 (cm/s). The fiber coated to a thickness by passing through a coating die is immediately cured by UV radiation. The coating thickness H_∞ is defined by $H_\infty = (R_0 - R_2)$, where $2R_0$ is the coated fiber diameter. For a photoinitiator having an activation energy ΔE_c (kcal/mol) and molar concentration $[c]$, the required UV power P_c (W) can be formulated as

$$P_c = \pi \, \Delta E_c \cdot [c](R_0^2 - R_2^2) \cdot V_2 \tag{3.64}$$

The concentration level of photoinitiator added to the prepolymer solution is 3–5% of which 4% concentration corresponds to $\sim 2 \times 10^{-4} \, \text{mol/cm}^3$ (see Table 3.6). Thus, the substitution of $2R_0 = 225 \, \mu\text{m}$ and $2R_2 = 125 \, \mu\text{m}$ with $[c]$ and ΔE_c into equation 3.64 gives the following result:

$$P_c = 0.016V_2 \tag{3.65}$$

For example, in case of a draw speed of 100 cm/s, the power is calculated to be 1.6 W. However, it must be noted that equation 3.65 is formulated based on the assumption that all the power supplied to the lamp is converted to UV light and then totally

TABLE 3.6   Data of Photoinitiator Irgacure 651 [60]

Activation energy, ΔE_c (kcal/mol)	70
Density, ρ_c (g/cm^3)	1.2
Molecular weight, M_w	256
Molar absorption coefficient, α_λ (L/(mol cm)) at $\lambda = 340 \, \text{nm}$	320

absorbed in the coating material. In a real system, there will be the component losses of the system, for example, in spectral power conversion of the lamp, in the reflector unit itself, and so on.

3.3.2   Impact of Drawing on Fiber Characteristics

It should be mentioned that fiber cooling and reheating for annealing influence the optical and mechanical characteristics of a fiber such as transmission loss and fiber strength [55,56]. The lightguiding fiber consists of the core and cladding that are two different kinds of materials. To produce a low-loss and high strength fiber, it is necessary to fully investigate the draw mechanics and to quantitatively identify the effects on the fiber performance. For example, the effect of draw tension on the fiber strength is directly related to the compositions of the core and cladding materials [61]. Certain combinations of the composition materials for core and cladding can change the state of axial component of the residual stresses in the cladding from the tensile stress to compressive stress or vice versa [39]. Many investigators [62,63] studied the tensile strength of fused silica fibers as a function of fiber diameter. They showed that the finer fibers tend to give higher strength than larger diameter fibers. In absence of gas blowing, the cooling rate is basically determined by the fiber diameter. First, let us examine the effect of fiber cooling rate on the fiber strength with a low draw tension (low draw speed). A quenching time estimated to be of order of the Maxwell relaxation time $\tau = \eta/G$ inhibits the formation of flaw in glass fibers, where η is the viscosity of glass and G is the shear modulus [64]. This implies higher strength value for the fiber than that for bulk glass. Hillig [65] measured the ultimate strength at liquid nitrogen temperature ($-196°C$) to be $140,000 \, kg/cm^2$ (2 million psi). In air and at room temperature the maximum strength is approximately $70,000 \, kg/cm^2$ (1 million psi). Griffith [26] also estimated the theoretical strength of a silica fiber based on the intrinsic bonding forces of molecular structure of a solid. He reached a conclusion that $0.1E$ is a good estimate of a reasonable value for the theoretical limit of the ultimate strength. Here, E is the Young's modulus of silica glass, which is $825,000 \, kg/cm^2$. In fact, the average strength of manufactured optical fibers falls far below this figure. We believe this is primarily due to surface flaws, internal defects, inclusions, minute bubbles, and so on.

Second, let us discuss the effect of fiber draw speed (draw tension) on the fiber strength in conjunction with the cooling rate. It was mentioned earlier that as the draw speed increases, the draw tension also increases in a linear fashion. Therefore, in the drawing process stresses are induced in the composite structure of a fiber, which will partially remain as residual stresses when the fiber cools down to room temperature. The residual stresses ultimately influence the fiber strength [66]. For example, the compressive state of the residual stress in the fiber cladding is considered to be favorable since it helps prevent a flaw or defect from growing. However, it is worthy of noting that the state of compression and tension of the residual stresses depends on the composition of fiber structure and the draw conditions [67,68]. Bouten et al. [69] investigated the strength of fibers with different residual tensile stresses at the surface in order to predict the lifetime of fibers. The strength and lifetime experiment determines the fatigue coefficient that is a key parameter in predicting the survival time of a fiber under a certain applied tension, when a fiber cable is installed for communications applications.

In addition, it is also important to note that the optical fiber loss is partly influenced by the drawing conditions. Hibino et al. [70] studied the optical characteristics of silica core with fluorine-doped silica cladding to clarify the effect of the drawing conditions on the optical transmission loss. They established the relationship between the draw-induced excess loss at the wavelength of 1.55 μm. The excess loss in silica core fibers decreases rapidly with increasing drawing temperature, reaching its minimum value near 2270 K. But the loss in germania-doped silica core fiber varies with drawing temperature. In this case, the increased loss is particularly pronounced in the long wavelength. This indicates that the increase in optical fiber loss depends on the draw tension that is determined by the furnace temperature, draw speed, and preform diameter.

Polymer coatings may not be impervious to moisture [71]. The moisture that permeates through the polymer coating degrades the fiber strength under a tension over a period of time due to fatigue. Considerable efforts have been undertaken for many different types of hermetic coatings. Particularly, metallic coatings such as copper, tin, silver, aluminum, and so on are attractive for hermetically sealing glass surfaces. An ideal hermetic coating material should be easily applied at a high draw speed and should not introduce any additional optical loss. Metallic coatings have unique advantages: (1) high fatigue parameter, (2) higher strength, (3) high temperature resistance, and (4) solderability [72–75]. Those advantages of metallic coatings over polymer-based coating are attractive to fiber manufacturers for the unique applications of fiber-based optical devices and components.

REFERENCES

[1] D. M. Dobkin and M. K. Zuraw (eds), "Principles of chemical vapor deposition," Kluwer Academic Press, 2003.

[2] C. K. Kao and G. A. Hockham, "Dielectric-fiber surface waveguides for optical frequencies," Proceedings of the IEE, vol. 133, pp. 1151–1158, 1966.

[3] F. C. Allard, "Fiber optics handbook for engineers and scientists," McGraw-Hill, 1990.

[4] S. E. Miller and A. G. Chynoweth (eds), "Optical fiber telecommunications," Chapter 8, Academic Press, 1979.

[5] S. R. Nagel, J. B. MacChesney, and K. L. Walker, "An overview of the modified chemical vapor deposition (MCVD) process and performance," IEEE Journal of Quantum Electronics, vol. QE-18, no. 4, pp. 459–476, 1982.

[6] T. J. Miller, D. A. Nicol, K. D. Pohl, and H. R. Clark, "Fluorine doping of pyrogenic silica," Material Research Society Symposium Proceedings, vol. 88, pp. 43–51, 1987.

[7] A. Marchall and K. R. Hallam, "Fluorine doping and etching reactions of Freon 12 in optical fiber manufacture," Journal of Lightwave Technology, vol. LT-4, no. 7, pp. 746–750, 1986.

[8] D. L. Wood, K. L. Walker, J. B. MacChesney, J. R. Simpson, and R. Csencsits, "Germanium chemistry in the MCVD process for optical fiber fabrication," IEEE Journal of Lightwave Technology, vol. LT-5, no. 2, pp. 277–285, 1987.

[9] K. L. Walker, F. T. Geyling, and S. R. Nagel, "Thermophoretic deposition of small particles in the modified chemical vapor deposition process," Journal of the American Ceramic Society, vol. 63 no. 9–10 pp. 552–558, 1979.

[10] K. L. Walker, J. H. Harvey, F. T. Geyling, and S. R. Nagel, "Consolidation of particulate layers in the fabrication of optical fiber preforms," Journal of the American Ceramic Society, vol. 63, no. 1–2 pp. 96–102, 1980.

[11] S. E. Miller and A. G. Chynoweth (eds), "Optical fiber telecommunications," Chapter 7, Academic Press, 1979.

[12] F. T. Geyling, K. L. Walker, and R. Csencsits, "The viscous collapse of thick walled tubes," Journal of Applied Mechanics, vol. 50, pp. 303–310, 1983.

[13] D. Marcuse, "Calculation of bandwidth from index profiles of optical fibers. 1: theory," Applied Optics, vol. 18, no. 12, pp. 2073–2080, 1979.

[14] D. H. Smithgall and R. E. Frazer, "Characterization of the preform stretching process," Journal of Lightwave Technology, vol. LT5, no. 12, pp. 1755–1762, 1987.

[15] H. R. Clark and M. Viriyayuthakorn, "The VAD preform stretching process: a numerical model for evaluation of internal distortion," Journal of Lightwave Technology, vol. LT-4, no. 8, pp. 1039–1047, 1986.

[16] T. J. Miller and F. T. Geyling, "One-dimensional models for the co-drawing of preform rods in tubes," Journal of Lightwave Technology, vol. LT-2, no. 4, pp. 349–354, 1984.

[17] A. Utsumi, T. Sagawa, and T. Shintani, "Modified rod-in-tube method for low loss step-index optical fiber," Optical Fiber Communications, Technical Digest, pp. 64–66, 1979.

[18] Herarus, "Application note, RIC preforms-jacketing process with 120 mm OD fused silica cylinders," AN0203-rev 1.0, July 8, 2002.

[19] F. P. Kapron, D. B. Keck, and R. D. Maurer, "Radiation losses in glass optical waveguide," Applied Physics Letters, vol. 17, pp. 423–425, 1970.

[20] U. C. Paek and C. M. Schroeder, "High speed coating of optical fibers with UV curable materials at a rate of greater than 5 m/s," Applied Optics, vol. 20, no. 23, pp. 4028–4034, 1981.

[21] U. C. Paek and R. B. Runk, "Physical behavior of the neck-down region during furnace drawing of silica fibers," Journal of Applied Physics, vol. 49, no. 8, pp. 4417–4422, 1978.

[22] U. C. Paek and C. M. Schroeder, "Fiber drawing from 7.5 cm diameter preform at high speed," Technical Digest of OFC/FS'88, New Orleans, LA, paper PD-3, 1988.

[23] Z. Yin and Y. Jaluria, "Thermal transport and material flow in high speed optical fiber drawing," ASME Journal of Heat Transfer, vol. 120, no. 4, pp. 916–930, 1998.

[24] J. F. Bacon, A. A. Hasapis, and J. W. Wholly, Jr., "Viscosity and density of molten silica and high silica content glass," Physics and Chemistry of Glasses, vol. 1, no. 3, pp. 90–98, 1960.

[25] D. Kalish and B. K. Tariyal, "Static and dynamic fatigue of a polymer-coated fused silica optical fiber," Journal of America Ceramic Society, vol. 61, no. 11–12 pp. 518–523, 1978.

[26] A. A. Griffith, "The phenomena of rapture and flow in solids," Philosophical Transactions of the Royal Society of London, vol. A221 pp. 163–198, 1920.

[27] R. B. Runk, "A zirconia furnace for precision silica waveguides," Digest of Topical Meeting on Optical Fiber Transmission, Williamsburg, VA, paper TuB5-1, 1977.

[28] J. Issa, Z. Yin, C. E. Polymeropoulos, and Y. Jaluria, "Temperature distribution in an optical fiber draw tower furnace," Journal of Material Processing & Manufacturing Science, vol. 4, pp. 221–232, 1996.

[29] S. Sakaguchi and T. Kimura, "A 1200 m/min speed drawing of optical fibers with pressurized coating," Technical Digest of OFC'85, San Diego, CA, paper MG2 1985.

[30] K. Kobayashi, K. Tsurusaki, Y. Sato, and S. Araki, "High-speed coating of optical fibers with UV curable resins," International Wire and Cable Symposium Proceedings, pp. 126–133, 1991.

[31] F. Hanawa, Y. Hibino, M. Shimizu, H. Sudo, and M. Horiguchi, "Influences of the drawing conditions on optical characteristics in undoped-silica-core single-mode fibers," Optics Letters, vol. 12, pp. 617–618, 1987.

[32] G. Scherer, "Thermal stresses in a cylinder: application to optical waveguide blanks," Journal of Non-Crystalline Solids, vol. 34, pp. 223–238, 1979.

[33] G. Scherer, "Stress-induced index profile distortion in optical waveguides," Applied Optics, vol. 19, no. 12, pp. 2000–2006, 1980.

[34] S. H.-K. Lee and Y. Jaluria, "The effect of geometry and temperature variations on the radiative transport during optical fiber drawing," Journal of Material Processing & Manufacturing Science, vol. 3, pp. 317–331, 1995.

[35] S. H.-K. Lee and Y. Jaluria, "Effects of variable properties and viscous dissipation during optical fiber drawing," ASME Journal of Heat Transfer, vol. 118, pp. 350–358, 1996.

[36] D. A. Kaminski, "Thermal transport in optical fiber manufacturing," Proceedings of the First International Symposium on Radiation Transfer, pp. 667–681, 1995.

[37] D. H. Smithgall, "Application of optimization theory to the control of the optical fiber drawing process," Bell System Technical Journal, vol. 58, pp. 1425–1435, 1979.

[38] A. Mulpur and C. Thompson, "Nonlinear control of optical fiber diameter variation," IEEE Control System Technology, vol. 4, no. 2, pp. 152–162, 1996.

[39] U. C. Paek and C. R. Kurkjian, "Calculation of cooling rate and induced stresses in drawing of optical fibers," Journal of American Ceramic Society, vol. 58, no. 7–8 pp. 330–335, 1975.

[40] U. C. Paek, "Free drawing and polymer coating of silica glass optical fibers," Journal of Heat Transfer, vol. 121, pp. 774–788, 1999.

[41] V. A. Petrov and V. Yu. Reznik, "The effect of thickness and temperature upon the total emissivity of quartz glass," Proceedings of 6th Symposium on Thermophysical Properties, pp. 301–304, 1973.

[42] S. R. Choudhury, Y. Jaluria, and S. H.-K. Lee, "Generation of neck-down profile for furnace drawing of optical fiber," National Heat Transfer Conference, ASME HTD, vol. 306 pp. 23–32, 1995.

[43] M. Rajala, K. Asikkala, M. Maekinen, T. Tuurnala, and E. Peltoluhta, "Combination furnace for drawing large optical fiber preforms at high speed," International Wire and Cable Symposium Proceedings, pp. 483–488, 1998.

[44] D. S. Vaidya and G. D. Mihalacopoulos, "Characterization of meltdown profile during fiber draw," International Wire and Cable Symposium Proceedings, pp. 73–80, 1998.

[45] O. L. Anderson, "Cooling time of strong glass fibers," Journal of Applied Physics, vol. 29, no. 1, pp. 9–12, 1958.

[46] R. G. C. Arridge and K. Prior, "Cooling time of silica fibers," Nature, vol. 203, pp. 386–387, 1964.

[47] L. G. Glicksman, "The cooling of glass fibers," Glass Technology, vol. 9, no. 5, pp. 131–138, 1968.

[48] M.-G. Gossiaux, J.-F. Bourhis, and G. Orcel, "Numerical simulation of optical fiber cooling during the fiber drawing process," International Wire and Cable Symposium Proceedings, pp. 81–84, 1998.

[49] T. Vaskopoulos, C. E. Polymeropoulos, and A. Zebib, "Heat transfer from optical fiber during the draw process," International Conference of Transport Phenomena in Processing, Honolulu, Hawaii, pp. 1–10, 1992.

[50] S. R. Choudhury, Y. Jaluria, T. Vaskopoulos, and C. E. Polymeropoulos, "Forced convective cooling of optical fiber during drawing process," ASME Journal of Heat Transfer, vol. 116, pp. 790–794, 1994.

[51] T. Vaskopoulos, C. E. Polymeropoulos, and A. Zebib, "Cooling of optical fiber in aiding and opposing forced gas flow," International Journal of Heat and Mass Transfer, vol. 18, no. 11, pp. 1933–1944, 1995.

[52] S. Sakaguchi and T. Kimura, "High-speed drawing of optical fibers with pressurized coating," IEEE Journal of Lightwave Technology, vol. LT-3, no. 3, pp. 669–673, 1985.

[53] H. Koaizawa, N. Orita, and T. Kamiya, "Study of coating diameter at high speed drawing of optical fiber with UV curable resin," Technical Digest of IOOC'95, FA1-2, vol. 4, pp. 4–5, 1995.

[54] K. Chida, S. Sakaguchi, M. Wagatsuma, and T. Kimura, "High-speed coating of optical fibers with thermally curable silicon resin using a pressure die," Electronics Letters, vol. 18, no. 16, pp. 713–715, 1982.

[55] S. Bateson, "Critical study of the optical and mechanical properties of glass fibers," Journal of Applied Physics, vol. 29, no. 1, pp. 13–21, 1958.

[56] A. Burgmann, "Liquid glass jets in the forming of continuous glass fibers," Glass Technology, vol. 11, no. 4, pp. 110–116, 1970.

[57] P. Walker, V. J. Webers, and G. A. Thommes, "Photopolymerizable reproduction systems—chemistry and applications," Journal of Photographic Science, vol. 18, pp. 150–157, 1970.

[58] T. Kokubun, Y. Katsuyama, K. Hogari, and S. Hatano, "Resin selection and high-speed coating of optical fibers with UV-curable materials," IEEE Journal of Lightwave Technology, vol. 7, no. 5, pp. 824–828, 1989.

[59] T. Ferington, "Kinetics of polymer formation by free radical mechanism," vol. 36, no. 4, pp. 174–181, 1959.

[60] U. C. Paek and C. M. Schroeder, "Calculation of photopolymerization energy required for optical fiber coating," Applied Optics, vol. 20, no. 7, pp. 1230–1233, 1981.

[61] D. A. Krohn and A. R. Cooper, "Strengthening of glass fibers; I. Cladding," Journal of American Ceramic Society, vol. 52, no. 12, pp. 661–664, 1969.

[62] F. O. Anderegg, "Strength of glass fibers," Industrial Engineering Chemistry, vol. 31, pp. 290–298, 1939.

[63] J. G. Moley, P. A. Andrew, and I. Whitney, "Strength of fused silica," Physical Chemistry and Glass, vol. 5, pp. 1–10, 1964.

[64] R. G. C. Arridge and K. Prior, "Cooling time of silica fibres," Nature, vol. 203, no. 4943, pp. 386–387, 1964.

[65] W. B. Hillig, "Strength of bulk fused quartz," Journal of Applied Physics, vol. 32, pp. 741, 1961.

[66] M. S. Maklad, A. R. Asam, and F. I. Akers, "Recent advances in high strength optical fibers having surface compression," International Wire and Cable Symposium Proceedings, pp. 340–343, 1979.

[67] R. K. Mohr, O. H. El-Bayoumi, and R. P. Ingel, "Static fatigue in glass optical fibers having surface compression," American Ceramic Society Bulletin, vol. 59, no. 11, pp. 1145–1150, 1980.

[68] C. R. Kurkjian and U. C. Paek, "Effect of drawing tension on residual stresses in clad glass fibers," Journal of American Ceramic Society, vol. 61, no. 3–4 pp. 137–138, 1978.

[69] P. C. Bouten, W. Hermann, C. M. Jochem, and D. U. Wiechert, "Drawing influence on the lifetime of optical fibers," IEEE Journal of Lightwave Technology, vol. 7, no. 3, pp. 555–559, 1989.

[70] Y. Hibino, F. Hanawa, and M. Horiguchi, "Drawing-induced residual stress effects on optical characteristics in pure-silica-core single-mode fibers," Journal of Applied Physics, vol. 65, no. 1, pp. 30–34, 1989.

[71] T. T. Wang, H. N. Varzirani, H. Schonhorn, and H. M. Zupko, "Effects of water and moisture on strength of optical glass (silica) fibers coated with a UV-cured epoxy acrylate," Journal of Applied Polymer Science, vol. 23, pp. 887–892, 1979.

[72] G. M. Bubel, J. T. Krause, B. J. Bickta, and R. T. Ku, "Mechanical reliability of metallized optical fiber for hermetic terminations," IEEE Journal of Lightwave Technology, vol. 7, no. 10, pp. 1488–1493, 1989.

[73] D. R. Biswas, C. R. Kurkjian, and H. H. Yuce, "Hermetic coating for optical glass fibers," International Wire and Cable Symposium Proceedings, pp. 317–324, 1995.

[74] P. G. Simkins, C. R. Kurkjian, and C. M. Schroeder, "Aluminum-coated silica fibers: strength and solderability," Electronics Letters, vol. 31, no. 9, pp. 747–749, 1995.

[75] V. A. Bogatyrjov, I. I. Cheremisin, E. M. Dianov, K. M. Golant, and A. L. Tomashuk, "Super-high-strength metal-coated low-hydroxyl low-chlorine all-silica optical fibers," IEEE Transactions on Nuclear Science, vol. 43, no. 3, pp. 1057–1060, 1996.

CHAPTER FOUR

Dispersion-Managed Single-Mode Fibers for Wavelength Division Multiplexing

4.1 WAVELENGTH ALLOCATIONS IN SINGLE-MODE FIBERS FOR WDM APPLICATIONS

In Chapter 2, standard specifications for single-mode fibers have been reviewed. In recent optical communications, signals in conventional band (C band) have expanded to neighboring optical bands to cover almost entire available optical window in single-mode optical fiber. The band allocation in International Telecommunication Union (ITU) is summarized in Table 2.8.

In transmission of signals through optical fiber, two major fundamental limitations were and have been optical loss and chromatic dispersion. With the advent of fiber amplifiers, the limitation imposed by optical loss has been virtually removed but within a finite range of gain bandwidth. In other words, previous loss limit has been changed to bandwidth limit of optical amplifiers. Among the above communication bands, S, C, and L bands are widely used due to availability of efficient fiber optical amplifiers.

The optical amplifiers in S, C, and L bands are schematically shown in Figure 4.1, along with an optical loss spectrum of conventional single-mode fiber. Since the successful deployment of EDFA for C band in both terrestrial and submarine cables, erbium ions in silica glass have been further studied to find that lower population inversion in a longer EDF length could provide a sufficient optical gain in L band [1].

Silica Optical Fiber Technology for Devices and Components: Design, Fabrication, and International Standards, First Edition. By Kyunghwan Oh and Un-Chul Paek.

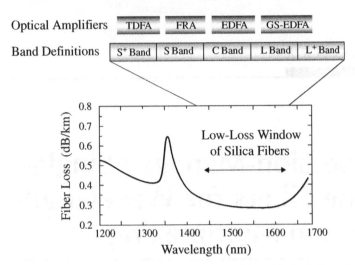

FIGURE 4.1   Band location and accompanying optical amplifiers. TDFA, thulium-doped fiber amplifier; FRA, fiber Raman amplifier; EDFA, erbium-doped fiber amplifier.

Instead of using erbium, thulium ions also have been attempted to obtain a gain and thulium-doped fiber amplifier (TDFA) in fluoride glasses [2] and multicomponent silicate glass [3] have been developed for S-band applications. Raman process in optical fiber has been well studied [4] but it is only recent for the process to be put into practical use with development of high power laser diode pumps. Fiber Raman amplifiers (FRA) can, in principle, provide gain at an arbitrary spectral location with an appropriate pump, but due to high pump power and long interaction requirements, they can hardly compete alone with rare earth–doped counterparts.

The other fundamental limitation in optical transmission, the chromatic dispersion, has been a major battleground for optical fiber design and fabrication for the last decade and it is still one of the major challenges to meet the demands for ever-increasing data rate, number of channel, and link distance. In order to control the chromatic dispersion, varieties of index profiles in optical fibers have been introduced with significant success in the shifting zero-dispersion wavelength and manipulating dispersion value and its slope in C bands. And these technical achievements have been arranged in international standards such as ITU-T G.653 [5], G.655 [6], and G.656 [7], whose optical attributes are summarized in Tables 2.10 and 2.11.

As the fiber compatible amplifiers further developed in L and S bands, the optimization of chromatic dispersion is subsequently demanded. There are two principal categories for dispersion control in wavelength division multiplexing (WDM) systems. One is to develop an optimal transmission fiber where all the necessary chromatic dispersion parameters are met within a band along the entire link distance. Dispersion-shifted fiber (DSF) and nonzero dispersion-shifted fiber

(NZ-DSF) belong to this category. The other is to provide a dispersion-compensating unit within an optical link where the accumulated dispersion in the preceding link distances is cancelled by the opposite sign of chromatic dispersion of the unit. Dispersion-compensating fiber (DCF) and tunable dispersion-compensating modules (DCM) of various kinds belong to the second category.

Dispersion control is indeed an essential step for modern WDM system in two notions: (1) maximizing the data rate by reducing time delay among signals and (2) minimizing nonlinear optical interaction such as four-wave mixing among WDM channels [8,9].

In this chapter, we will review the waveguide design parameters for dispersion control in single-mode fiber along with the latest specifications and requirements.

In WDM systems in an optical band, optical signals are multiplexed and amplified to meet the optical power budget. In order for the dispersion to be tailored for WDM channels, therefore, spectral location of channels is one of the most important prerequisites in fiber design.

ITU recently finalized the WDM spectral grids in their standards, ITU-T G.694.1 for dense WDM (DWDM) and 694.2 for coarse WDM (CWDM), respectively [10,11].

According to ITU-T G.694.1, the frequency grid, anchored to 193.1 THz, supports a variety of channel spacings ranging from 12.5, 25.0, and 50 to 100 GHz [10]. For channel spacings of 12.5 GHz, the allowed channel frequencies (in THz) are defined by

$$193.1 + n \times 0.0125$$

where n is a positive or negative integer including 0.

For wider channel spacings, the allowed channel frequencies (in THz) are defined by

$$193.1 + n \times 0.025 \quad \text{(for 25 GHz)}$$

$$193.1 + n \times 0.050 \quad \text{(for 50 GHz)}$$

$$193.1 + n \times 0.100 \quad \text{(for 100 GHz)}$$

Table 4.1 lists some central frequencies and corresponding wavelength in vacuum within the C and L bands based on the above channel spacings for DWDM.

Note that we need to use the value of speed of light in vacuum, 2.99792458×10^8 m/s, in order to convert ITU grid frequencies into wavelengths.

In ITU-T G.694.2 the spectral grids for CWDM are defined [11]. CWDM technology is proposed to utilize the whole optical communication bands by adopting a wider channel spacing of 20 nm to realize cost-effective applications. In CWDM flexible system design can be achieved through a combination of uncooled lasers, relaxed laser wavelength selection tolerances, and wide passband MUX/DEMUXs. Table 4.2 summarizes some of the central wavelengths of CWDM grids.

TABLE 4.1   Some of Central Frequencies of DWDM Channels in C and L Bands and Their Corresponding Wavelengths in Vacuum [10]

Nominal Central Frequencies (THz) for Spacings				Wavelengths (nm)
12.5 GHz	*25 GHz*	*50 GHz*	*100 GHz*	
195.9375	–	–	–	1530.04
195.9250	195.925	–	–	1530.14
195.9125	–	–	–	1530.24
195.9000	195.900	195.90	195.9	1530.33
195.8875	–	–	–	1530.43
195.8750	195.875	–	–	1530.53
195.8625	–	–	–	1530.63
195.8500	195.850	195.85	–	1530.72
195.8375	–	–	–	1530.82
195.8250	195.825	–	–	1530.92
195.8125	–	–	–	1531.02
195.8000	195.800	195.80	195.8	1531.12
195.7875	–	–	–	1531.21
195.7750	195.775	–	–	1531.31
195.7625	–	–	–	1531.41
195.7500	195.750	195.75	–	1531.51
195.7375	–	–	–	1531.60
195.7250	195.725	–	–	1531.70
195.7125	–	–	–	1531.80
193.2375	–	–	–	1551.42
193.2250	193.225	–	–	1551.52
193.2125	–	–	–	1551.62
193.2000	193.200	193.20	193.2	1551.72
193.1875	–	–	–	1551.82
193.1750	193.175	–	–	1551.92
193.1625	–	–	–	1552.02
193.1500	193.150	193.15	–	1552.12
193.1375	–	–	–	1552.22
193.1250	193.125	–	–	1552.32
193.1125	–	–	–	1552.42
193.1000	193.100	193.10	193.1	1552.52
193.0875	–	–	–	1552.62
193.0750	193.075	–	–	1552.73
193.0625	–	–	–	1552.83
193.0500	193.050	193.05	–	1552.93
193.0375	–	–	–	1553.03
193.0250	193.025	–	–	1553.13
193.0125	–	–	–	1553.23
184.7750	184.775	–	–	1622.47
184.7625	–	–	–	1622.58
184.7500	184.750	184.75	–	1622.69

Table 4.1 *Continued*

Nominal Central Frequencies (THz) for Spacings				Wavelengths (nm)
12.5 GHz	*25 GHz*	*50 GHz*	*100 GHz*	
184.7375	–	–	–	1622.80
184.7250	184.725	–	–	1622.91
184.7125	–	–	–	1623.02
184.7000	184.700	184.70	184.7	1623.13
184.6875	–	–	–	1623.24
184.6750	184.675	–	–	1623.35
184.6625	–	–	–	1623.46
184.6500	184.650	184.65	–	1623.57
184.6375	–	–	–	1623.68
184.6250	184.625	–	–	1623.79
184.6125	–	–	–	1623.90
184.6000	184.600	184.60	184.6	1624.01
184.5875	–	–	–	1624.12
184.5750	184.575	–	–	1624.23
184.5625	–	–	–	1624.34
184.5500	184.550	184.55	–	1624.45
184.5375	–	–	–	1624.56
184.5250	184.525	–	–	1624.67
184.5125	–	–	–	1624.78

TABLE 4.2   Central Wavelengths (nm) of CWDM Grids [11]

Central Wavelengths (nm)
1270
1290
1310
1330
1350
1370
1390
1410
1430
1450
1470
1490
1510
1530
1550
1570
1590
1610

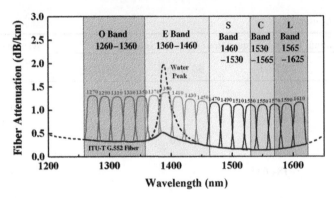

FIGURE 4.2 CWDM bands over optical communication windows [13].

The locations of these CWDM grids over ITU-T bands are overlaid in Figure 4.2. Note that the CWDM band centered at 1390 nm would be operable only in low water peak single-mode fibers as defined in ITU-T G.652.C and D [12].

The CWDM grids have the uniform channel spacing of 20 nm in wavelength in contrast to the frequency-based DWDM grids. Total wavelength variation in current uncooled signal light sources of the order of ± 6−7 nm is expected to be compatible with current filter technologies in mass production scale. As for the guard band, one third of the minimum channel spacing is sufficient. Therefore, in order to maximize the number of channels, 20 nm has been chosen. CWDM can, therefore, take advantage of low-cost uncooled lasers and wide passband filters to make the total system cost compatible to access network environment.

With these spectral locations of optical channels in each band, the chromatic dispersion should have optimal parameters to guarantee a certain data rate over a certain distance with a certain level of bit error rate (BER). In Chapter 2, various types of fibers such as DSF in ITU-T G.653.A and B [5] and NZ-DSFs in ITU-T G.655.A, B, C, D, and E [6] and ITU-T G.656 [7] have been introduced with modified chromatic dispersion properties as listed in Table 2.11.

Chromatic dispersion of these types of fibers is shown in Figure 4.3 along with fiber attenuation spectra of conventional SMF and low OH peak fiber.

In comparison with DSF, NZ-DSFs can have zero-dispersion wavelengths outside C band so that the zero-dispersion wavelength is shifted toward either the shorter or the longer edge of the band of interest. By locating the zero-dispersion wavelength out of the spectral range of interests, we can suppress four-wave mixing among the WDM channels. Most of currently commercial products are covering C and L bands and further expansion into S-band coverage has been reported by various vendors and some of their products are listed in Table 4.3.

Recent development of low OH fibers can open another practical window, E band, along with fiber Raman amplifier. Further evolution of NZ-DSFs is being expected in E band and this adaptation of optical bands could continue till the full optical bands can be exploited with corresponding amplifiers in long-haul transmission.

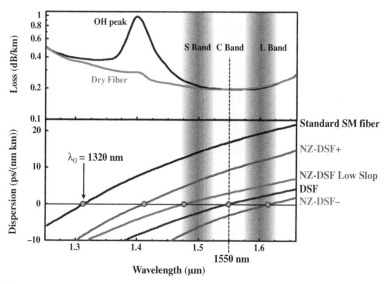

FIGURE 4.3 Typical chromatic dispersion curves of dispersion-shifted fibers, NZ-DSF+ and NZ-DSF− stand for nonzero dispersion-shifted fiber with positive dispersion and negative value at 1550 nm, respectively.

4.2 OPTIMIZATION OF WAVEGUIDE PARAMETERS FOR DISPERSION CONTROL

Chromatic dispersion of an optical fiber has contributions from material dispersion and waveguide dispersion, the latter of which depends on the fiber refractive index profile. In order to achieve an optimum chromatic dispersion, various types of index profiles have been attempted. In this section, we will review a few of recently reported index profiles for dispersion-controlled fibers such as DSF, dispersion-flattened fiber (DFF), DCF, and NZ-DSF in S, C, and L bands, whose dispersion properties were analyzed by a computer simulation program OPACAD based on the cascaded boundary matrix (CBM) method described in Chapter 9.

Schematic refractive index profiles of conventional SMF and dispersion-controlled fibers are listed in Figure 4.4.

In the design of refractive index profiles of conventional SMF and dispersion-controlled fibers, the dispersion value over a transmission band is optimized along with its slope to balance the trade-off between the linear intersymbol interference (ISI) penalty at a high chromatic dispersion value and the nonlinear penalties at zero dispersion. There are two major system penalties associated with chromatic dispersion: (1) ISI penalty in the linear regime and (2) four-wave mixing (FWM) penalty in the nonlinear regime. These penalties are to be fully understood for optimization of dispersion control.

The fiber chromatic dispersion introduces a phase delay among the different spectral components, temporally broadening the transmitted signals. If the pulse train

TABLE 4.3 Commercially Available NZ-DSFs

Fiber Brands	C-Band Dispersion (ps/(nm km)) (1530–1565 nm)	L-Band Dispersion (ps/(nm km)) (1565–1625 nm)	Dispersion (D) (ps/(nm km)) at 1550 nm	Dispersion Slope (S) (ps/(nm^2 km)) at 1550 nm	Relative Dispersion Slope (RDS: S/D) (nm^{-1}) at 1550 nm
Lucent TrueWave RS fiber	2.6–6.0	4.0–8.9	4.4	0.045	0.010
Corning LEAF	2.0–6.0	4.5–11.2	4.2	0.085	0.020
Lucent TrueWave REACH	5.5–8.9	6.9–11.4	7.5	0.042	0.0056
Alcatel TeraLight Ultra	5.5–10.0	7.5–13.4	8.0	0.052	0.0065
Sumitomo Pure Guide	5.0–10.0	12.4 at 1625 nm	8.0	0.058	0.0073

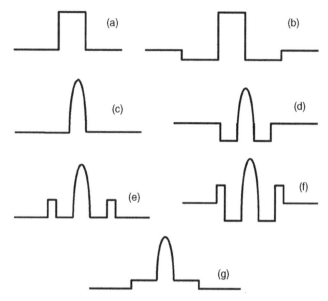

FIGURE 4.4 Schematic refractive index profiles: (a and b) conventional SMFs and (c–g) DSF, NZ-DSF, and DCF [14].

of optical signal is broadened to overlap with adjacent time slots, interference between the neighboring bits will enhance ISI to significantly increase BER. In order to understand the dispersion penalty due to ISI, let us consider propagation of Gaussian input pulse along a single-mode optical fiber.

In frequency domain, the input pulse in the fundamental mode is centered at ω_0 and each spectral component at ω will have the propagation constant $\beta(\omega)$ to result in the electric field expressed as [15,16]

$$\vec{E}(\vec{r}, t) = \hat{x}F(x,y)A(z,t)e^{j\omega_0 t}\,e^{-j\beta_0 z} \tag{4.1}$$

Here $F(x, y)$ is the field distribution of the fundamental LP_{01} mode as discussed in Chapter 2. $A(z, t)$ represents the signal envelope function whose shape is affected by the chromatic dispersion as the pulse propagates along z direction. For quasi-monochromatic field, whose spectral width is significantly narrow, we can expand the propagation constant β near the central frequency ω_0 in a Taylor series:

$$\beta(\omega) = \beta_0 + \beta_1(\omega - \omega_0) + \frac{1}{2}\beta_2(\omega - \omega_0)^2 + \frac{1}{6}\beta_3(\omega - \omega_0)^3 + O((\omega - \omega_0)^4) \tag{4.2}$$

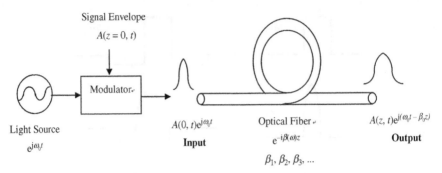

FIGURE 4.5 The schematic of pulse distortion along an optical fiber.

Here $\beta_i = (d^i\beta/d\omega^i)|_{\omega=\omega_0}$ and they are related to group velocity v_g, chromatic dispersion $D(\lambda)$, and dispersion slope $S = dD(\lambda)/d\lambda$ as follows:

$$\beta_1 = \frac{1}{v_g}$$

$$\beta_2 = -\frac{\lambda^2}{2\pi c}D(\lambda)$$

$$\beta_3 = \frac{\lambda^4}{4\pi^2 c^2}S(\lambda) + \frac{\lambda^3}{\pi^2 c^2}D(\lambda)$$

(4.3)

The model of Gaussian pulse response of an optical fiber is schematically shown in Figure 4.5. For a quasi-monochromatic light source, $e^{j\omega_0 t}$, which is initially modulated by signal $A(z=0, t)$, we have the input of the fiber in the time domain as $A(z = 0, t)e^{j\omega_0 t}$. This input signal will propagate along the fiber that is characterized by the phase of $e^{-j\beta(\omega)z}$ and dispersion parameters $\beta_1, \beta_2,$ and β_3. The pulse after propagation at the fiber position z, $A(z, t)$, will be affected by these propagation and dispersion parameters to result in distortion of the initial signal envelop function.

It can be shown that the envelope function obeys the following nonlinear partial differential equation [17]:

$$\frac{\partial A}{\partial z} + \beta_1 \frac{\partial A}{\partial t} - \frac{i}{2}\beta_2\frac{\partial^2 A}{\partial t^2} - \frac{1}{6}\beta_3\frac{\partial^3 A}{\partial t^3} + \frac{\alpha}{2}A = 0$$

(4.4)

where α is the power attenuation coefficient and it is assumed to be negligible for the following analysis. As indicated in the above propagation equation, the envelope function and its Fourier component in spectral domain are directly affected by dispersion parameters, β_i. In conventional single-mode fiber, therefore, the light signal whose wavelength is over the zero-dispersion wavelength near 1310 nm will experience a pulse broadening due to positive chromatic dispersion and its slope.

For a Gaussian input pulse

$$A(0, t) = e^{-[t^2/((\tau_0/2)^2)]} \tag{4.5}$$

it is shown that the output is given [18,19] by successive Fourier transformations as follows:

$$A(z, t) = 2\sqrt{\pi}|B|^{-(1/3)} \times \exp\left(\frac{2 - 3AB - 6C^2}{3B^2} - jC\frac{3AB + 2C^2 - 6}{3B^2}\right)$$

$$\times A_i\left[(1 - AB - C^2 + j2C)B^{-(4/3)}\right] \tag{4.6}$$

where A, B, and C are related to dispersion parameters

$$A = \frac{4(t - \beta_1 z)}{\tau_0}$$

$$B = \frac{32\beta_3 z}{\tau_0^3} \tag{4.7}$$

$$C = \frac{8\beta_2 z}{\tau_0^2}$$

Here A_i is the Airy function [20] and β_i are evaluated at $\omega = \omega_0$.

In the case when β_3 can be neglected compared with lower order β_i, the output pulse can be also approximated as Gaussian:

$$A(z, t) \approx \frac{1}{(1 + C^2)^{1/4}}\exp\left[\frac{A^2}{4(1 + C^2)}\right] \tag{4.8}$$

Marcuse has analyzed the variation of root mean square (RMS) width of signal pulse to obtain the broadening factor in an analytic form as follows [15,16,21]:

$$\frac{\sigma_{\text{out}}}{\sigma_{\text{in}}} = \left[1 + (1 + V_\omega^2)\left(\frac{\beta_2 L}{2\sigma_{\text{in}}^2}\right)^2 + \frac{1}{2}(1 + V_\omega^2)\left(\frac{\beta_3 L}{4\sigma_{\text{in}}^3}\right)^2\right]^{1/2} \tag{4.9}$$

The RMS width of the pulse is defined by

$$\sigma \equiv \left[\langle t^2 \rangle - \langle t \rangle^2\right]^{1/2}$$

$$\langle t^k \rangle = \frac{\displaystyle\int_{-\infty}^{+\infty} t^k |A(z, t)|^2 dt}{\displaystyle\int_{-\infty}^{+\infty} |A(z, t)|^2 dt}, \quad k = 1, 2 \tag{4.10}$$

The parameter V_ω is defined by

$$V_\omega = \sigma_{in}\sigma_{in}^{\bar\omega} \tag{4.11}$$

where $\sigma_{in}^{\bar\omega}$ is the spectral RMS width assuming an input Gaussian source spectrum and σ_{in} is the temporal RMS width for the Gaussian input equation 4.4.

The broadened pulse width, σ_{out}, should be compared with the allocated bit period, $T_B = 1/B$, for a given signal data rate, B. Commonly used criterion for ISI penalty is such that at least 95% of Gaussian pulse is retained in a bit period and it is expressed by:

$$\sigma_{out} \leq \frac{T_B}{4} \quad \text{or} \quad \sigma_{out}B \leq \frac{1}{4} \tag{4.12}$$

Now we will consider how the above criterion casts the dispersion penalties in transmission of data rate, B, in terms of chromatic dispersion, $D(\lambda)$, and its slope, $S(\lambda)$. The dispersion penalty is highly affected by the spectral width of optical source, $\sigma_{in}^{\bar\omega}$, or equivalently V_ω [19,21]:

(1) Optical source with a broad spectral width, $V_\omega \gg 1$. This case corresponds to directly modulated LED and Fabry–Perot type LD.

   (a) Far away from zero-dispersion wavelength, λ_o: Let us consider an optical communication system whose operating wavelength is away from the zero-dispersion wavelength, λ_o, which satisfies $D(\lambda = \lambda_o) = 0$. Then β_3 can be neglected compared with lower order β_i. Using equations 4.3 and 4.8, the output pulse width is obtained as

$$\sigma_{out} \approx \sigma_{in}\left[1 + \left(\frac{\beta_2 L \sigma_{in}^\omega}{\sigma_{in}}\right)^2\right]^{1/2} = [\sigma_{in}^2 + (DL\sigma_{in}^\lambda)^2]^{1/2} \tag{4.13}$$

   where σ_{in}^λ is the RMS source spectral width in the wavelength domain.

   In most of high–data rate, long-distance optical transmission, we will have $\sigma_{in} \ll DL\sigma_{in}^\lambda$ and therefore $\sigma_{out} \approx |D|L\sigma_{in}^\lambda$. Then by equation 4.11, we will have the dispersion penalty as follows:

$$BL|D|\sigma_{in}^\lambda \leq \frac{1}{4} \tag{4.14}$$

   It is noted that the data rate B is proportional to $|D|^{-1}$, L^{-1}, and $(\sigma_{in}^\lambda)^{-1}$.

(b) At the zero-dispersion wavelength, λ_o: When the communication system is operated at the dispersion wavelength λ_o, we will have $\beta_2 = 0$ and equation 4.8 is governed only by β_3. Using equation 4.3 we have

$$\sigma_{out} \approx \sigma_{in} \left[1 + \frac{1}{2} \left\{ \frac{\beta_3 L (\sigma_{in}^\omega)^2}{\sigma_{in}} \right\}^2 \right]^{1/2} = \left[\sigma_{in}^2 + \frac{1}{2} \{ SL(\sigma_{in}^\lambda)^2 \}^2 \right]^{1/2}$$

(4.15)

Note that dispersion broadening is now affected by dispersion slope, S, instead of D compared with equation 4.12. When the initial RMS width, σ_{in}, is negligible compared with dispersion broadening factor, equation 4.11 will lead to the criterion as follows:

$$BL|S|(\sigma_{in}^\lambda)^2 \leq \frac{1}{\sqrt{8}}$$

(4.16)

and B scales as L^{-1}, $(\sigma_{in}^\lambda)^{-2}$, and $|S|^{-1}$.

(2) Optical source with a narrow spectral width, $V_\omega \ll 1$. This case corresponds to single longitudinal distributed feed back (DFB) LD, which is externally modulated.

(a) Far away from zero-dispersion wavelength, λ_o: In equation 4.8, we once again neglect β_3 to obtain:

$$\sigma_{out} \approx \sigma_{in} \left[1 + \left(\frac{\beta_2 L}{2\sigma_{in}^2} \right)^2 \right]^{1/2} = \left[\sigma_{in}^2 + \left(\frac{\beta_2 L}{2\sigma_{in}} \right)^2 \right]^{1/2}$$

$$\approx \left[\sigma_{in}^2 + \left(\frac{\lambda^2 DL}{4\pi c \sigma_{in}} \right)^2 \right]^{1/2}$$

(4.17)

In the above equation, it is noted that RMS output pulse width, σ_{out}, can be minimized by proper choice of σ_{in}. Minimum of σ_{out} is given by

$$\sigma_{out} \approx (|\beta_2|L)^{1/2} = \left(\frac{\lambda^2 L}{2\pi c} |D| \right)^{1/2}$$

(4.18)

at the optimal $\sigma_{in} = (|\beta_2|L/2)^{1/2}$.

Applying the same criterion for ISI penalty, we obtain the following condition:

$$B(|\beta_2|L)^{1/2} \leq \frac{1}{4}$$

(4.19)

$$B\left(\frac{\lambda^2 L}{2\pi c}|D|\right)^{1/2} \leq \frac{1}{4} \tag{4.20}$$

It is apparent that the dispersion penalty does have different dependence such that B scales with $|D|^{-1/2}$ and $L^{-1/2}$ rather than $|D|^{-1}$ and L^{-1} as in equation 4.13.

(b) At the zero-dispersion wavelength, λ_o: We will have again $\beta_2 = 0$ and the output pulse width is given by

$$\sigma_{\text{out}} \approx \left[\sigma_{\text{in}}^2 + \frac{1}{2}\left(\frac{\beta_3 L}{4\sigma_{\text{in}}^2}\right)^2\right]^{1/2} \tag{4.21}$$

Similar to the above situation, the output pulse width can be minimized by an appropriate input pulse width. The minimum pulse width is given by

$$\sigma_{\text{out}} \approx \left(\frac{3}{2}\right)^{1/2}\left(\frac{|\beta_3|L}{4}\right)^{1/3} \tag{4.22}$$

at the optimal $\sigma_{\text{in}} = (|\beta_3|L/4)^{1/3}$.

The ISI penalty is expressed in the conditions as follows:

$$B(|\beta_3|L)^{1/3} \leq 0.324 \tag{4.23}$$

$$B\left(\frac{|S(\lambda)|\lambda^4 L}{4\pi^2 c^2}\right)^{1/3} \leq 0.324 \tag{4.24}$$

Note that B scales with $|S|^{-1/3}$, which is a major disparity compared with the case in equation 4.15.

According to above discussions, an optical communication system operating near zero dispersion with a low dispersion slope would be most preferred. In fact, for the time domain multiplexing (TDM) system operating at a single wavelength channel, DSF has been developed to provide a very low chromatic dispersion at the operating wavelength near 1.55 μm. In the case of current WDM system, however, zero dispersion within the band will induce a nonlinear dispersion penalty.

In addition to linear responses in optical fibers such as attenuation and chromatic dispersion, transmission of multiple light signals at a close-spaced channel spacing over a long distance is affected by a nonlinear response of the refractive index. This nonlinearity in silica glass is expressed as [17]

$$n = n_0 + n_2 \frac{P}{A_{\text{eff}}} \tag{4.25}$$

where P is the optical power (W) carried through the optical fiber and A_{eff} is the effective area of the guided optical mode where the optical power is confined. The coefficient n_2, for silica, is $\sim 2.6 \times 10^{-20}\,\text{m}^2/\text{W}$ [17].

Due to the nonlinear term in equation 4.25, the envelope propagation equation 4.4 is modified as

$$\frac{\partial A}{\partial z} + \beta_1 \frac{\partial A}{\partial t} - \frac{i}{2}\beta_2 \frac{\partial^2 A}{\partial t^2} - \frac{1}{6}\beta_3 \frac{\partial^3 A}{\partial t^3} + \frac{\alpha}{2}A = -i\frac{2\pi}{\lambda}\bar{n}_2|A|^2 A \qquad (4.26)$$

where $\bar{n}_2 = (n_0/2Z_0)n_2$ with vacuum impedance $Z_0 = 376.7\,\Omega$. Note that the nonlinear term on the right-hand side of equation 4.26 will affect the phase of the envelope, not the magnitude because it is an imaginary number. The results of nonlinear phase modulation of optical signals in silica fiber can be classified in general into self-phase modulation (SPM), cross-phase modulation (CPM), and FWM. Among these, FWM was found to be the most significantly affecting the WDM systems [8]. Through the FWM process, three signals of frequencies, f_i, f_j, and f_k $(j \neq k)$, generate spurious signal at the frequency $f_{ijk} = f_i + f_j - f_k$ as schematically shown in Figure 4.6. For unequal spacing between the incoming signals, f_1, f_2, and f_3, we have nine FWM frequencies as shown in Figure 4.6a. In the case of equal spacing such that $f_1 = f_2 - \Delta f$ and $f_3 = f_2 + \Delta f$, the FWM frequencies overlap and they do show the same spacing of Δf as shown in Figure 4.6b.

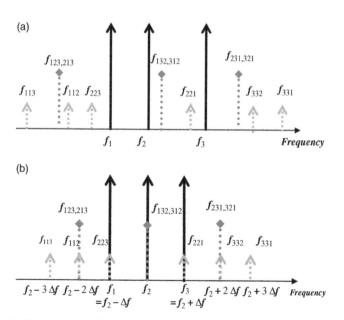

FIGURE 4.6 Frequency combinations for three optical signals with (a) unequal frequency spacing and (b) the equal spacing of Δf.

Therefore, WDM signals with an equal spacing will be more severely affected by FWM than the unequally spaced channels due to cross-talk between the signal and spurious FWM outputs.

For three input signals with the same polarization state, the time-averaged optical power at the frequency component f_{ijk} is given as [22–24]:

$$P_{ijk} = \left(\frac{D_{ijk}}{3}\frac{2\pi}{\lambda}\frac{n_2}{A_{eff}}\right)^2 P_i P_j P_k \left|\frac{e^{(i\,\Delta\beta-\alpha)L}-1}{i\,\Delta\beta-\alpha}\right|^2 \quad \text{(in MKS unit)} \qquad (4.27)$$

or

$$P_{ijk} = \left(\frac{D_{ijk}}{n^2}\frac{32\pi^3}{c}\frac{\chi_{1111}}{A_{eff}}\right)^2 P_i P_j P_k \left|\frac{e^{(i\,\Delta\beta-\alpha)L}-1}{i\,\Delta\beta-\alpha}\right|^2 \quad \text{(in cgs unit)} \qquad (4.28)$$

where L is the fiber length, D_{ijk} is the degeneracy factor ($D_{ijk}=3$ for $i=j$, otherwise 6), n is the refractive index of the core, c is the speed of light, A_{eff} is the effective mode area for the guided HE_{11} mode, n_2 is defined in equation 4.25, and χ_{1111} is the third-order nonlinear susceptibility of silica and the value is 6×10^{-15} cm^3/erg. $\Delta\beta$ represents the phase mismatch among the incident signals and FWM signal, and it is expressed in terms of the signal frequency differences and chromatic dispersion D and its slope $dD/d\lambda$ [23]:

$$\Delta\beta = \beta_i + \beta_j - \beta_k - \beta_{ijk} = \frac{2\pi\lambda_k^2}{c}|f_i-f_k||f_j-f_k|$$

$$\times\left[D(\lambda_k)+\frac{dD(\lambda_k)}{d\lambda}\frac{\lambda_k^2}{2c}(|f_i-f_k|+|f_j-f_k|)\right] \qquad (4.29)$$

Equation 4.27 can be further parameterized as

$$P_{ijk} = \eta\left(\frac{D_{ijk}}{3}\gamma L_{eff}\right)^2 P_i P_j P_k\, e^{-\alpha L} \qquad (4.30)$$

where the nonlinear coefficient γ is given by

$$\gamma = \frac{2\pi n_2}{\lambda A_{eff}} \qquad (4.31)$$

along with the effective fiber length L_{eff} defined by

$$L_{eff} = \frac{1-e^{-\alpha L}}{\alpha} \qquad (4.32)$$

And the FWM efficiency η is expressed as

$$\eta = \frac{\alpha^2}{\alpha^2 + \Delta\beta^2} \left\{ 1 + \frac{4\,e^{-\alpha L}\sin^2(\Delta\beta L/2)}{(1 - e^{-\alpha L})^2} \right\} \tag{4.33}$$

It is noted that FWM efficiency, η, takes its maximum value of 1 for the phase-matched condition, $\Delta\beta = 0$, and it decreases with a larger $\Delta\beta$, which makes chromatic dispersion and its slope critical in fiber design to minimize FWM penalty for optical communications.

From the FWM power descriptions, in equations 4.30–4.33, we can deduce the optimal directions of fiber design to minimize the nonlinear penalty. First of all, we need a larger effective mode area, A_{eff}, to reduce the nonlinear coefficient given in equation 4.31. At the same time we need to have a large magnitude of chromatic dispersion with its slope of the same sign as in equation 4.29 to reduce the FWM efficiency in equation 4.33. Note that the last condition is directly contradicting to low dispersion requirement in the linear regime as described in equations 4.14 and 4.20. Chromatic dispersion control, therefore, requires highly delicate optimization process for the choice of optical waveguide and material to give the minimal penalty for WDM systems in both linear and nonlinear regimes.

The effects of chromatic dispersion on FWM efficiency have been analyzed by Maeda et al. [25] for conventional single-mode fiber and dispersion-shifted fiber near 1550 nm. As shown in Figure 4.7, the FWM efficiency drops sharply for the larger magnitude of chromatic dispersion with increasing signal frequency separation.

The fiber length dependence on FWM efficiency is analyzed for various chromatic dispersion values in Figure 4.8. It is once again confirmed that the larger fiber chromatic dispersion results in a lower FWM power accumulation along the propagation length.

Besides these general optimization directions for optical fiber design, there is another critical condition to avoid FWM penalty, the location of zero-dispersion

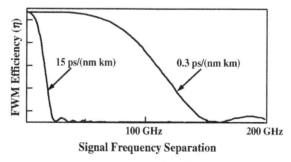

FIGURE 4.7 FWM efficiency in linear scale for various chromatic dispersion values. The effective signal frequency difference is defined as $\sqrt{|f_i - f_k||f_j - f_k|}$. Fiber length: $L = 12$ km; fiber attenuation $\alpha = 0.2$ dB/km; $dD/d\lambda = 0.09$ ps/(km nm^2) [25].

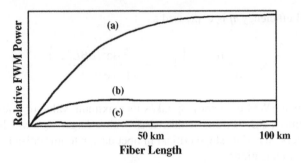

FIGURE 4.8 Relative FWM power at the eighth channel in a 16-channel WDM system. Here the launched signal power per channel is assumed to be equal with 10-GHz channel spacing. (a) $D = 0.3$ ps/(nm km), (b) $D = 1$ ps/(nm km), and (c) $D = 15$ ps/(nm km) [25].

wavelength, $\lambda_0 = c/f_0$, where $D(\lambda_0) = 0$. The behavior of FWM near the zero-dispersion wavelength has been studied by Inoue [26] and Marcuse et al. [24].

If all the input signals are located near the zero-dispersion frequency f_0, then the phase mismatch $\Delta\beta$ is given in terms of only the dispersion slope neglecting the chromatic dispersion, $D(\lambda_0) = 0$ [26]:

$$\Delta\beta \approx -\frac{\pi\lambda^4}{c^2}\frac{dD(\lambda_0)}{d\lambda}(f_i - f_k)(f_j - f_k)\{(f_i - f_0) + (f_j - f_0)\} \qquad (4.34)$$

Now let us assume a case where the zero-dispersion frequency f_0 is located at the middle between two signals f_i and f_j as shown in Figure 4.9a.

Then we have $(f_i - f_0) + (f_j - f_0) = 0$ and FWM is therefore inherently phase matched, $\Delta\beta = 0$, to give the maximum efficiency. In this particular case, the phase-matching condition is satisfied irrespective of the dispersion parameters to result in growth of spurious FWM signal with the maximum FWM efficiency $\eta = 1$. FWM signal would be located at the opposite side of f_k with respect to f_0 as shown in Figure 4.9a.

When two of the three input signals are degenerate as $f_i = f_j$, then the phase mismatch is modified as

$$\Delta\beta \approx -2\frac{\pi\lambda^4}{c^2}\frac{dD(\lambda_0)}{d\lambda}(f_i - f_k)^2(f_i - f_0) \qquad (4.35)$$

If the zero-dispersion frequency f_0 of the fiber coincides with the degenerate signal at f_i, then we have $f_i = f_0$ to make $\Delta\beta = 0$ again, resulting in the maximum FWM efficiency $\eta = 1$. The frequency locations are illustrated in Figure 4.9b and FWM signal will grow at the symmetric position to f_k with respect to f_0.

Marcuse et al. [24] further analyzed degenerate FWM signal evolution for the case when one of signals is located at the zero-dispersion wavelength to find that one of FWM signals grows proportional to the length L while the other FWM at the opposite side oscillates.

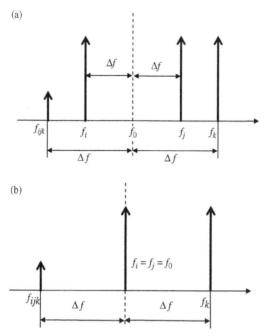

FIGURE 4.9   FWM phase matching where the zero-dispersion frequency f_0 is located (a) in the middle of the two channels, f_i and f_j, and (b) at the degenerated channel $f_i = f_j$.

For the two incoming signals at the frequencies f_1 and f_2, spurious FWM channels are generated at f_3 and f_4. More specifically, when we assume that the fiber loss is negligible ($\alpha = 0$) and one of the signal frequencies, for example, f_1, coincides with the zero chromatic dispersion frequency, f_0, the FWM efficiency of the power accumulation in the sideband at frequencies f_3 and f_4 is obtained from equation 4.33 as

$$\eta \propto \frac{\sin^2(\Delta\beta L/2)}{L_{\text{eff}}^2 \, \Delta\beta^2} \tag{4.36}$$

The equation shows that FWM signal at f_3 continually grows as L increases ($\Delta\beta = 0$) but the signal at f_4 is in sinusoidal behavior ($\Delta\beta \neq 0$). This is shown in Figure 4.10.

To summarize the requirement of fiber dispersion for WDM transmission within a certain range of optical communication band, we need chromatic dispersion low enough to avoid linear regime ISI penalty, yet high enough to suppress FWM. We also require large effective mode area to suppress fiber nonlinearity. Furthermore, the zero-dispersion wavelength should be located far enough from the spectral positions of the WDM signal channels. In addition, the dispersion slope should be kept low to encompass wider bandwidth for WDM channels. These requirements are schematically expressed in Figure 4.11.

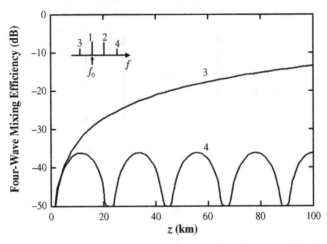

FIGURE 4.10   Evolution of degenerate FWM efficiency for when one of the signal frequencies, f_1, coincides with zero chromatic dispersion frequency, f_0 [24].

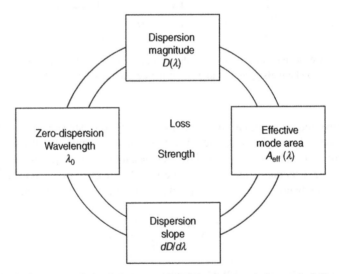

FIGURE 4.11   Design parameters for optimal transmission optical fiber.

4.3   REFRACTIVE INDEX PROFILE ANALYSIS FOR DISPERSION-SHIFTED FIBERS

In this section, some of refractive index profiles for dispersion-controlled transmission fibers are studied in terms of the impacts of waveguide parameters over the chromatic dispersion and their tolerances. These refractive index profiles are from the prior reports where the numerical waveguide parameters are available such as in a review paper by Ainslie and Day [27] along with recent reports [14]. The analysis was carried out using the CBM method, which will be discussed in Chapter 9.

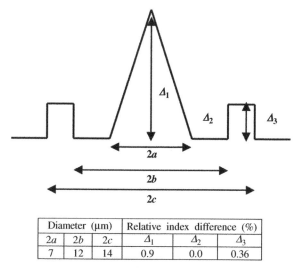

Diameter (μm)			Relative index difference (%)		
$2a$	$2b$	$2c$	Δ_1	Δ_2	Δ_3
7	12	14	0.9	0.0	0.36

FIGURE 4.12   Refractive index profile and waveguide parameters for DSF1.

4.3.1 Dispersion-Shifted Fiber Design

Figure 4.12 shows a DSF refractive index profile of a triangular core along with a segmented ring that has been reported by Paek et al. [28] and Bhagavatula et al. [29].

The mode field diameter (MFD) and effective mode area (A_{eff}) for the fundamental mode are 10.18 μm and 81.48 μm², respectively, at 1550 nm. The LP_{11} mode cutoff (λ_c) is at 1320 nm. The central triangular core mainly shifts the dispersion near 1550 nm and the segmented ring in the inner cladding improves the bending loss and increases mode field area.

The waveguide parameter in Figure 4.12 can affect the chromatic dispersion and the effects of the radii of the core and relative index differences and are analyzed in Figures 4.13 and 4.14, respectively, along with the corresponding changes.

The solid square in Figure 4.13 is the dispersion curve for the parameters given in Figure 4.12. The open circles are for the radius a variation within ±5% and the open triangle is for variation in b within ±5%. The variation in the core radius a did show a significant impact over the zero-dispersion wavelength, dispersion slope, and chromatic dispersion value at 1550 nm. The effect of b was minimal compared to the same magnitude of variation in a.

The refractive index between the triangular core ($r = a$) and the segmented ring ($r = b$) also has a significant impact over the dispersion and the results are shown in Figure 4.14.

It is found that the relative index differences of the triangular core center (Δ_1), inner clad (Δ_2), and the segmented ring (Δ_3) do affect the chromatic dispersion behavior. It is of particular interest to note the cases (2) and (4) in Figure 4.14, where the inner clad index, Δ_2, and the segment ring index, Δ_2, are kept at a negative value, respectively.

The case (2) has the depressed inner clad similar to W-type fibers [30–32] that shows the negative dispersion over a wide wavelength range. In fact, this is the very

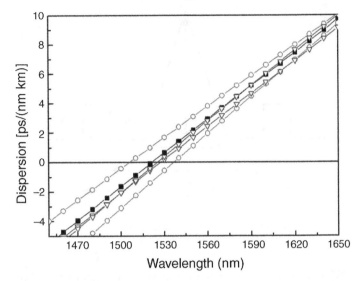

FIGURE 4.13 Impact of radius parameters on chromatic dispersion of DSF1.

principle of getting a negative dispersion in dispersion-compensating fiber, which will be explained in detail Section 4.4.

The case (4) has the refractive index with a pedestal surrounding the triangular core whose impact is the shift of zero-dispersion wavelength to a shorter wavelength along

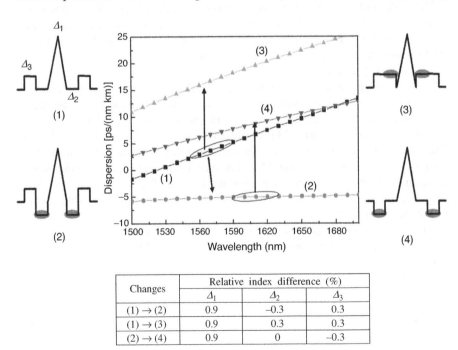

FIGURE 4.14 Impacts of refractive index parameters over dispersion of DSF1.

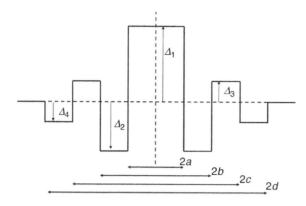

Diameter [μm]				Index difference (%)			
$2a$	$2b$	$2c$	$2d$	Δ_1	Δ_2	Δ_3	Δ_4
6.2	10.2	18	20	0.55	0.55	0.055	0.055

FIGURE 4.15   Refractive index profile and waveguide parameters for DFF1.

with a reduced dispersion slope. This is the very principle of designing an NZ-DSF and in fact the pedestal played a key role in TrueWave™ fiber design [33].

In Figure 4.15, a refractive index profile of DFF is shown [34,35]. It is noted that DFF has more index ring structure in the cladding to tailor the dispersion slope compared with DSF.

For the given waveguide parameters in Figure 4.15, MFD and A_{eff} for the fundamental mode are 6.94 μm and 37.8 μm², respectively, at 1550 nm. The LP_{11} mode cutoff is at 900 nm. The DFF showed a very flat chromatic dispersion from 1450 to 1650 nm within ±1 ps/(nm km) with zero dispersion near 1450 and 1550 nm.

The impact of the core radii, a and b, is analyzed in Figure 4.16. The central core radius, a, showed more prominent impact than the inner cladding radius b, similar to the DSF1.

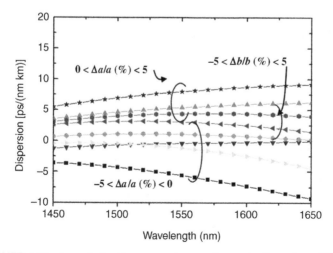

FIGURE 4.16   Impact of radius parameters on chromatic dispersion of DFF1.

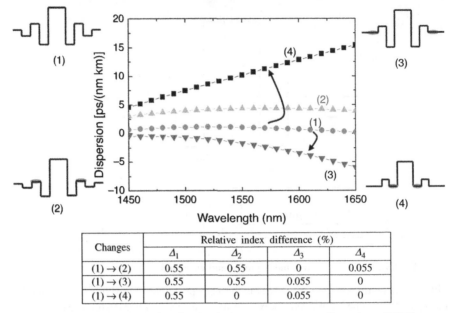

Changes	Relative index difference (%)			
	Δ_1	Δ_2	Δ_3	Δ_4
$(1) \rightarrow (2)$	0.55	0.55	0	0.055
$(1) \rightarrow (3)$	0.55	0.55	0.055	0
$(1) \rightarrow (4)$	0.55	0	0.055	0

FIGURE 4.17 Impacts of refractive index parameters over dispersion of DFF1.

The effects of segmented rings are compared in Figure 4.17. It is found that the relative index differences of the central core (Δ_1), first ring (Δ_2), second ring (Δ_3), and fourth ring (Δ_4) do affect the chromatic dispersion behavior. It is of particular interest to note the cases (2) and (3) in Figure 4.17, where Δ_3 and $\Delta 4$ are kept at zero, respectively.

Case (2) indicates that we could have a finite small dispersion over a wide wavelength region to suppress FWM penalty. Case (3) is once again W-type fiber profile with a negative dispersion value in most of C and L bands. In comparison with the case (2) in Figure 4.14, the amount of suppressed inner cladding index is found to affect the magnitude of negative dispersion, which is important information for DCF. The case (4) in Figure 4.17 shows the clear contrast to triangular core and step-index core in terms of zero-dispersion location. In the case of step-index core, the zero dispersion is no longer near 1500 nm and triangular core would be most efficient choice to shift the zero-dispersion wavelength for a given index difference. However, DFF suffers from relative small A_{eff} and subsequent nonlinear penalties.

In Figure 4.18, the refractive index profile of an NZ-DSF is shown where the core is composed of concentric rings, which have been applied to LEAF™ fibers with an increased effective area [36,37].

The intensity distribution around the central rings in this profile will have a flat-top distribution for the fundamental HE_{11} mode compared with conventional solid core fibers, to result in a larger effective area with appropriate dispersion characteristics.

The fiber has MFD of 10.7 µm and A_{eff} of 90.4 µm^2 at 1550 nm along with λ_c of 1210 nm. The fiber showed nonzero dispersion of 2–5 ps/(nm km) in the entire C band with zero-dispersion wavelength outside the band. The impacts of core radius

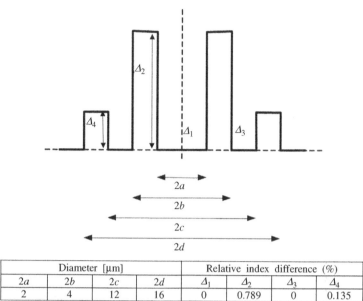

Diameter [μm]				Relative index difference (%)			
$2a$	$2b$	$2c$	$2d$	Δ_1	Δ_2	Δ_3	Δ_4
2	4	12	16	0	0.789	0	0.135

FIGURE 4.18 Refractive index profile and waveguide parameters for NZ-DSF1.

parameters are analyzed for a and b in Figure 4.19 keeping the other parameters the same as in Figure 4.18, where the structure showed a good robustness against the variations of the radius parameters.

For the given central ring core parameters, a and Δ_2, other refractive index parameters were varied and the results are summarized in Figure 4.20. It is found that the dispersion and its slope can be flexibly controlled with these refractive index parameters. It is noted that the outer ring index, Δ_4, can directly change the dispersion

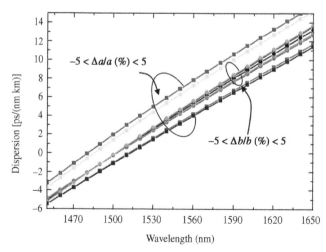

FIGURE 4.19 Impact of radius parameters on chromatic dispersion of NZ-DSF1.

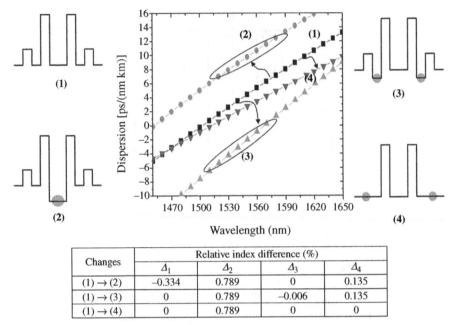

Changes	Relative index difference (%)			
	Δ_1	Δ_2	Δ_3	Δ_4
(1) → (2)	−0.334	0.789	0	0.135
(1) → (3)	0	0.789	−0.006	0.135
(1) → (4)	0	0.789	0	0

FIGURE 4.20 Impacts of refractive index parameters over dispersion of NZ-DSF1.

slope as in the case (4) of Figure 4.20, which can be further applied to NZ-DSF for wider bandwidth including S, C, and L bands.

The transmission characteristics of single-mode fibers having a trapezoidal index profile were first investigated in 1983. It was demonstrated that the zero chromatic dispersion at a wavelength greater than 1.5 μm could be readily achieved with a small dispersion slope [38,39]. Figure 4.21 is another type of NZ-DSF with a smaller effective area, with a trapezoidal core and segmented ring or pedestal [40]. The fiber showed an effective area of 63 μm² with MFD of 8.8 μm at 1550 nm and λ_c was near 1200 nm.

The fiber showed finite dispersion of 2–6 ps/(nm km) in the entire C band with zero-dispersion wavelength near 1470 nm. The impacts of core radius parameters are analyzed for *a* and *b* in Figure 4.22 keeping the other parameters the same as in Figure 4.21. In this case *b* played a major role than *a* in modification of dispersion characteristics in contrast to prior results in NZ-DSF1.

For the given central trapezoidal core parameters, *a* and Δ_1, other refractive index parameters were varied and the results are summarized in Figure 4.23. It is found that the dispersion and its slope can be flexibly controlled with these refractive index parameters.

For the case (2), the fiber shows the characteristics of DSF with the zero dispersion near 1550 nm, from which we can deduce the role of segmented ring or pedestal to shift the zero dispersion to a shorter wavelength. For the case (4), we have pedestal of inner cladding rather than segmented ring as in (3) or (1); the results are further reduction of chromatic dispersion value within C and L bands along with more or less similar dispersion slope.

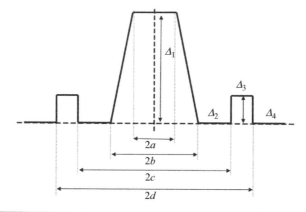

Diameter [µm]				Index difference (%)			
$2a$	$2b$	$2c$	$2d$	Δ_1	Δ_2	Δ_3	Δ_4
3	7	11	14	0.90	0	0.288	0

FIGURE 4.21   Refractive index profile and waveguide parameters for NZ-DSF2.

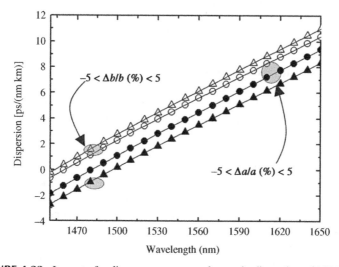

FIGURE 4.22   Impact of radius parameters on chromatic dispersion of NZ-DSF2.

4.4 DISPERSION-COMPENSATING FIBERS USING THE FUNDAMENTAL MODE

In previous sections, we have reviewed fiber designs to reduce chromatic dispersion especially in the C band, where the optical gain in EDFA exists for WDM applications. The designs came up with development of DSF, DFF, and NZ-DSF that will reduce dispersion penalty in long-distance telecommunications compared with conventional single-mode fibers (SMFs) that have a high anomalous dispersion

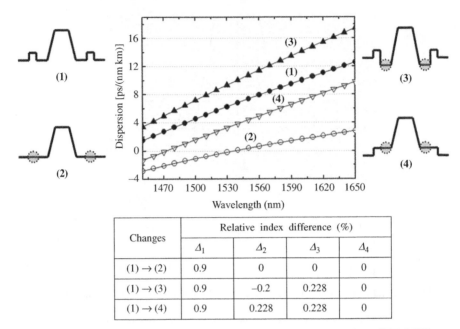

Changes	Relative index difference (%)			
	Δ_1	Δ_2	Δ_3	Δ_4
(1) → (2)	0.9	0	0	0
(1) → (3)	0.9	−0.2	0.228	0
(1) → (4)	0.9	0.228	0.228	0

FIGURE 4.23   Impacts of refractive index parameters over dispersion of NZ-DSF2.

of ~17 ps/(nm km) at 1550 nm. Nevertheless, the residual dispersions of these fibers still linearly accumulate over the transmission distance. Therefore, to control the chromatic dispersion we should have twofold strategies: one is to reduce dispersion in optimal fiber design and the other is to compensate accumulated dispersion with the opposite sign of chromatic dispersion and its slope.

In this section, we will review the dispersion-compensating techniques and especially focus on DCF utilizing the fundamental mode.

A dispersion management scheme is illustrated in Figure 4.24; the chromatic dispersion at a certain wavelength in a transmission fiber accumulates to $+D$ ($=D_{Tr} \times L$) for the length L and is then compensated by the negative dispersion at a DCF by $-D$ ($=D_{DC} \times L'$) to result in total dispersion of zero end-to-end.

The net dispersion D_{net} and dispersion slope S_{net} of a composite link consisting of transmission fiber span length L with DCF length L' can be defined as

$$D_{net} = D_{Tr} \times L + D_{DC} \times L' \qquad (4.37)$$

$$S_{net} = S_{Tr} \times L + S_{DC} \times L' \qquad (4.38)$$

where D_{Tr} and S_{Tr} are the chromatic dispersion in the unit of ps/(nm km) and its slope in ps/(nm^2 km) of the transmission fiber, respectively, and D_{DC} and S_{DC} are the chromatic dispersion and its slope of a DCF. Here dispersion slope is defined in the wavelength domain such that $S(\lambda) = dD(\lambda)/\lambda$.

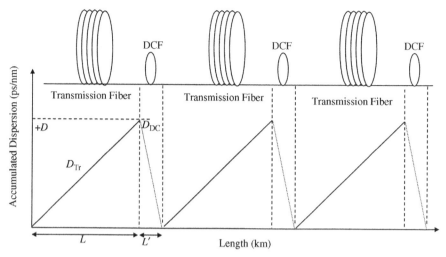

FIGURE 4.24 Schematic diagram for a dispersion-managed system.

In WDM application, dispersion compensation devices are specified by two important parameters, the relative dispersion slope (RDS) and the figure of merit (FOM), and they are defined as follows:

$$\text{RDS}(\lambda) = \frac{S(\lambda)}{D(\lambda)} \tag{4.39}$$

$$\text{FOM}(\lambda) = \frac{|D(\lambda)|}{\alpha(\lambda)} \tag{4.40}$$

where α is the attenuation of DCF. Usually RDS and FOM are defined at 1550 nm for C-band applications.

In ideal compensation, a DCF should provide an RDS value same as the transmission fiber to be compensated and a highest possible FOM for longer span compensation.

RDS of the commercially available transmission fibers are listed in Table 4.4.

RDS of transmission fibers do show a variety of ranges from 0.00341 of SMF to 0.202 of LEAF®. It is noted that a DCF should adapt its dispersion properties, especially RDS, so that they match those of the transmission fiber types for applications in wideband multichannel communications. Commercially available DCF shows FOM in the range of 50–250.

Besides RDS and FOM, there are other requirements in DCF, or equivalently in DCM, for practical applications. The requirements are namely [41,42]

- low dispersion variation, or ripple
- low insertion loss
- minimum multiple path interference
- low nonlinearity

TABLE 4.4 Transmission Fiber Dispersion Properties

Fiber Types	Dispersion (ps/(nm km)) (C Band)	Dispersion (ps/(nm km)) (L Band)	D (ps/(nm km)) at 1550 nm	S (ps/(nm^2 km)) at 1550 nm	RDS (m^{-1}) at 1550 nm
Conventional SMF	16–18	18–22	17	0.058	0.00341
TrueWave RS	2.6–6.0	4.0–8.9	4.4	0.045	0.010
LEAF	2.0–6.0	4.5–11.2	4.2	0.085	0.0202
TrueWave REACH	5.5–8.9	6.9–11.4	7.5	0.042	0.0056
TeraLight Ultra	5.5–10.0	7.5–13.4	8.0	0.052	0.0065
Pure Guide	5.0–10.0	12.4 at 1625 nm	8.0	0.058	0.0073

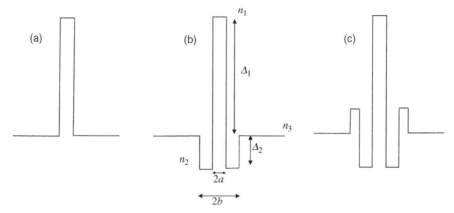

FIGURE 4.25 Schematic refractive index profiles for the fundamental mode DCFs.

- low polarization-dependent loss (PDL) and low polarization mode dispersion (PMD)
- good temperature stability
- high mechanical strength

Among available technologies for dispersion compensation, the fundamental mode DCF provides the most of above requirements for practical field applications.

The fundamental mode DCF utilized both material and waveguide dispersion to generate a large negative dispersion and negative dispersion slope for the HE_{11} mode. The refractive index profiles are schematically shown in Figure 4.25.

The profiles have two common factors, high core index ($\Delta_1 > 1.5\%$) and small core ($2a < 5\,\mu m$), to increase the waveguide dispersion contribution to result in a large negative dispersion value. Type (a) is a matched clad structure providing only positive dispersion slope and it is no longer used in WDM systems. Types (b) and (c) are based on W-type fiber [30,31] and they provide negative dispersion slope and large negative dispersion, well suited for current WDM applications. Type (c) has additional segmented ring to further tailor dispersion slope and improve bending performances.

The fundamental advantage of W-type fiber in dispersion control can be explained by Figure 4.26.

At a short wavelength, say $\lambda < \lambda_1$, the HE_{11} mode would experience a step-index profile within the inner cladding, whose index difference is given by $n_1 - n_2$. The trace of the effective index evolution, therefore, can be expressed as a solid line in Figure 4.26. As the wavelength increases, however, the mode will realize that there exists the outer cladding with refractive index n_3, and changes its route to the dotted line. Within a relatively narrow region of wavelength, $\lambda_1 < \lambda < \lambda_2$, the effective index $n_{eff}(\lambda)$ shows a rapid change in its slope, which results in a large negative dispersion along with negative dispersion slope [43].

In W-type fiber, therefore, the ratio between inner cladding parameters and core parameters will play a major role in dispersion control. For example, the relative index

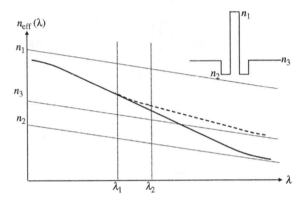

FIGURE 4.26 Schematic diagram for change of effective index in W-type fiber.

ratio, $R_\Delta = \Delta_2/\Delta_1$ along with Δ_1, as shown in Figure 4.27, has a strong impact over the chromatic dispersion [44].

In Figure 4.27 it is noted that for Δ_1 of +0.6%, which is almost the same as that of conventional single-mode fiber, and R_Δ of −0.285, the dispersion is negative but with a small magnitude and the dispersion slope near 1550 nm is positive. As R_Δ gets a larger negative value of −0.800, for the same Δ_1, the negative dispersion significantly increases with a negative dispersion slope. However, the most optimal dispersion properties are obtainable for higher core index Δ_1 of 2.0% and an appropriate R_Δ of −0.285.

In the following discussions, we will review three types of fundamental mode DCFs and detailed analysis will be given in terms of waveguide parameter impact analysis and optimization based on prior published papers.

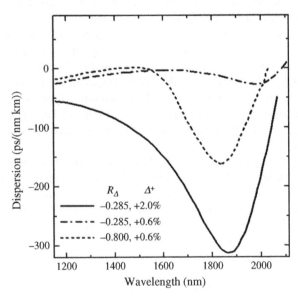

FIGURE 4.27 Impact of inner cladding parameters over chromatic dispersion [44].

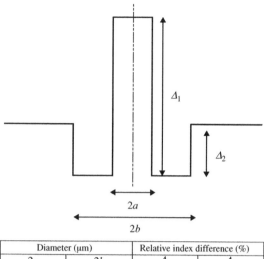

Diameter (μm)		Relative index difference (%)	
$2a$	$2b$	Δ_1	Δ_2
2.8	6.2	1.5	−0.6

FIGURE 4.28 Refractive index profile and waveguide parameters for DCF1.

Figure 4.28 shows a refractive index profile of a typical step-index W-type DCF [45]. In this case $R_\Delta = -0.400$ and the DCF1 has an MFD of 4.9 μm and an A_{eff} of 20 μm^2.

The impacts of core radius parameters, a and b, are summarized in Figure 4.29 for their relative variations within ±5%. In Figure 4.29, it is noted that the impact of core radius, a, variation is significantly greater in chromatic dispersion than that of inner cladding radius, b. Smaller values of a were found to provide larger negative

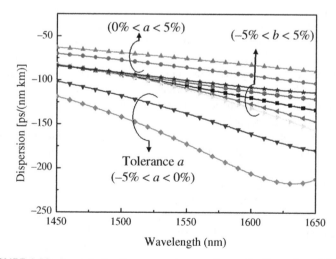

FIGURE 4.29 Impact of radius parameters on chromatic dispersion of DCF1.

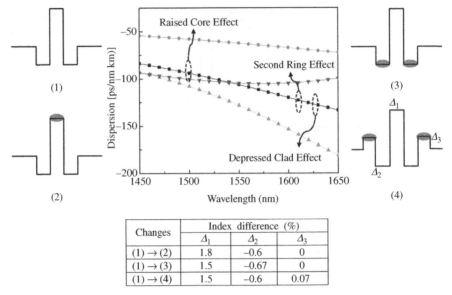

Changes	Index difference (%)		
	Δ_1	Δ_2	Δ_3
$(1) \to (2)$	1.8	−0.6	0
$(1) \to (3)$	1.5	−0.67	0
$(1) \to (4)$	1.5	−0.6	0.07

FIGURE 4.30 Impacts of refractive index parameters over dispersion of DCF1.

dispersions in the W-type fiber in accordance with prior reports, which should be balanced by the requirements for larger effective area. The variation in b was found to be more effective in overall dispersion slope control.

Maintaining the radius parameters, a and b as in Figure 4.28, the refractive index parameters were varied and the results are summarized in Figure 4.30. The case (2) corresponds to raised core index Δ_1, and the case (3) to depressed inner clad index Δ_2. The case (4) corresponds to addition of segmented ring with relative index Δ_3, whose width is the same as the depressed inner cladding. It is noted that in this particular set of refractive index parameters, further increase of Δ_1 did not provide a larger negative dispersion in C and L bands but Δ_2 did. Addition of segmented ring in the outer cladding changed the dispersion slope, as expected from the results of DSF in previous section.

As another type of DCF, we will consider the parabolic or so-called α profile refractive index core, along with the depressed inner cladding and the raised segmented ring. This profile has been fabricated by outside vapor deposition (OVD) process to compensate SMF dispersion [46]. Later on similar profiles have been adopted in vapor axial deposition (VAD) process, which is referred as reverse dispersion fiber (RDF) that was initially designed to compensate both dispersion and dispersion slope of DSF [47]. The profile further developed to achieve dispersion compensation over S, C, and L bands [48].

The refractive index profile is given in Figure 4.31.

Here we have assumed the parabolic core index. For the given parameters, the DCF2 has the MFD of 6 μm and A_{eff} of 28.3 μm^2 at 1550 nm. The impacts of radius parameters are analyzed in Figure 4.32. There are two notable observations that can be made. First, the inner cladding variation, $\Delta b/b$, showed more significant impacts than core radius variation, $\Delta a/a$, in contrast to DCF1 as shown in Figure 4.29. Second, this profile showed almost the same or larger variation in dispersion value for less variation

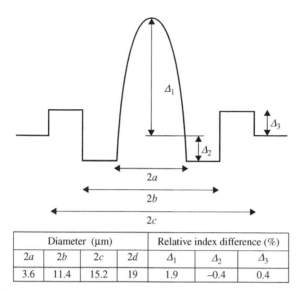

Diameter (μm)				Relative index difference (%)		
$2a$	$2b$	$2c$	$2d$	Δ_1	Δ_2	Δ_3
3.6	11.4	15.2	19	1.9	−0.4	0.4

FIGURE 4.31 Refractive index profile and waveguide parameters for DCF2.

of radius parameters, ±3%, compared with previous DCF1. Therefore, tight control of radius parameters would be required for DCF2 in the fabrication process.

Maintaining the radius parameters, a, b, c, and d, and the core refractive index, Δ_1, as in Figure 4.31, other refractive index parameters were varied and the results are summarized in Figure 4.33. The case (2) corresponds to elimination of segmented ring index Δ_3, and the case (3) to matched inner clad index $\Delta_2 = 0$. The case (4) corresponds to addition of another depressed index ring with relative index Δ_4, whose width is the same as the segmented raised ring.

It is noted that as in the cases (2) and (3) the profiles require the raised segment ring ($\Delta_3 > 0$) and the depressed inner cladding ($\Delta_2 < 0$) in order to have a large negative

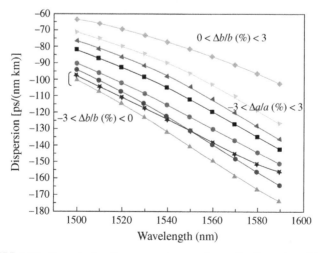

FIGURE 4.32 Impact of radius parameters on chromatic dispersion of DCF2.

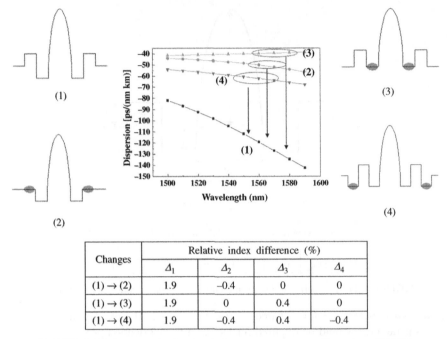

Changes	Relative index difference (%)			
	Δ_1	Δ_2	Δ_3	Δ_4
(1) → (2)	1.9	−0.4	0	0
(1) → (3)	1.9	0	0.4	0
(1) → (4)	1.9	−0.4	0.4	−0.4

FIGURE 4.33 Impact of refractive index parameters over dispersion of DCF2.

dispersion. Additional depressed ring ($\Delta_4 < 0$) as in case (4) did not improve the negative dispersion.

The core refractive index profile can be further varied to a triangular shape in DCF structure assuming the depressed inner cladding and the raised segmented ring. This profile design has been mainly fabricated using modified chemical vapor deposition (MCVD) process for dispersion and dispersion slope compensation of conventional single-mode fibers [49].

The profile of the fiber is shown in Figure 4.34 and DCF3 has MFD of 8.6 μm and A_{eff} of 57.7 μm² for the given parameters. The impacts of radius parameters are analyzed in Figure 4.35 keeping the refractive index parameters, Δ_1, Δ_2, and Δ_3 as in Figure 4.34.

Similar to DCF2 with a parabolic core index, triangular core DCF3 also showed more significant impact by the inner cladding radius, $\Delta b/b$, than the core radius, $\Delta a/a$, yet the dispersion took a larger value in general for a smaller core radius and a smaller inner cladding radius.

Maintaining the radius parameters, a, b, and c, and the core refractive index, Δ_1, as in Figure 4.34, other refractive index parameters were varied and the results are summarized in Figure 4.36. The case (2) corresponds to matched inner clad index $\Delta_2 = 0$, and the case (3) to elimination of segmented ring index Δ_3. The case (4) corresponds to addition of another depressed index ring with relative index Δ_4, whose width is the same as the segmented raised ring.

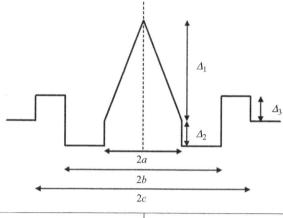

Diameter [μm]			Index difference (%)		
$2a$	$2b$	$2c$	Δ_1	Δ_2	Δ_3
4.6	12.4	20.8	2	−0.4	0.4

FIGURE 4.34   Refractive index profile and waveguide parameters for DCF3.

For the given waveguide parameters, it is noted that DCF3 is more affected by the raised segmented ring, Δ_3, than the depressed inner cladding, Δ_2, which is quite a contrast to the parabolic DCF2 as in Figure 4.33. Other behavior is in general in a good accordance with previous two profiles.

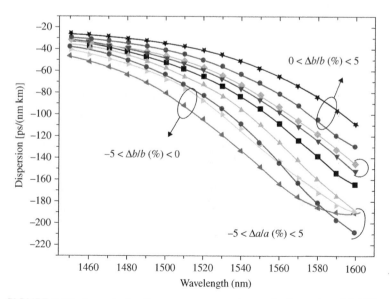

FIGURE 4.35   Impact of radius parameters on chromatic dispersion of DCF3.

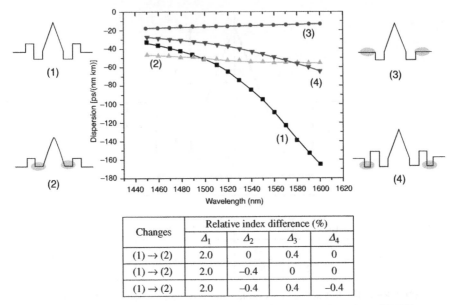

FIGURE 4.36 Impact of refractive index parameters over dispersion of DCF3.

4.5 DISPERSION COMPENSATION USING HIGH-ORDER MODES

Chromatic dispersion is uniquely determined by the wavelength-dependent effective index and its derivatives of a guided mode. Higher order modes such as LP_{11} and LP_{02} modes have been utilized to exploit their inherently large negative dispersion near their cutoffs. In contrast to the fundamental LP_{01} mode, these higher order modes have cutoff wavelength where the effective index of the mode takes the refractive index of silica cladding and the mode extends out to the cladding. Therefore, near the cutoff there exist abrupt changes in the effective index along with mode confinement to result in a large negative dispersion [50–57].

Basic schematics for higher order mode dispersion compensation are illustrated in Figure 4.37. The optical pulse from a transmitter is experiencing pulse broadening due

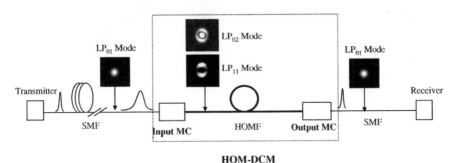

HOM-DCM

FIGURE 4.37 Schematic of high-order mode dispersion compensation technique.

to the accumulated chromatic dispersion along the single-mode fiber, where the light is guided in the fundamental LP_{01} mode. This dispersion is compensated by a high-order mode (HOM) DCM, which is composed of input and output mode converter (MC) and high-order mode fiber (HOMF).

At the input MC, the LP_{01} mode is converted to HOM such as the LP_{11} or LP_{02} modes. These HOMs will travel through HOMF where these modes will experience a large negative dispersion and dispersion slope to compensate the dispersion accumulation in preceding SMF. After appropriate dispersion compensation along HOMF, HOMs are converted into the LP_{01} mode in SMF by the output MC.

Here SMF and HOMF are single-mode fiber and high-order mode fiber, respectively. MC and HOM-DCM are mode converter and high-order mode dispersion-compensating module.

Among the high-order modes, the LP_{11} and LP_{02} modes have been intensively investigated due to high mode conversion efficiency from the LP_{01} mode and high figure of merit. In Table 4.5, the mode characteristics are summarized.

It is noted that the LP_{11} mode consists of TE_{01}, TM_{01}, and doubly degenerate HE_{21} modes with their distinctive polarization directions, while the LP_{02} mode shows the same twofold degeneracy as the fundamental LP_{01} mode. For a step-index profile optical fiber the chromatic dispersion of LP_{11}, LP_{21}, and LP_{02} modes is plotted in Figure 4.38 assuming a perfectly circular symmetry. The dispersion was evaluated near cutoff in a step-index fiber with $\Delta = 0.02$ and the LP_{11} mode cutoff $\lambda_{cutoff} = 1620$ nm. The core radius was adjusted for each curve to make the cutoff wavelength the same.

Considering usual values of the LP_{01} mode DCF in the range of ~ -100 ps/(nm km) as discussed in Chapter 3, these HOMs, especially LP_{11} and LP_{02}, can give several-fold larger negative dispersion to make HOM-DCM compact and efficient.

The large negative dispersion near the cutoff can be qualitatively understood by considering mode power overlap with the cladding. The mode power overlap is

TABLE 4.5 Characteristics of Important High-Order Modes Used in Dispersion Compensation

Exact solution	HE_{11}	TE_{01}	TM_{01}	HE_{21}	EH_{11}	HE_{12}	HE_{31}
Weakly guiding	LP_{01}	LP_{11}			LP_{21}		LP_{02}
V_{cutoff}	0	2.405			3.832		

The arrows indicate the direction of the electric field. The cutoff V numbers are for matched step-index fibers and they change as a function of refractive index profiles.

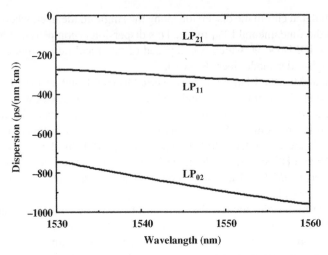

FIGURE 4.38 Comparison of dispersion versus wavelength for the first few high-order LP modes [50].

usually represented by a ratio, P_{clad}/P_{total}, and is given in a weakly guiding approximation [32] as

$$\frac{P_{clad}}{P_{total}} = \frac{U}{V}\frac{dU}{dV} \tag{4.41}$$

As the wavelength increases toward cutoff and mode power extends further into the cladding, the effective index of the mode rapidly approaches to that of silica cladding. In the case of normal dispersion ($d^2\beta/d\omega^2 > 0$), the group velocity will increase as the wavelength increases. Thus, the group delay time, which is the reciprocal of the group velocity, decreases with wavelength and, therefore, the chromatic dispersion becomes a negative value.

The larger magnitude of the dispersion in the higher order modes results from the rapid transition with wavelength of mode power from the core to the cladding near cutoff, which is directly related to waveguide dispersion.

Assuming the weakly guiding approximation [32], the waveguide dispersion in ideal circular step-index fiber can be expressed as

$$D_w = \frac{\Delta n}{c\lambda}\frac{\Psi(V)}{V}, \quad \Psi(V) = V^2\frac{d^2}{dV^2}\left(\frac{U^2}{V}\right) \tag{4.42}$$

Here Δn is the core–cladding index difference in a step-index fiber. V is the normalized frequency and U is the core transverse waveguide parameter as defined in Chapter 2.

Figure 4.39 shows the normalized dispersion, $\Psi(V)/V$, plotted versus cladding overlap given by equation 4.42, for the first four LP modes.

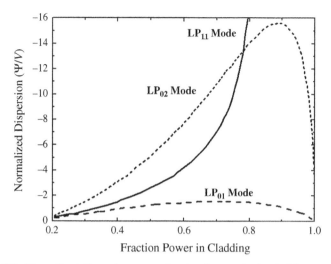

FIGURE 4.39 Normalized dispersion versus fractional power in the cladding, P_{clad}/P_{total}, for the first few LP modes of a step-index fiber [50].

It is noted that in the range of 0.5–0.75 fractional power in the cladding, the LP_{02} mode provides the largest magnitude dispersion. In this region, the LP_{02} and LP_{11} waveguide dispersions are as much as an order of magnitude larger than the LP_{01} mode. In contrast, the LP_{21} mode is limited to a maximum of ∼0.45 fractional power in the cladding at the cutoff and the waveguide dispersion, subsequently, is also limited.

From equation 4.42 and Figure 4.39, we can figure out the optimal direction of fiber design for HOM-DCM. First, the magnitude of the chromatic dispersion scales with the core–cladding index difference so that high Δ fiber would be preferable for a larger negative dispersion. V-value ranges are 2.5–2.7 for the LP_{11} mode fiber and 4.1–4.4 for the LP_{02} mode fiber for optimal fractional power in the cladding of 0.5–0.75.

In this section, we will review HOM-DCM based on the LP_{11} and the LP_{02} modes, discussing merits and demerits of system applications. We will, however, limit the discussion on the MC in this section, which is based on periodic mode coupling such as fiber grating and will be referred in the later section on fiber gratings.

Thus far, the analysis was based on a weakly guiding approximation along with scalar wave equation, where the degeneracy of the HOM is assumed. In real physical world, however, the HOM, especially LP_{11} mode, will show highly polarization-dependent mode dispersion so that the amount of dispersion compensation in HOMF will significantly depend on the incoming polarization state. For perfect circular symmetric optical waveguides, the LP_{01} mode is degenerate showing the same effective indices for two orthogonal polarizations, which are the same for the LP_{02} mode. But in the case of LP_{11} mode, the constituent TE_{01}, TM_{01}, and doubly degenerate HE_{21} modes will have their own dispersion properties even in a perfect cylindrical symmetric optical fiber as long as the index difference between the core and cladding is high enough, say over 1% of Δ [53].

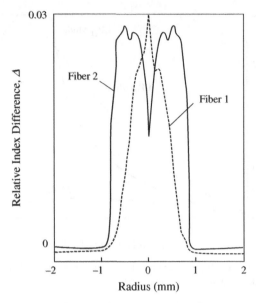

FIGURE 4.40 LP$_{11}$ mode dispersion-compensating fiber characteristics [50].

Poole et al. [50,51] first experimentally demonstrated the potential of HOM-DCM using the LP$_{11}$ mode in elliptic core two-mode fibers and they reported polarization dependence in dispersion compensation. The fiber characteristics are summarized in Figure 4.40.

Figure 4.41 shows the experimental measurement of LP$_{11}$ mode dispersion in C band. Measurements of dispersion in two elliptic core fibers operated in the LP$_{11}$ mode near cutoff showed dispersion values as large as -770 ps/(nm km) at 1555 nm. It is noted that the scalar theory prediction shows a rough trend but differs

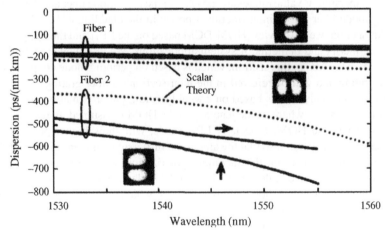

FIGURE 4.41 Dispersion versus wavelength for the higher order modes of the two test fibers. Arrows indicate polarization [50].

significantly from experimental measurements. The direction of polarization does change the response in chromatic dispersion as shown in the mode diagram along with the arrows. Poole et al. further proposed a polarization nulling scheme [51] by using a polarization rotating mirror to cancel out polarization dependence in these LP_{11} modes and simultaneous cancellation of the dispersion and the dispersion slope in a single-mode fiber span has been demonstrated, allowing compensation over the entire C band. Application of the dispersion compensation technique to digital transmission has been also demonstrated in a 2.5 Gb/s system experiment over 50 km of conventional single-mode fiber that had been completely compensated by 720 m of compensating fiber in terms of chromatic dispersion and its slope.

The effects of core refractive index profiles in a matched cladding fiber over the waveguide dispersion have been theoretically investigated by Eguchi et al. [52] using finite element method (FEM). Their analysis is summarized in Figure 4.42, where the total chromatic dispersion spectra are obtained by concatenating a conventional 1.3 μm zero-dispersion single-mode optical fiber of 40 km length. In the analysis the dispersion compensation was analyzed for α-power refractive index profile dual-mode optical fiber whose length is set to cancel the total chromatic dispersion of conventional single-mode fiber at 1.55 μm. The α-power refractive index profile is

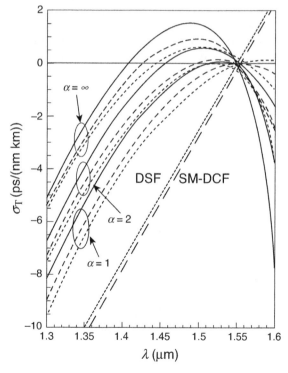

FIGURE 4.42 Total chromatic dispersion obtained by concatenating conventional single-mode fiber of 40 km length with compensating dual-mode fiber. The solid line, $D = 3\%$; the dashed line, $\Delta = 2\%$; the dotted line, $\Delta = 1.5\%$ [52].

discussed in Chapter 5 and $\alpha = 1$ and 2 correspond to triangular and parabolic index profiles, respectively. Step index is represented by $\alpha = \infty$.

Three cases have been analyzed at $\alpha = \infty$ for step, $\alpha = 2$ for parabolic, and $\alpha = 1$ for triangular core refractive index profile. In Figure 4.42, the solid, the dashed, and the dotted lines correspond to the results for $\Delta = 3, 2,$ and 1.5%, respectively. DSF and SM-DCF correspond to the results for a step-index profile dispersion-shifted fiber ($\Delta = 0.7\%$, $2a = 4.449\,\mu m$) and an LP_{01} mode dispersion-compensating fiber with step-index profile core ($\Delta = 3\%$, $2a = 2\,\mu m$), respectively. It is noted that the net chromatic dispersion obtained by using the dual-mode fibers is smaller than that of LP_{01} mode DCF for a span of conventional single-mode fiber in a wideband including S, C, and L bands keeping the dispersion slope small.

Oh et al. [53] have further developed theoretical analysis of LP_{11} mode DCF using a full vector solution of wave equation for depressed inner cladding optical fibers with various a profiles. The waveguide parameters are shown in Figure 4.43 that is a W-type fiber [30] and the cutoff of higher order modes can be flexibly controlled with the inner cladding parameters, Δ^- and b.

The principal difference between the matched cladding dual-mode fiber (DMF) and depressed cladding DMF is schematically illustrated in Figure 4.44. The LP_{01} mode in SMF transforms to LP_{11} mode in DMF by a mode converter. In cylindrical DMFs, the degeneracy of the LP_{11} mode can be removed intrinsically if very high index difference is provided even without external geometrical ellipticity. In order to analyze the time delay between the polarization modes, the following parameters were introduced for the TE_{01}, the TM_{01}, and the HE_{21} modes:

$$\frac{\Delta \tau_1}{L} = \frac{n_{\text{eff}}^{TE_{01}} - n_{\text{eff}}^{TM_{01}}}{c} - \frac{\lambda}{c} \frac{d(n_{\text{eff}}^{TE_{01}} - n_{\text{eff}}^{TM_{01}})}{d\lambda} \tag{4.43}$$

$$\frac{\Delta \tau_2}{L} = \frac{n_{\text{eff}}^{TE_{01}} - n_{\text{eff}}^{HE_{21}}}{c} - \frac{\lambda}{c} \frac{d(n_{\text{eff}}^{TE_{01}} - n_{\text{eff}}^{HE_{21}})}{d\lambda} \tag{4.44}$$

Note that TE_{01}, TM_{01}, and HE_{21} modes are allowed in the matched cladding DMF with their cutoffs located nearby.

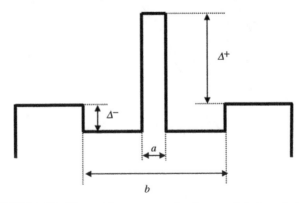

FIGURE 4.43   Waveguide parameters for depressed dual-mode fiber.

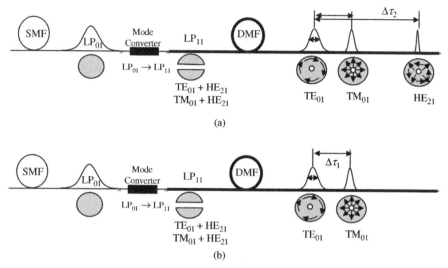

FIGURE 4.44 Schematic comparison of matched cladding dual-mode fiber (DMF) and depressed cladding DMF. In the depressed cladding DMF, the HE_{21} mode is cutoff and only TE_{01} and TM_{01} modes are allowed [53]. (a) Cylindrical DMF with the matched cladding; (b) cylindrical DMF with the depressed cladding.

In the case of depressed inner cladding DMF, however, the cutoff behavior of the high-order modes can be further tailored such that the HE_{21} mode can be made cutoff and only TE_{01} and TM_{01} modes are allowed in DMF as shown in Figure 4.44b. In this case the modal delay can be significantly improved to enable dispersion compensation for high–data rate, long-distance communications.

Further waveguide design can result in the same value of chromatic dispersion for the TE_{01} and TM_{01} modes in the depressed inner cladding DMF, as shown in Figure 4.45.

Note that as Δ^- takes a larger negative value, the HE_{21} mode is no longer defined for a certain range of Δ^- and only the TE_{01} and the TM_{01} modes are guided. The arrows in Figure 4.45 indicate the conditions for which the TE_{01} and TM_{01} modes show the same chromatic dispersion value. The waveguide parameters in these cases are summarized in Table 4.6.

Considering inherent polarization degeneracy for a circular fiber and large negative dispersion as shown in Figure 4.39, dispersion compensation scheme based on the LP_{02} mode would have significant advantages over those based on the LP_{11} mode, which suffer from polarization mode dispersion as discussed earlier. Ramachandran et al. [54] have reported application of an LP_{02} mode dispersion compensator for 40 Gb/s transmission over 1000 km.

The design principle of LP_{02} mode fiber is schematically shown in Figure 4.46. Note that the fiber has the depressed inner cladding and the raised segmented ring outside. As wavelength increases, the entire mode profile of the LP_{01} mode expands across the core, while in the LP_{02} mode, only the side lobe expands especially around the depressed inner cladding and the raised segmented ring. The depressed inner cladding confines the side lobe of the LP_{02} mode in the shorter wavelength in a very

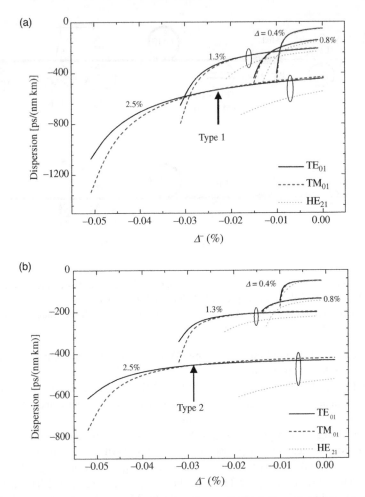

FIGURE 4.45 Chromatic dispersion of high-order modes in dual-mode fibers with the step-index core and depressed inner cladding structures. The dispersion values are plotted as a function of Δ^-, for b/a ratio of 5.0 and 7.5 as shown in (a) and (b), respectively, at 1550 nm [53].

TABLE 4.6 Waveguide Parameters and Dispersion Properties of Two Types of DMF with Step-Index Core and Depressed Inner Cladding Structure

Optical Parameters	Type 1	Type 2
Δ^+ (%)	2.5	2.5
b/a ratio	5.0	7.5
Δ^- (%)	−0.024	−0.029
Dispersion (ps/(nm km))	−538	−453
Modal delay, τ_1 (ns/km)	3.36	3.30

These fibers have the same chromatic dispersion for the TE_{01} and the TM_{01} modes.

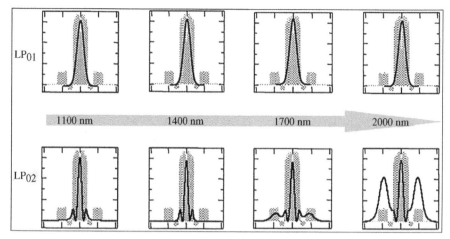

FIGURE 4.46 Schematic evolution of LP_{01} and LP_{02} modes as a function of wavelength. Mode evolution with wavelength. Gray background is the refractive index profile. Different modes expand at different rates. Modal expansion is intimately related to propagation properties [54].

similar manner as a W-type fiber does for the LP_{01} mode as described earlier in Figures 4.25 and 4.26. Utilizing this unique wavelength-dependent mode expansion, we can flexibly design an LP_{02} mode fiber with a large negative dispersion and negative dispersion slope.

The dispersion of the reported LP_{02} mode DCM is shown in Figure 4.47. Chromatic dispersion at 1550 nm was over $-420\,\text{ps/(nm km)}$ and RDS was $0.008\,\text{nm}^{-1}$. The dispersion curve was smooth and continuous over entire C band. Ripples at $\lambda < 1510\,\text{nm}$ indicate multiple path interference–related degradations due to inadequate mode conversion between the LP_{01} and LP_{02} modes.

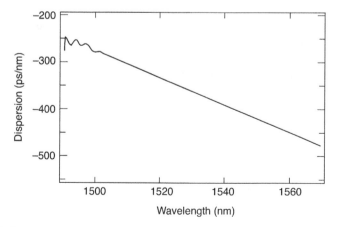

FIGURE 4.47 Dispersion versus wavelength in LP_{02} mode DCM [54].

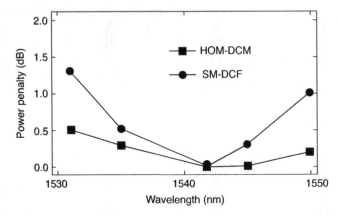

FIGURE 4.48   Power penalty for 40 Gb/s RZ signals. Broader compensation with LP_{02}-DCM is observed compared with the LP_{01}-DCF showing a bandwidth over 20 nm for 0.5 dB penalty [54].

The LP_{02}-DCM have been evaluated in a 100 km span of a nonzero dispersion-shifted fiber, TrueWave RS, with signals at a bit rate of 40 Gb/s. The performance was compared with a similar span that has a commercial LP_{01} mode DCF. The commercial LP_{01}-DCF had an RDS of 0.0065 nm^{-1}, while that for the LP_{02}-DCM was 0.008 nm^{-1}. The broadband performance of the module is quantified by measuring BER power penalties in a 40 Gb/s return-to-zero (RZ) transmission system. Figure 4.48 shows the BER power penalty at a BER of 10^{-9} compared to the back-to-back signal.

TrueWave RS has the dispersion of 4.5 ps/(nm km) and its slope 0.045 at 1550 nm, to result in RDS of 0.010. The LP_{02}-DCM had a closer RDS of 0.008 than the LP_{01}-DCF to provide a broader wavelength range of dispersion compensation.

The LP_{02}-DCM had an effective area over 65 μm^2, which is at least four times larger than that of the LP_{01}-DCFs. In addition, the length of fiber in the module is a factor of two less in the LP_{02}-DCM. Thus, simulations indicate that up to 11 dB higher signal powers may be introduced into the HOM-DCM without incurring optical nonlinear distortions. This has been experimentally confirmed by raising input power levels to as high as 10 dBm into the LP-DCM while not detecting any nonlinearity-related penalties.

Based on these dispersion evaluations for a single LP_{02}-DCM, Ramachandran et al. further demonstrated a discretely tunable dispersion compensation in C band [55].

The schematic structure of the adjustable higher order mode dispersion compensator (AHOM-DC) is shown in Figure 4.49, where mechanically switchable long-period grating (SLPG) spectra are also shown for cross and bar states. SLPG was introduced to convert the LP_{01} mode to the LP_{02} mode, functioning as a binary 2×2 mode switch. By concatenating the LP_{02} HOMF with different lengths via SLPGs, various combinations of dispersion compensation and slope compensation have been experimentally demonstrated as shown in Figure 4.50.

Thus far, mode converters in HOMF-DCM have been based on the period coupling technique, where periodic perturbations along a fiber provide the phase-matching condition between the LP_{01} mode and higher order modes. The most widely used periodic coupling techniques are stress-induced perturbation [58], microbending [59,60],

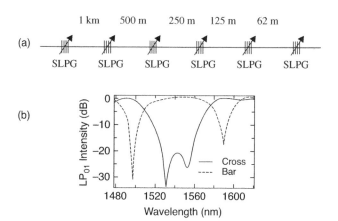

FIGURE 4.49   (a) The adjustable higher order mode dispersion compensator (AHOM-DC) schematic. (b) Spectrum of SLPG in the cross and bar states—broadband in both states [55].

and photoinduced index changes [57,61]. These mode converters inherently suffer from spectral location and bandwidth of operation, due to stringent requirements in the phase-matching conditions. In other words, the mode converter works as it is only for a range of wavelength that satisfies the phase-matching condition:

$$\beta_{01} - \beta_{lm} = \frac{2\pi}{\Lambda} \tag{4.45}$$

where β_{01} and β_{lm} are the propagation constants for the LP_{01} and higher order mode LP_{lm}. Λ is the pitch of the periodic structure in optical fiber. In most of prior mode converters the bandwidth of operation has been limited to less than 30 nm.

An attempt to obviate this phase-matching limitation has been recently reported by Choi and Oh [56] by using a unique hollow optical fiber (HOF) [62]. HOF has a three-layered structure composed of the central air hole, GeO_2–SiO_2 ring core,

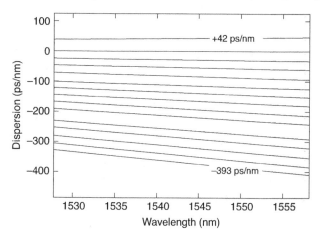

FIGURE 4.50   Dispersion versus wavelength. Tuning range = 435 ps/nm in 14 ps/nm steps at 1550 nm. Wavelength-continuous response over 30 nm was demonstrated [55].

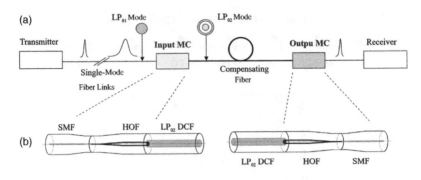

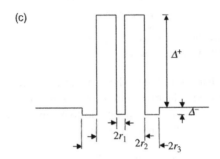

FIGURE 4.51 Proposed dispersion compensation scheme. (a) Transmission links cascaded with SMF-HOF mode converter LP_{02} mode DCF and (b) a pair of mode converters. (c) LP_{02} mode DCF, with $r_1 = 0.6\,\mu m$, $r_2 = 3.7\,\mu m$, and $r_3 = 4.7\,\mu m$, respectively. Δ^+ and Δ^- were 1.968 and 0.064%. The LP_{02} mode cutoff was 1.60 μm [56].

and SiO_2 cladding. When one of the ends of HOF segment is sealed to a solid core SMF, while the other end is left with the air hole, it can provide a unique adiabatic mode conversion from the LP_{01} mode to a ring-shaped mode with a very low loss [63]. Because of its adiabatic nature, the HOF mode converter provides a very wide band of operation, in principle unlimited as long as the light signal is guided in the LP_{01} mode in the SMF side. Detailed description of HOF is given in Chapter 8.

Choi and Oh [56] have reported a broadband LP_{02} mode dispersion-compensating technique based on an HOF mode converter and mating ring core HOM-DCFs for C band applications in 1.53–1.57 μm. The schematics of the proposed technique are shown in Figure 4.51. The input mode converter converts the incident LP_{01} mode into an annulus mode, which couples efficiently to the LP_{02} mode in HOM-DCF. The LP_{02} mode DCF was designed in a concentric ring core structure to yield a high coupling efficiency between its LP_{02} mode and the annulus fundamental mode of HOF.

Chromatic dispersion of the LP_{02} mode in DCF and LP_{01} mode in SMF is overlayed in C band in Figure 4.52. At 1.55 μm LP_{01} mode in SMF has dispersion of ~ 16 ps/(nm km) while the LP_{02} mode in the DCF has dispersion of ~ -770 ps/(nm km).

RDS, the ratio of dispersion slope to dispersion, was 0.00356 for SMF and 0.00695 nm^{-1} for the LP_{02} mode DCF. In order to reduce chromatic dispersion and its slope in the entire C band, 60 km SMF and 1.31 km DCF were combined to predict the total dispersion less than ± 1 ps/(nm km), and dispersion slope between -0.09 and

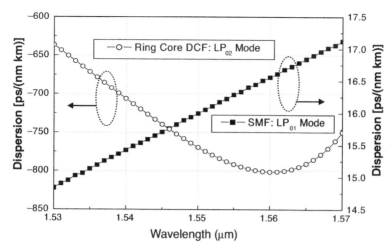

FIGURE 4.52 Chromatic dispersion versus wavelength for the LP_{01} and LP_{02} modes [56].

TABLE 4.7   Dispersion Parameters for LP_{02} Mode Based on HOF

Parameters	Unit	SMF	LP_{02} HOF DCF	TrueWave RS
Dispersion (D)	ps/(nm km)	16	-770	4.5
Dispersion slope (S)	ps/(nm^2 km)	0.057	-5.364	0.045
RDS $= S/D$	nm^{-1}	0.00356	0.0069	0.010

$0.3\,ps/(nm^2\,km)$ in the entire C band. The main characteristics of LP_{02} mode dispersion compensation techniques are summarized in Table 4.7.

REFERENCES

[1] H. Ono, M. Yamada, T. Kanamori, S. Sudo, and Y. Ohishi, "1.58-µm band gain-flattened erbium-doped fiber amplifiers for WDM transmission systems," Journal of Lightwave Technology, vol. 17, no. 3, pp. 490–496, 1999.

[2] T. Komukai, T. Yamamoto, T. Sugawa, and Y. Miyajima, "Upconversion pumped thulium-doped fluoride fiber amplifier and laser operating at 1.47 µm," IEEE Journal of Quantum Electronics, vol. 31, no. 11, pp. 1880–1889, 1995.

[3] J. Minelly and A. Ellison, "Applications of antimony-silicate glasses for fiber optic amplifiers," Optical Fiber Technology, vol. 8, no. 2, pp. 123–138, 2002.

[4] J. Bromage, "Raman amplification for fiber communications systems," Journal of Lightwave Technology, vol. 22, no. 1, pp. 79–93, 2004.

[5] ITU-T G.653, "Characteristics of a dispersion-shifted single-mode optical fibre and cable," 2010.

[6] ITU-T G.655, "Characteristics of a non-zero dispersion-shifted single-mode optical fibre and cable," 2009.

[7] ITU-T G.656, "Characteristics of a fibre and cable with non-zero dispersion for wideband optical transport," 2010.

[8] A. R. Chraplyvy, "Limitations on lightwave communications imposed by optical fiber nonlinearities," Journal of Lightwave Technology, vol. 8, no. 10, pp. 1548–1557, 1990.

[9] M. Wu and W. I. Way, "Fiber nonlinearity limitations in ultra-dense WDM systems," Journal of Lightwave Technology, vol. 22, no. 6, pp. 1483–1498, 2004.

[10] ITU-T G.694.1, "Spectral grids for WDM applications: DWDM frequency grid," 2002.

[11] ITU-T G.694.2, "Spectral grids for WDM applications: CWDM wavelength grid," 2003.

[12] ITU-T G.652, "Characteristics of a single-mode optical fiber cable," 2009.

[13] http://www.exfo.com/en/Library/WaveReview/2006-February/WRarticle2.asp.

[14] M. J. Lee and D. A. Nolan, "Optical transmission fiber design evolution," Journal of Lightwave Technology, vol. 26, no. 9, pp. 1079–1092, 2008.

[15] D. Marcuse, "Pulse distortion in single mode fibers," Applied Optics, vol. 19, no. 10, pp. 1653–1660, 1980.

[16] D. Marcuse, "Pulse distortion in single mode fibers. 3. Chirped pulses," Applied Optics, vol. 20, no. 20, pp. 3573–3579, 1981.

[17] G. P. Agrawal, "Nonlinear fiber optics," Academic Press, 1989.

[18] M. Miyagi and S. Nishida, "Pulse spreading in a single mode fiber due to third order dispersion," Applied Optics, vol. 18, no. 5, pp. 678–682, 1979.

[19] M. Miyagi and S. Nishida, "Pulse spreading in a single mode fiber due to third order dispersion: effect of optical source bandwidth," Applied Optics, vol. 18, no. 13, pp. 2237–2240, 1979.

[20] G. B. Arfken and H.-J. Weber, "Mathematical methods for physicists," Harcourt/Academic Press, 2001.

[21] D. Marcuse, "Pulse dispersion in single mode fibers, part 2," Applied Optics, vol. 20, no. 17, pp. 2969–2974, 1981.

[22] K. O. Hill, D. C. Johnson, B. S. Kawasaki, and I. R. MacDonald, "CW three wave mixing in single mode optical fibers," Journal of Applied Physics, vol. 49, pp. 5098–5106, 1978.

[23] N. Shibata, R. P. Braun, and R. G. Waarts, "Phase mismatch dependence of efficiency of wave generation through four-wave mixing in a single mode optical fiber," IEEE Journal of Quantum Electronics, vol. QE-23, no. 7, pp. 1205–1210, 1987.

[24] D. Marcuse, A. R. Chraplyvy, and R. W. Tkach, "Effect of fiber nonlinearity on long-distance transmission," Journal of Lighwave Technology, vol. 9, no. 1, pp. 121–128, 1991.

[25] M. W. Maeda, W. B. Sessa, W. I. Way, A. Yi-Yan, L. Curtis, R. Spicer, and R. I. Raming, "The effect of four-wave mixing in fibers on optical frequency division multiplexed systems," Journal of Lightwave Technology, vol. 8, no. 9, pp. 1402–1408, 1990.

[26] K. Inoue, "Four wave mixing in an optical fiber in the zero dispersion wavelength region," Journal of Lightwave Technology, vol. 10, no. 11, pp. 1553–1561, 1992.

[27] B. J. Ainslie and C. R. Day, "A review of single mode fibers with modified dispersion characteristics," Journal of Lightwave Technology, vol. LT-4, no. 8, pp. 967–979, 1986.

[28] U. C. Paek, G. E. Peterson, and A. Carnevale, "Dispersionless single-mode lightguides with a-index profiles," Bell System Technology Journal, vol. 60, pp. 583–598, 1981.

[29] V. Bhagavatula, M. Spotz, W. F. Love, and D. B. Keck, "Segmented-core single-mode fibers with low loss and low dispersion," Electronics Letters, vol. 19, pp. 317–318, 1983.

[30] S. Kawakami and S. Nishida, "Characteristics of a doubly clad optical fiber with a low-index inner cladding," IEEE Journal of Quantum Electronics, vol. QE-11, no. 10, pp. 879–887, 1974.

[31] S. I. Onoda, T. P. Tanaka, and M. Sumi, "W fiber design considerations," Applied Optics, vol. 15, no. 8, pp. 1930–1935, 1976.

[32] D. Gloge, "Weakly guiding fibers," Applied Optics, vol. 10, no. 10, pp. 2252–2258, 1971.

[33] D. DiGiovanni, D. P. Jablonowski, and M. F. Yan, "Advances in fiber design and processing," Chapter 4, Optical Fiber Communications IIIA, I. P. Kaminow and T. L. Koch (eds), Academic Press, 1997.

[34] H. Hatayama, T. Kato, M. Onishi, E. Sasaoka, and M. Nishimura, "Dispersion flattened fiber with large-effective-core area more than $50\,\mu m^2$," Optical Fiber Communication Conference, 1998 OSA Technical Digest Series, vol. 2, Optical Society of America, 1998, paper ThK4.

[35] P. K. Bachmann, D. Leers, H. Wehr, D. U. Wiechert, J. A. van Steenwijk, D. L. A. Tjaden, and E. R. Wehrhahn, "Dispersion-flattened single-mode fibers prepared with PCVD: performance, limitations, design optimization," Journal of Lightwave Technology, vol. LT-4, no. 7, pp. 858–863, 1986.

[36] Y. Liu, A. J. Antos, and M. A. Newhouse, "Large effective area dispersion shifted fibers with dual-ring index profiles," Optical Fiber Communications, Technical Digest, pp. 165–167, 1996, paper WK15.

[37] M. J. Yadlowsky, E. M. Deliso, and V. L. Da Silva, "Optical fibers and amplifiers for WDM systems," Proceedings of the IEEE, vol. 85, no. 11, pp. 1765–1779, 1997.

[38] U. C. Paek, "Dispersionless single-mode fibers with trapezoidal-index profiles in the wavelength region near 1.5 µm", Applied Optics, vol. 22, no. 15, pp. 2363–2369, 1983.

[39] U. C. Paek, "A single mode fiber lightguide fiber having a trapezoidal refractive index profile," U.S. Patent No. 4,516,826, 1985.

[40] P. Nouchi, "Maximum effective area for non-zero dispersion shifted fiber," Optical Fiber Communication Conference, Technical Digest, pp. 303–304, 1998, paper ThK3.

[41] M. J. Li, "Recent progress in fiber dispersion compensator," Proceedings of 27th European Conference on Optical Communications (ECOC), Technical Digest, pp. 486–489, 2001, paper Th.M.1.1.

[42] A. Willner and B. Hoanca, "Fixed and tunable management of fiber chromatic dispersion," Chapter 14, Optical Fiber Telecommunications, vol. IVB, I. Kaminow and T. Li (eds), Academic Press, 2002.

[43] M. Monerie, "Propagation in doubly clad single-mode fibers," IEEE Journal of Quantum Electronics, vol. QE-18, no. 4, pp. 535–542, 1982.

[44] Y. Akasaka, R. Sugizaki, S. Arai, Y. Suzuki, and T. Kamiya, "Dispersion flat compensation fiber for dispersion shifted fiber," 22nd European Conference on Optical Communications (ECOC), Technical Digest, pp. 2.221–2.224, 1996, paper TuP.01.

[45] Y. Akasaka, R. Sugizaki, A. Umeda, and T. Kamiya, "High dispersion compensating ability and low nonlinearity of W-shaped DCF," Optical Fiber Communication Conference, Technical Digest, pp. 201–202, 1996, paper ThA3.

[46] A. J. Antos and D. K. Smith, "Design and characterization of dispersion compensating fiber based on the LP01 mode," Journal of Lightwave Technology, vol. 12, no. 10, pp. 1739–1745, 1994.

[47] K. Mukasa and T. Yagi, "Dispersion flat and low non-linear optical link with new type of reverse dispersion fiber (RDF-60)," Optical Fiber Communication Conference, 2001, paper TuH7.

[48] M. Hirano, A. Tada, T. Kato, M. Onishi, Y. Makio, and M. Nishimura, "Dispersion compensating fiber over 140 nm bandwidth," Proceedings of 27th European Conference on Optical Communications (ECOC), Technical Digest, pp. 494–495, 2001, paper ThM1.4.

[49] A. M. Vengsarkar and W. A. Reed, "Dispersion-compensating single-mode fibers: efficient designs for first- and second-order compensation," Optics Letters, vol. 18, pp. 923–926, 1993.

[50] C. D. Poole, J. M. Wiesenfeld, D. J. DiGiovanni, and A. M. Vengsarkar, "Optical fiber-based dispersion compensation using higher order modes near cutoff," Journal of Lightwave Technology, vol. 12, no. 10, pp. 1746–1758, 1994.

[51] C. D. Poole, J. M. Wiesenfeld, A. R. McCormick, and K. T. Nelson, "Broadband dispersion compensation by using the higher-order spatial mode in a two-mode fiber," Optics Letters, vol. 17, no. 14, pp. 985–987, 1992.

[52] M. Eguchi, M. Koshiba, and Y. Tsujji, "Dispersion compensation based on dual-mode optical fiber with inhomogeneous profile core," Journal of Lightwave Technology, vol. 14, no. 10, pp. 2387–2394, 1996.

[53] K. Oh, H. S. Seo, J. K. Lee, and U. C. Paek, "Polarization dependent dispersion characteristics of high order modes in a cylindrical dual mode fiber with an arbitrary index profile," Optics Communications, vol. 159, pp. 139–148, 1999.

[54] S. Ramachandran, B. Mikkelsen, L. C. Cowsar, M. F. Yan, G. Raybon, L. Boivin, M. Fishteyn, W. A. Reed, P. Wisk, D. Brownlow, R. G. Huff, and L. Gruner-Nielsen, "All-fiber grating-based higher order mode dispersion compensator for broad-band compensation and 1000-km transmission at 40 Gb/s," IEEE Photonics Technology Letters, vol. 13, no. 6, pp. 632–634, 2001.

[55] S. Ramachandran, S. Ghalmi, S. Chandrasekhar, I. Ryazansky, M. F. Yan, F. V. Dimarcello, W. A. Reed, and P. Wisk, "Tunable dispersion compensators utilizing higher order mode fibers," IEEE Photonics Technology Letters, vol. 15, no. 5, pp. 727–729, 2003.

[56] S. Choi and K. Oh, "A new LP_{02} mode dispersion compensation scheme based on mode converter using hollow optical fiber," Optics Communications, vol. 221, nos. 4–6, pp. 307–312, 2003.

[57] S. Ramachandran, "Dispersion-tailored few mode fibers: a versatile platform for in-fiber photonic devices," Journal of Lightwave Technology, vol. 23, no. 11, pp. 3426–3433, 2005.

[58] R. C. Youngquist, J. L. Brooks, and H. J. Shaw, "Two-mode fiber modal coupler," Optics Letters, vol. 9, no. 5, pp. 177–179, 1984.

[59] J. N. Blake, B. Y. Kim, and H. J. Shaw, "Fiber-optic modal coupler using periodic microbending," Optics Letters, vol. 11, no. 3, pp. 177–179, 1986.

[60] C. D. Poole, C. D. Townsend, and K. T. Nelson, "Helical-grating two-mode fiber spatial-mode coupler," Journal of Lightwave Technology, vol. 9, no. 5, pp. 598–604, 1991.

[61] F. Bilodeau, et al., "Efficient, narrowband $LP_{01} \leftrightarrow LP_{02}$ mode converters fabricated in photosensitive fibre: spectral response," Electronics Letters, vol. 27, no. 8, pp. 682–685, 1991.

[62] K. Oh, S. Choi, Y. Jung, and J. W. Lee, "Novel hollow optical fibers and their applications in photonic devices for optical communications," Journal of Lightwave Technologies, vol. 23, no. 2, pp. 524–532, 2005.

[63] S. Choi, et al., "A low loss mode converter based on the adiabatically tapered hollow optical fiber," Electronics Letters, vol. 37, pp. 823–825, 2001.

CHAPTER FIVE

Multimode Fibers for Large-Bandwidth Applications

5.1 HISTORY AND RECENT APPLICATION TRENDS OF MULTIMODE OPTICAL FIBERS

In comparison to single mode fibers' (SMFs) dynamic, rapid, and variegated development, multimode optical fibers (MMFs) have shown three distinctive evolution steps in the past three decades, from a step index (SI) MMF to graded index (GI) MMF and then toward most recent laser optimized (LO) MMF as shown in Figure 1.2. This MMF development is closely related with demands in optical networks where they use different data rates, link distances, and light sources in order to take fullest advantage of MMFs' easy connection to economic light sources and detectors.

MMF applications include fiber distributed data interface (FDDI) or Ethernet-based local area network (LAN) [1,2], fiber channel (FC)–based storage area network (SAN) [3], and synchronous digital hierarchy (SDH)–based very-short-reach (VSR) networks [4]. Recently MMFs furthermore expanded applications in digital image communication (DICOM)–based picture archiving communication systems (PACS) [5] and media-oriented system transport (MOST)–based automobile intranetwork [6]. These various short-haul optical systems require ever-increasing transmission capacity as in the case of long-haul applications, and intensive efforts have been focused on development of high-bandwidth MMF in terms of material choice, waveguide structure optimization, and reliable characterization methods (Figure 5.1).

The demand for large capacity in short-haul optical transmission rapidly increases in recent years as shown in Figure 5.2. Especially gigabit Ethernet has been a driving force for development of MMF for larger bandwidth.

Silica Optical Fiber Technology for Devices and Components: Design, Fabrication, and International Standards, First Edition. By Kyunghwan Oh and Un-Chul Paek.
© 2012 John Wiley & Sons, Inc. Published 2012 by John Wiley & Sons, Inc.

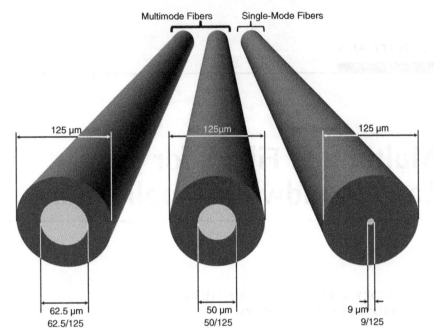

FIGURE 5.1 Comparison of multimode optical fibers and single-mode fiber.

The higher data rates required in the recent short haul transmission systems have forced light emitting diode (LED) to be replaced with other light sources such as Fabry-Perot laser diode (FP-LD) and vertical cavity surface emitting laser (VCSEL), which have a higher modulation speed. Schematic comparison of these light sources are summarized in Table 5.1.

It would have been a very simple upgrade if only the light source could be replaced from LED to lasers on the existing MMF links. However, replacing LED with lasers has brought about more fundamental issues in MMF design such as the light source

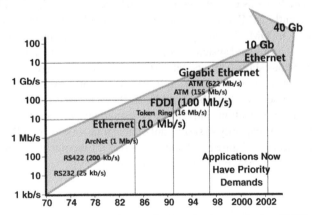

FIGURE 5.2 Recent data rate requirement in short-haul transmission. FDDI, fiber distributed data interface.

TABLE 5.1 Comparison of Light Sources for Short-Haul Transmission

	LED		VCSEL		FP-LD	
Cost	Cheap		Relatively cheap		Expensive	
Wavelength	850 nm multimode	1300 nm multimode	850 nm multimode	1300 nm single mode	1310 nm single mode	1550 nm single mode
Data rate	~200 Mbps		~1 Gbps		~10 Gbps	
Link distance	2 km		Hundreds of meters to a few kilometers		Tens of kilometers	
Launching condition	Overfilled launch		Selective mode launch		Selective mode launch	

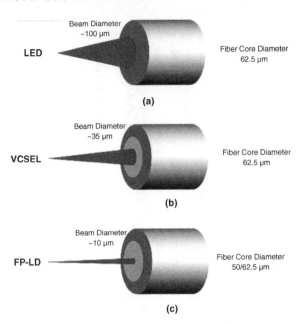

FIGURE 5.3 Comparison of launching conditions for different light sources: (a) overfilled launch in FDDI; (b) gigabit Ethernet, short wavelength; (c) gigabit Ethernet, long wavelength.

launching conditions, selective mode excitation, and differential modal delay to result in revival of optimal refractive index design issue for new high speed laser light sources.

In Figure 5.3, the launching conditions for three different types of light sources are schematically shown. In the case of LED, the light from light source has much larger beam diameter than that of the core in MMF to excite all the available modes. This is called overfilled launch condition and has been widely used up to the speed of $\sim$100 Mbps and within the link distance of a few kilometers.

In the case of VCSEL and single-mode laser diodes, the beam diameter is significantly smaller than the core diameter—only a fraction of the MMF core area—and the excited modes are highly localized. This change in launching condition has raised a fundamental issue in modal delays between the excited modes in GI-MMFs for large-bandwidth applications.

MMF, as the name stands for, carries hundreds to thousands of modes along the fiber, and the modal delay among them plays the key role in defining overall performances. In contrast to SMF, the most important attributes of MMF is "the bandwidth of transmission," which is given in the unit of MHz km [7]. For low speed MMF systems using LED light sources the bandwidth has been in the range of 200–400 MH km. For large capacity MMF optical communication systems based on high speed LD transmitters, a new attribute called "differential modal delay" (DMD) has become a critical issue [8]. Highly confined LD spot size and its local excitation of finite number of guided modes resulted in a certain DMD response imposing a fundamental limit for transmission capacity, which is highly dependent on refractive index profile of MMFs. Recent laser-optimized MMF is the result of intensive efforts to minimize DMD by precisely controlling the refractive index profile for the LD or VSCEL transmitters [9,10].

Laser-optimized 50 µm multimode fiber is the latest line of MMF but in fact this is a kind of revival and modification of standard 50 µm fiber initially produced in 1970s (designated as OM2 in ISO/IEC 11810 standard) [11]. In 1970s, 50 µm fiber was the most popular of the early fiber types available, and was used for both long-haul and short-reach applications. After the introduction of single-mode fiber in the 1980s for long-haul applications, multimode fibers were mainly applied to short-reach interconnects within a few kilometers. Fifty micrometer fiber was used with 850 nm wavelength LEDs, yet this combination was not satisfactory in terms of optical power budget. The 50 µm core was too small to collect the output from the LEDs. Obvious solution to this power coupling problem was to increase the core diameter and/or numerical aperture. These approaches to increase core diameter and numerical aperture 62.5 µm multimode fiber (designated as OM1 in ISO/IEC 11810 standard) were introduced in 1985 to couple more light from LEDs supporting 2 km link distances at 10 Mb/s [11]. Over two decades, 62.5 µm core MMF became a *de facto* standard among the vast majority of LAN installations. Despite continual upgrades in LAN bandwidth requirements, FDDI grade fiber remained a workhorse for backbone fiber installations for many years, and is still present in legacy systems.

While 62.5 µm fiber dominated the premises market for more than a decade, changing market conditions have reestablished 50 µm fiber as the best solution for applications whose data rate is over 10 Mbps. The 100 Mbps Fast Ethernet standards, published in 1995, called for the use of LEDs that take advantage of lower fiber attenuation at 1300 nm wavelength [12]. This offset the LED coupling loss with 50 µm fiber caused by its smaller core diameter. Therefore, 50 µm MMF was able to support the same 2 km reach at 100 Mbps as 62.5 µm fiber. Only 3 years later, the IEEE Gigabit Ethernet standard published in 1998 specified low-cost 850 nm wavelength vertical cavity surface-emitting lasers (VCSELs) that can reach 1000 m over 50 µm fiber, compared to 220–275 m on standard 62.5 µm fiber. As data rates rise to multi-gigabit speeds, it is apparent that 62.5 µm fiber is stretched beyond its performance limit, due to its lower bandwidth at 850 nm. By comparison, 50 µm fiber can provide as much as 10 times the bandwidth of the 62.5 µm fiber, enabling more robust support of 1 and 10 Gbps applications. Because 1 and 10 Gbps transmitters use small spot size lasers, concerns over power coupling efficiencies into 50 µm fiber are no longer an issue [13].

The 10 Gb/s Ethernet standard, published in 2002, takes advantage of laser-optimized 50 µm fiber (designated as OM3 in ISO/IEC 11810 standard) that supports 300 m reach using 850 nm VCSELs [14].

In this chapter we will review the fundamental principles of GI silica MMF design and fabrication process, identifying key parameters of large-bandwidth applications. Bandwidth and DMD will be then discussed for further optimization of waveguide structure along with recent standard activities. We will continue discussion over most updated international standards.

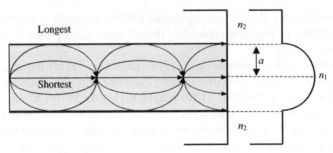

FIGURE 5.4   Schematic meridional ray trajectories for a graded-index multimode fiber.

5.2   PRINCIPLE OF MULTIMODE OPTICAL FIBER DESIGN

A schematic diagram for a GI-MMF is shown in Figure 5.4, where meridional ray trajectories are drawn for more intuitive explanation. Among the rays, the one passing through the center of the core will travel the shortest path, while those passing the core–cladding boundaries will experience longer paths. In order to compensate this difference in the geometrical paths, graded-index structure has been implemented, where the highest index is allocated for the ray with shortest path, while the other skew rays take a lower refracted index. In GI-MMF, therefore, the time of flight for each mode was made to be equalized by varying the refractive index profile in the core.

In contrast to an SI-MMF, GI-MMF can provide a significantly reduced time delay among the modes and subsequently a larger bandwidth. GI-MMF has been extensively used in a relatively short distance optical communication network such as in LAN environment.

As depicted in Figure 5.4, the refractive index has the maximum at the center of the core, taking a decreasing value for radially outward position. In an ideal mathematical form, the core refractive index profile of a GI-MMF has been expressed in "α-profile":

$$n(r) = \begin{cases} n_1 \left[1 - 2\Delta \left(\dfrac{r}{a} \right)^{\alpha} \right]^{1/2} & (0 \le r \le a) \\[3mm] n_2 = n_1 (1 - 2\Delta)^{1/2} & (a < r) \end{cases} \tag{5.1}$$

Here n_1 is the refractive index at the center of the core, $r = 0$, and Δ is given by

$$\Delta = \frac{n_1^2 - n_2^2}{2n_1^2} \tag{5.2}$$

Note that for $\alpha = 1$, the refractive index assumes a triangular profile, and for $\alpha = 2$ a parabolic one as shown in Figure 5.5. The profile converges to the step-index profile for $\alpha \to \infty$.

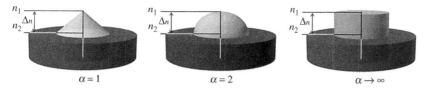

FIGURE 5.5   Refractive index profiles for various α.

Given the distribution of refractive index n or equivalent ε, the mode analysis for GI-MMF starts from the Maxwell's equations as described in Chapter 2:

$$\vec{\nabla} \cdot \vec{D} = 0 \tag{2.1}$$

$$\vec{\nabla} \cdot \vec{B} = 0 \tag{2.2}$$

$$\vec{\nabla} \times \vec{E} = -\frac{\partial \vec{B}}{\partial t} = -\mu_0 \frac{\partial \vec{H}}{\partial t} \tag{2.3}$$

$$\vec{\nabla} \times \vec{H} = \frac{\partial \vec{D}}{\partial t} = \varepsilon \frac{\partial \vec{E}}{\partial t} \tag{2.4}$$

Here we have assumed the fiber material, silica, is nonmagnetic such that we set $\mu = \mu_0$. Similar to step-index fibers described in Chapter 2, the transverse components (E_t, H_t) of electric/magnetic fields can be expressed in terms of longitudinal components (E_z, H_z):

$$E_t = -j \frac{\beta}{\omega^2 \varepsilon \mu_0 - \beta^2} \left[\nabla E_z - \frac{\omega \mu_0}{\beta} k \times \nabla H_z \right] \tag{5.3}$$

$$H_t = -j \frac{\beta}{\omega^2 \varepsilon \mu_0 - \beta^2} \left[\nabla H_z - \frac{\omega \mu_0}{\beta} k \times \nabla E_z \right] \tag{5.4}$$

The above equations can be further developed to lead us to vectorial differential equations for the transverse components:

$$\nabla^2 E_t + (\omega^2 \varepsilon \mu_0 - \beta^2) E_t + \nabla \left[\frac{\nabla \varepsilon}{\varepsilon} . E_t \right] = 0 \tag{5.5}$$

$$\nabla^2 H_t + (\omega^2 \varepsilon \mu_0 - \beta^2) H_t + \frac{\nabla \varepsilon}{\varepsilon} \times (\nabla \times H_t) = 0 \tag{5.6}$$

In the weakly guiding approximation, $|\nabla \varepsilon / \varepsilon| \ll 1$, the equation can be further simplified as follows:

$$\nabla^2 E_t + [\omega^2 \varepsilon(r) \mu_0 - \beta^2] E_t = 0 \tag{5.7}$$

$$\nabla^2 H_t + [\omega^2 \varepsilon(r) \mu_0 - \beta^2] H_t = 0 \tag{5.8}$$

A scalar wave equation can be derived for $E_t = E_x$ or $E_y = R(r)e^{-jl\theta}$:

$$\frac{1}{r}\frac{d}{dr}\left(r\frac{dR}{dr}\right) + \left[\omega^2\varepsilon(r)\mu_0 - \beta^2 - \frac{l^2}{r^2}\right]R = 0 \tag{5.9}$$

$$l = \begin{cases} 1, & \text{for TE and TM modes } (n = 0) \\ n+1, & \text{for EH modes } (n \geq 1) \\ n-1, & \text{for HE modes } (n \geq 1) \end{cases} \tag{5.10}$$

The above scalar differential equation can be transformed very similar to Schrodinger equation by introducing a new variable:

$$\hat{R}(r) = \sqrt{r}R(r) \tag{5.11}$$

$$\frac{d^2\hat{R}}{dr^2} + [E - U(r)]\hat{R} = 0 \tag{5.12}$$

$$E = k^2 n_1^2 - \beta^2 \tag{5.13}$$

$$U(r) = \left[k^2 n_1^2 - k^2 n^2(r)\right] + \frac{l^2 - (1/4)}{r^2} \tag{5.14}$$

The above differential equations are commonly encountered in quantum mechanics problems to describe a particle motion with a given energy of E and potential $U(r)$. Fairly accurate solutions have been obtained using WKB method by imposing boundary conditions as follows:

$$R(a) = R_{\text{clad}}(a)$$
$$\left[\frac{dR}{dr}\right]_{r=a} = \left[\frac{dR_{\text{clad}}}{dr}\right]_{r=a} \tag{5.15}$$

In GI-MMFs' modal analysis WKB method has been one of most widely used techniques and early classical works of Gloge and Marcatili [7] and Olshansky and Keck [15] are based on the method.

In this section, we will briefly review the principal results of WKB modal analysis for a GI-MMF with the α-profile.

We start with the expression for the propagation constant of an LP$_{lp}$ mode. For an LP$_{lp}$ mode guided in a general GI-MMF, the propagation constant β can be found from an integral equation [7]:

$$\int_{r_1}^{r_2} dr\sqrt{k^2 n^2(r) - \beta^2 - \frac{l^2 - (1/4)}{r^2}} = \left(p - \frac{1}{2}\right)\pi \tag{5.16}$$

Here r_i are the turning points at which the following equation is satisfied [15]:

$$E = U(r) \tag{5.17}$$

5.2.1 Scalar Equation for Wave Propagation

For the case of α-profile GI-MMF, it is shown that the propagation constant, β, can be expressed in a closed analytic form:

$$\beta = kn_1 \left[1 - 2\Delta \left(\frac{\nu}{N} \right)^{\alpha/(\alpha+2)} \right]^{1/2} \tag{5.18}$$

Here ν is the number of mode whose propagation constant is in the range between β and $\beta_{max} = kn_1$. N is the total number of propagating modes. These parameters are given by

$$\nu(\beta) = \alpha^2 k^2 n_1^2 \Delta \frac{\alpha}{\alpha+2} \left(\frac{k^2 n_1^2 - \beta^2}{2k^2 n_1^2 \Delta} \right) \tag{5.19}$$

$$N = \nu(\beta = kn_2) = a^2 k^2 n_1^2 \Delta \frac{\alpha}{\alpha+2} \tag{5.20}$$

If we introduce normalized frequencies, u and v, as in Chapter 2:

$$
\begin{aligned}
u^2 &= (k^2 n_1^2 - \beta^2) a^2 \\
v^2 &= k^2 n_1^2 a^2 (2\Delta) \\
x &= \frac{u^2}{v^2} \\
\frac{\nu}{N} &= x^{(\alpha+2)/\alpha}
\end{aligned}
\tag{5.21}
$$

we can further simplify the above expressions for the propagation constant as follows:

$$\beta = kn_1 [1 - 2x\Delta]^{1/2} \tag{5.22}$$

Group delay, τ, is defined as the time required by an impulse to propagate over a unit length of fiber and is given by

$$\tau = \frac{d\beta}{d\omega} = \frac{1}{c} \frac{d\beta}{dk} \tag{5.23}$$

For the case of α-profile GI-MMF, the group delay is expressed in terms of Δ as follows:

$$\tau = \frac{N_1}{c}\left[1+\Delta\frac{\alpha-2-y}{\alpha+2}x+\frac{\Delta^2}{2}\frac{3\alpha-2-2y}{\alpha+2}x^2+O(\Delta^3)\right] \qquad (5.24)$$

Here N_1 is group index defined as

$$N_1 = \frac{d(kn_1)}{dk} = n_1 - \lambda\frac{dn_1}{d\lambda} \qquad (5.25)$$

And y is a parameter introduced by Olshansky and Keck to express the difference between the material dispersion in the core and cladding and is given by [15]

$$y = \frac{2n_1}{N_1}k\Delta\frac{d\Delta}{dk} = -\frac{2n_1}{N_1}\lambda\Delta\frac{d\Delta}{d\lambda} \qquad (5.26)$$

Assuming weakly guiding structure, $\Delta \ll 1$, it is noted that group delay becomes nearly constant for the condition:

$$\alpha = 2+y \qquad (5.27)$$

$$\tau \approx \frac{N_1}{c}\left[1+\frac{x^2}{2}\Delta^2+O(\Delta^3)\right] \approx \frac{N_1}{c} \qquad (5.28)$$

When the index profile parameter α is designed to a certain value as in equation 5.27, the group delay becomes constant, that is, the difference in delay time between modes vanishes to result in ideal multimode fiber.

For the given propagation constants, β, and group delay, τ, we can furthermore estimate the total root mean square (RMS) pulse width, σ_{tot}, in GI-MMFs, which is composed of intermodal dispersion contribution, σ_{inter}, and intramodal dispersion contribution, σ_{intra}. σ_{intra} is composed of material dispersion and waveguide dispersion contributions as described in Chapter 2:

$$\sigma_{inter} = \frac{LN_1\Delta}{2c}\frac{\alpha}{\alpha+1}\left(\frac{\alpha+2}{3\alpha+2}\right)^{1/2}\times\left[C_1^2+\frac{4C_1C_2\Delta(\alpha+1)}{2\alpha+1}+\frac{4\Delta^2C_2^2(2\alpha+2)^2}{(5\alpha+2)(3\alpha+2)}\right]^{1/2}$$

$$(5.29)$$

$$C_1 = \frac{\alpha-2-y}{\alpha+2}, \qquad C_2 = \frac{3\alpha-2-2y}{2(\alpha+2)} \qquad (5.30)$$

$$\sigma_{intra} = \frac{\delta\lambda_s}{\lambda}\left[\left(-\lambda^2\frac{d^2n_1}{d\lambda^2}\right)^2-2\lambda^2\frac{d^2n_1}{d\lambda^2}(N_1\Delta)\left(\frac{\alpha-2-\varepsilon}{\alpha+2}\right)\left(\frac{2\alpha}{2\alpha+2}\right)\right.$$

$$\left.+(N_1\Delta)^2\left(\frac{\alpha-2-\varepsilon}{\alpha+2}\right)^2\left(\frac{2\alpha}{3\alpha+2}\right)\right]^{1/2} \qquad (5.31)$$

where $\delta\lambda_s$ is the root mean square spectral width of the light source:

$$\sigma_{tot} = \sqrt{\sigma_{inter}^2 + \sigma_{intra}^2} \tag{5.32}$$

It is noted that dispersion is heavily dependent on the refractive index profile parameter, α, and there exists an optimal value, α_{opt} as given in equation 5.33, which will give the minimum total dispersion D_{TOTAL}:

$$\alpha_{opt} = 2 + y - \Delta\frac{(4+y)(3+y)}{5+2y} \tag{5.33}$$

For the optimal refractive index profile, $\alpha = \alpha_{opt}$, we have a negligible intermodal dispersion and intramodal dispersion becomes pure material dispersion to make the minimum total dispersion:

$$\sigma_{inter} \approx 0$$

$$\sigma_{intra} \approx \frac{\delta\lambda_s}{\lambda}\left(-\lambda^2\frac{d^2n_1}{d\lambda^2}\right)^2 \approx \sigma_{TOTAL} \tag{5.34}$$

The total rms pulse widths for α-profiles can be predicted from equation 5.32. In Figure 5.6, the rms pulse width is shown as a function of α for three types of light sources operating at $\lambda = 0.9\,\mu m$, but having different spectral bandwidths [15]. The sources are taken to be an LED, a gallium arsenide injection laser, and a distributed feedback laser having rms spectral widths, $\delta\lambda_S$, of 15, 1, and 0.2 nm, respectively.

Here the GI-MMF is assumed to have 3.4 wt% TiO_2-doped silica core, $n_1(\lambda)$, and fused silica cladding $n_2(\lambda)$. The curves were computed with the following choice of parameters: $n_1 - n_2 = 0.02$, $\Delta = 0.0135$, $n_2 = 1.457$, and $a = 30\,\mu m$ for the core radius. All modes are assumed to carry equal power. A dashed curve represents the rms width predicted if material dispersion and intramodal broadening are ignored.

In Figure 5.6 it is clearly noted that at the optimal refractive index parameter, $\alpha = \alpha_{opt} \approx 2.25$, the total rms pulse width, σ_{tot}, takes the minimum values. The effects of pulse width reduction near α_{opt} are more significant for narrow linewidth lasers than LED. It is also noted that material and intramodal contribution should be taken into account to find realistic α_{opt}.

Neglecting the material dispersion, α_{opt} is located near 2 as in the dotted curve in Figure 5.6. In this case theoretical limit of MMF 3 dB bandwidth for 1 km fiber length was estimated by Marcuse and coworkers [16] and the results are shown in Figure 5.7. The curves were computed with the following choice of parameters: $n_1 - n_2 = 0.02$, $\Delta = 0.0135$, $n_2 = 1.457$, and $a = 30\,\mu m$ for the core radius. For the optimal α_{opt} located near 2, it is predicted that the bandwidth could exceed 10 GHz km.

The optimal α is heavily dependent on the core material and it varies with wavelength as shown in Figure 5.8. Depending on the core glass dopants, GeO_2,

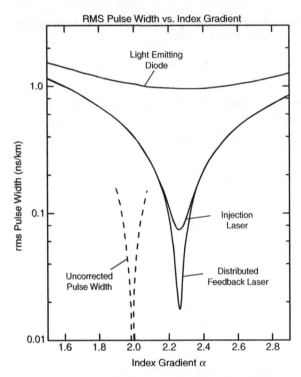

FIGURE 5.6   Assuming equal power in all modes, the RMS pulse width is shown as a function of α for three different sources, all operating at $0.9\,\mu m$ [15].

P_2O_5, and B_2O_3, the optimal α_{opt} varies in the range of 1.6–2.1 in the spectral domain from 800 to 1300 nm [17]. Except P_2O_5, other glass compositions show significant variation of α_{opt} over wavelength. For the case of GeO_2–P_2O_5–SiO_2 glass, which is most widely used in commercial GI-MMFs, α_{opt} takes values of $\sim$2 and 1.85 for 850 and 1310 nm, respectively. Therefore, very precise index control and material composition are critical to provide large bandwidth at a given operating wavelength. In present technology, it is very difficult to maintain high bandwidth at both 850 and 1310 nm. Recent laser-optimized MMF is using 850 nm as the operating wavelength, adapting low-cost VSCEL light sources.

The relation between α and bandwidth at 850 and 1300 nm is shown in Figure 5.9. The vertical and horizontal axes are for 1300 and 850 nm, respectively. The plots are two-dimensional plots of theoretical bandwidth for various α. There are three segments of line plots. In plot (a), as α increases from 1.68 to 1.80, the bandwidth at 1300 nm rapidly increases but 850 nm bandwidth maintains the same value. For the range of α from 1.89 to 1.96, plot (b), the 1300 nm bandwidth decreases and 850 nm bandwidth rapidly increases. In plot (c), the situation is opposite to plot (a) such that 850 nm bandwidth rapidly increases with more or less fixed 1300 nm bandwidth for increasing α. The dashed lines are bandwidth at 1000 MHz km and the crossing

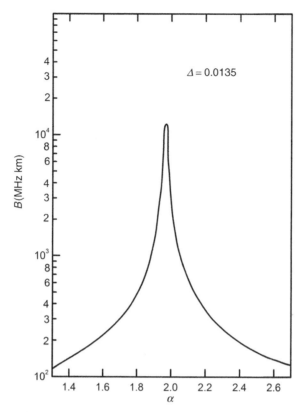

FIGURE 5.7 Calculation of bandwidth for GI-MMF with an optimal α-profile [16].

point lies near $\alpha \approx 1.9$ range. However, this region is rather unstable to result in severe fluctuation of bandwidth for small variations in α. In recent laser-optimized GI-MMFs the bandwidth requirement exceeds 10 GHz km and optimization of α at the 850 nm operating wavelength has been established for 50 μm core diameter MMF along with VSCEL light source.

5.3 IMPACTS OF NONIDEAL α-REFRACTIVE INDEX PROFILE ON TRANSMISSION BANDWIDTH

The basic principles of waveguide design in the α-profile and choice of core material in GI-MMF have been reviewed in Section 5.2. The optimal waveguide structure is given in equation 5.1 for the refractive index profile with optimal α_{opt} given in equation 5.33. Depending on the operation window (850 or 1310 nm), the core size (50 or 62.5 μm), and the transmission bandwidth, GI-MMFs of various kinds have been fabricated using chemical vapor deposition (CVD) techniques described in Section 2.2.

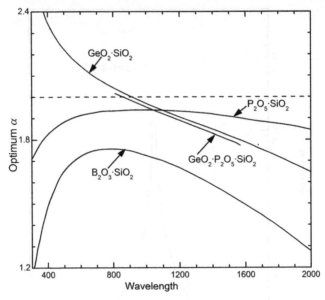

FIGURE 5.8 Variation of optimum *a* for different core glass compositions as a function of wavelength [17].

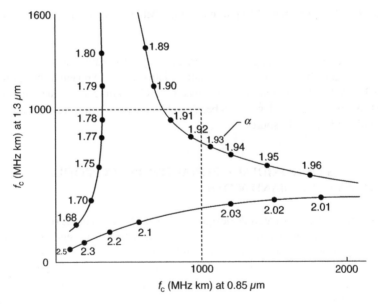

FIGURE 5.9 Relation between α and 6 dB bandwidth (f_c in MHz km) at 0.85 and 1.3 μm [18].

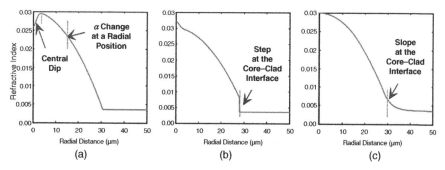

FIGURE 5.10 Typical deviations from ideal α refractive index profile in GI-MMFs. (a) Central dip and change of α within the core; (b) central peak and step-like distortion at the core/cladding interface; (c) gradual distortion at the core/cladding interface by dopant diffusion.

There exist practical limits in optical fiber fabrication process to result in deviation from the ideal refractive index profiles. Typical deviations are

- Distortion at core center (depression or peak)
- Distortion at core/cladding interface (diameter variation)
- Varying profile parameter (α) within the core

These deviations are schematically shown in Figure 5.10.

The primary impact of these deviations from the ideal α profile is significant reduction of transmission bandwidth. Marcuse and coworkers have theoretically analyzed the impact of nonideal α-profile over GI-MMF bandwidth in their classical paper [16], where three types of deviations widely met in CVD process were described:

(a) Central refractive index dip,

(b) Composite index profile made up of two sections of power-law curves with different exponents, $\alpha_1 \neq \alpha_2$,

(c) Sinusoidal perturbation superimposed over optimum α.

In the early days of GI-MMF, due to transformation of GeO_2 into gaseous GeO in the final collapse process CVD, the central dip in the index profile often existed as in Figure 5.11a. The modes whose intensity is confined near the central dip, which is the fundamental modes and a few symmetric excited modes, will experience significant pulse distortion to result in bandwidth reduction.

The refractive index with a Gaussian central dip is expressed as

$$n(r) = n_2 + \left[1 - \exp\left(\frac{-r^2}{w^2}\right)\right]\left[n_1\left(1 - 2\Delta\left(\frac{r}{a}\right)^\alpha\right)^{1/2} - n_2\right] \quad (5.35)$$

For the GI refractive index profile with $\alpha = 1.968$, $\Delta = 0.0135$, and $a = 30\,\mu m$, a Gaussian central dip was assumed whose depth reaches the cladding index $n_2 = 1.457$

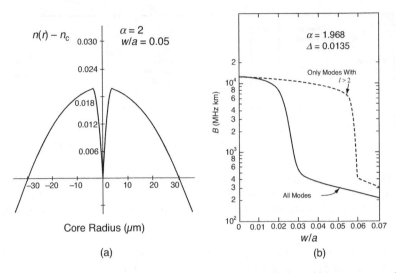

FIGURE 5.11 (a) Power-law profile with Gaussian central dip, width parameter of dip $w/a = 0.05$, and $\alpha = 1.968$, and (b) its bandwidth. The solid curve is for all modes equally excited; the dotted curve resulted from omitting modes with azimuthal mode numbers $l = 0$ and 1 [19].

and width, w, is given in the range from zero to $0.07a$. Here we assumed that GeO_2 is totally depleted to leave behind only SiO_2 glass at the central dip.

Calculated bandwidth neglecting material dispersion is shown in Figure 5.11b. When all the modes are equally excited, the maximum bandwidth decreases by orders of magnitude when the central dip expands to $w > 0.02a$. It is also noted that when the modes are selectively excited excluding the modes with azimuthal number $l = 0$ and 1, the bandwidth is not affected until the central dip width reaches $w > 0.06a$. Central refractive index dip, therefore, affects the bandwidth significantly and its impacts could be reduced by selective excitation of modes. The latter finding is widely applied in "offset launching" techniques to enhance the bandwidth.

Even when the central dip is absent, practical fibers do not always approach the optimal index profile. The index profile can be composed of several different α as in Figure 5.12a. In this case the profile is divided into two segments: the upper segments corresponding to $\alpha = 1.35$, while the lower segment is with $\alpha = 2.5$. The dotted curve represents a single best fit with the optimal $\alpha = 1.968$. Note that even though the fitting resulted in the optimal index profile, there exists a fitting error E, which is defined by

$$E = 100 \left[\left\{ \sum_{i=1}^{N} (n(r_i) - n_{\text{fit}}(r_i))^2 \right\} \left\{ \sum_{i=1}^{N} n^2(r_i) \right\}^{-1} \right]^{1/2} \quad (5.36)$$

Here $n(r)$ and $n_{\text{fit}}(r)$ correspond to the actual and fitted refractive index profiles, respectively. The bandwidth of MMF with composite profiles is shown in Figure 5.12b for the fitting error range from 0 to 10%. For 1% fitting error, the

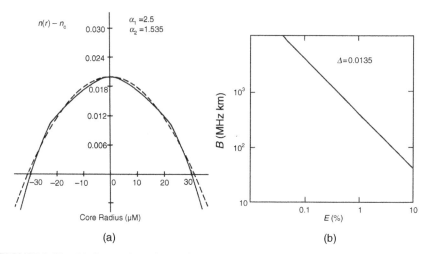

FIGURE 5.12   (a) Composite index profile made up of two sections of power-law curves with different exponents: $\alpha = 1.535$ for the upper part of the curve and 2.5 for the lower part. The dotted curve is a best-fit power-law curve with $\alpha = 1.968$. (b) Bandwidth of fibers with composite power-law index profiles as a function of the fitting error [19].

bandwidth drops from $\sim$10 GHz km to $\sim$100 MHz km. Even if a measured index profile seems to approximate the optimum power law closely, the bandwidth of the corresponding fiber could be significantly lower than the designed value. Therefore, it is absolutely important to optimize the fabrication process to keep the single a value through the whole cross-section of the core.

In most of CVD processes the core of MMF is formed by layer-by-layer deposition of SiO_2–GeO_2 glass, which can result in refractive index ripple as shown in Figure 5.13. The ripple can be represented by a sinusoidal perturbation over the optimal refractive index profile as equation 5.37 [19]:

$$n(r) = n_1 \left[1 - 2\Delta \left(\frac{r}{a} \right)^{\alpha_{\mathrm{opt}}} \right]^{1/2} + A \sin\left(\frac{2\pi N r}{a} \right) \qquad (5.37)$$

Here A is the amplitude of perturbation given relative to the index difference, $n_1 - n_2$, such that 1% of $n_1 - n_2 = 0.02$ corresponds to $A = 0.0002$. For the given amplitude, A, the bandwidth of the fiber is calculated in Figure 5.13b for the perturbation period in the range from 1 to 25. There exists a rapid drop of bandwidth from $\sim$10 GHz km for the optimal profile to $\sim$100 MHz km with increasing spatial frequency of the perturbation.

These imperfections in refractive index profiles fabricated by CVD processes could result in significant reduction of bandwidth in GI-MMF. In the early days of LAN deployment most of MMF links were operated at low data rates in short distances using overfilled launch condition for LED exciting all of allowed modes, and impacts of these defects were not critical issues. However, as the demand of transmission capacity in short-haul systems rises rapidly, these imperfections in

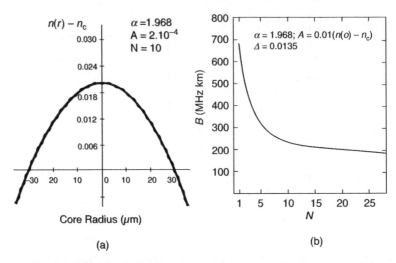

FIGURE 5.13   (a) Refractive index profile with optimum α and superimposed sinusoidal perturbation of amplitude $A = 0.0002$ and spatial frequency of $10/a$. (b) Bandwidth of fiber with sinusoidally perturbed optimum index profile as a function of the number N of sinusoidal periods for the amplitude $A = 0.01$ [19].

refractive index profiles have played critical role. Recently index profiling technology in CVD process has developed and can produce large-bandwidth GI-MMFs on a large scale that could be used with high speed LD and VCSEL light sources [9,10].

In Figure 5.14, the spectral distributions of bandwidth given in MHz km are plotted for 50 and 62.5 µm core GI-MMFs. It is noted that the bandwidth peaks are located near 1030 and 1230 nm for 50 and 62.5 µm core GI-MMFs, respectively. These wavelengths, however, suffer from high attenuation and 850 and 1310 nm windows are used. Fifty micrometer core GI-MMFs have a higher bandwidth at 850 nm and 62.5 µm MMFs have advantages near 1310 nm. The most recent laser-optimized MMF has 50 µm core diameter such that 850 nm window is targeted to combine low-cost light source, large bandwidth, and low attenuation.

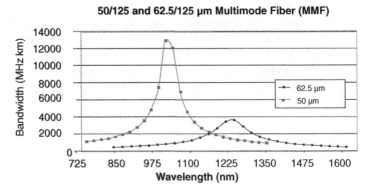

FIGURE 5.14   Bandwidth of standard GI-MMFs.

5.4   MAIN ATTRIBUTES OF GI-MMFS-BANDWIDTH AND DIFFERENTIAL MODAL DELAY

As discussed in Sections 5.1 and 5.2, MMF inherently suffers from modal dispersion, which results in different times of flight for excited modes. Elaborated refractive index profile was, therefore, required to minimize the average delays among the modes as discussed in Section 5.3. When LED light source was used, the large Lambertian beam overfills the MMF core to excite all the available modes. In this case a slight deviation from the optimal index profiles, especially the central index dip, was acceptable to a certain degree. The impact of central dip is schematically shown in Figure 5.15. In comparison to an ideal refractive index profile in (a), the profile with the central dip will cause significant amount of modal delay between the modes represented as the central straight line and other meridional rays. This modal delay will spread the output pulse to increase rms pulse width, which will subsequently reduce the bandwidth of MMF, especially for the case of small spot size VCSELs or LDs.

The differences in delays among the modes excited at different locations in the core are called DMD. DMD is highly dependent on the light source beam diameter and its launching position within the core. A typical scenario for two contrasting cases in the MMF with the central refractive index dip is schematically shown in Figure 5.16. In the case of LED light source we assume all the modes are excited, while in LD incident at the central refractive index dip, two modes will be assumed for simple comparison. In the case of LED, the amount of light power carried by the mode excited near the central dip is outnumbered by other modes as shown in Figure 5.16a.

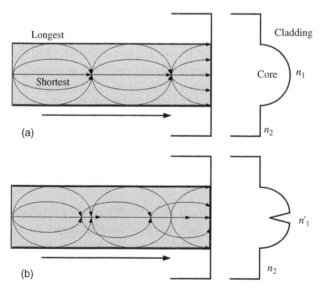

FIGURE 5.15   Impact of the central dip in the refractive index profile over differential modal delay. (a) Minimal delay between guided modes for an ideal refractive index profile; (b) modal delay significantly develops between the modes excited at the center and neighboring modes.

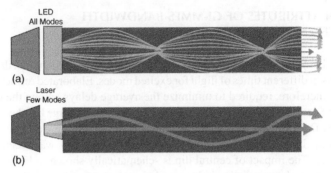

FIGURE 5.16 Impacts of differential modal delay (DMD) in a GI-MMF with the central refractive index dip: (a) overfilled launch condition with LED and (b) restricted mode launch with LD.

The secondary pulse from the mode excited near the central dip will have significantly lower amplitude than the major pulse contributed from all the rest of modes to give acceptable performances at the receiver. In the case of LD, however, the number of modes excited near the central dip becomes very comparable to the other modes. The secondary pulse becomes notable and the output appears to split into two pulses as in Figure 5.16b, which will give rise to serious bit errors at the receiver.

In this section, we will first review characterization of MMF bandwidth based on international standards IEC 60793-1-41, "Measurement methods and test procedures—bandwidth." Then we will discuss DMD based on IEC 60793-1-49, "Measurement methods and test procedures—differential mode delay."

The 3 dB bandwidth of an MMF is defined as follows [20]:

The value numerically equal to the lowest modulation frequency (in Hz) at which the magnitude of the baseband transfer function of an optical fiber (in 1 km) decreases to a specified fraction, generally to one half of the zero frequency value.

The baseband frequency response of an optical fiber can be measured by launching a light source into the fiber core. The input mode power distribution excited by a laser beam is largely dependent on the launching condition and is localized within the narrow region. This causes a deviation in the frequency response of a graded-index fiber. The steady-state mode excitation routine should be systematically established so that the baseband frequency response of the GI-MMFs can be measured reproducibly.

The baseband frequency response of a given MMF can be directly measured in the frequency domain by taking the RF spectrum from the fiber output for a sinusoidally modulated light source. The baseband response can also be measured by observing the broadening of a narrow pulse of light using a sampling scope in time domain. Each method can be performed using one of the following two launches: an overfilled launch (OFL) condition or a restricted mode launch (RML) condition.

OFL is the condition where the source launches light uniformly into all modes of the multimode fiber. The launch condition of this measurement is similar to that of an

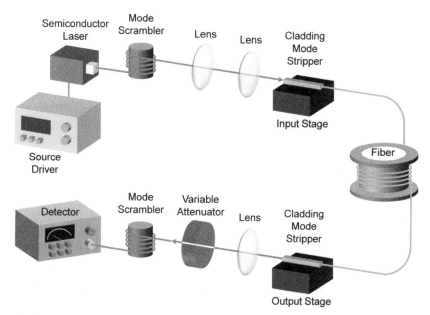

FIGURE 5.17   Schematic diagram for OFL condition for bandwidth measurement.

LED source. Therefore, this measurement method gives a good indication of system performance when using legacy protocols utilizing LED sources.

Figure 5.17 depicts the OFL bandwidth measurement test setup defined by the IEC 60793-1-41 standard [20]. Light launch conditions are consistently established through the use of a mode scrambler, which is positioned between the light source and test fiber to produce a radiation distribution overfilling the test fiber core and numerical aperture, irrespective of the spatial radiation properties of the light source.

Structures of the mode scrambler are schematically shown in Figure 5.18. In the step–graded–step structure, a series combination of 1 m lengths of step-, graded-, and step-index fibers is spliced together to efficiently scramble the modes [21,22]. In the step-with-bend structure, microbending is applied along a segment of long step-index MMF to excite and couple the modes in the MMF.

The output from the mode scrambler is focused with a larger numerical aperture and larger beam diameter than the test fiber core using imaging lenses. This imaging optics usually excites cladding modes and it is necessary to provide means to remove cladding light from the test fiber. Often the fiber coating is sufficient to perform this function. Otherwise, it will be necessary to use cladding mode strippers, usually the fiber immersed in index matching liquid, near both ends of the test sample. The fibers may be retained on the cladding mode strippers with small weights, but care must be taken to avoid microbending-induced loss at these sites.

The traditional OFL method of bandwidth measurement has accurately predicted fiber bandwidth for LED applications. However, high speed lasers excite only a few modes in a position-dependent manner, which made OFL bandwidth measurement no longer valid to predict what the fiber's bandwidth would be if the fiber were to be used

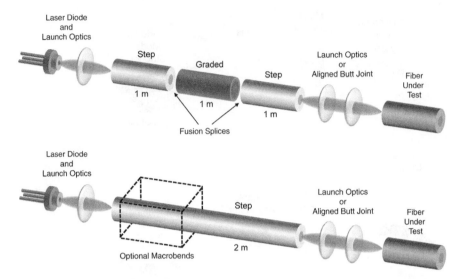

FIGURE 5.18   Two types of mode scrambler for MMF measurement procedure [20].

with a laser light source. In order to overcome this uncertainty, an RML method was developed as shown in Figure 5.19, which is defined by the IEC 60793-1-41 standard [20]. The main difference between OFL and RML is that in the latter, the overfilled launch condition is further filtered with a special RML fiber. The OFL for RML fiber needs to be only large enough to overfill the RML fiber both angularly and spatially. The RML fiber has a core diameter of $23.5 \pm 0.1\,\mu\text{m}$ and a numerical

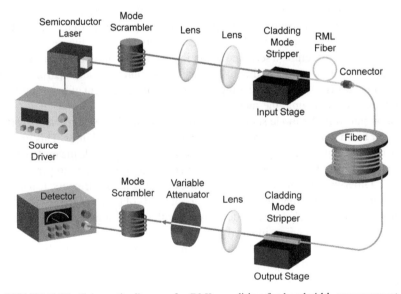

FIGURE 5.19   Schematic diagram for RML condition for bandwidth measurement.

aperture of 0.208 ± 0.01. The fiber must have a graded-index profile with an α of approximately 2 and an OFL bandwidth greater than 700 MHz km at 850 and 1300 nm. The RML fiber should be at least 1.5 m in length to eliminate leaky modes, and it should be less than 5 m in length to avoid transient loss effects. The launch exiting the RML fiber is then coupled into the fiber under test.

This RML is intended to replicate a VCSEL. However, test results can still vary depending on exactly how the light source is coupled with the fiber. In order to achieve the highest measurement reproducibility, tight alignment tolerances are required in the connection between the RML fiber and the fiber under test to ensure the RML fiber is centered to the fiber under test.

For the above launching setups in Figure 5.18 or 5.19, the bandwidth of the test fiber is measured by pulse distortion method (optical time domain) or frequency domain method.

The time domain measurement begins with characterization of the input pulse, $a(t)$, and the fiber output pulse, $b(t)$, using a pulsed light source and sampling scope. The Fourier transforms of the input and output pulses are calculated and normalized to the zero frequency value using the following formula:

$$A(f) = \frac{\int_{-\infty}^{+\infty} a(t)e^{-j2\pi ft}\, dt}{\int_{-\infty}^{+\infty} a(t)dt} \tag{5.38}$$

$$B(f) = \frac{\int_{-\infty}^{+\infty} b(t)e^{-j2\pi ft}\, dt}{\int_{-\infty}^{+\infty} b(t)dt} \tag{5.39}$$

where $a(t)$ is the temporal input pulse, $b(t)$ is the temporal output pulse out of fiber, $A(f)$ is the normalized input pulse Fourier transform, and $B(f)$ is the normalized fiber output pulse Fourier transform.

For the time domain method, the fiber transfer function is calculated as

$$H(f) = \frac{B(f)}{A(f)} \tag{5.40}$$

For the frequency domain method, the frequency response, $H(f)$, is given as

$$H(f) = \log_{10}\left[\frac{P_{\text{out}}(f)}{P_{\text{in}}(f)}\right] \tag{5.41}$$

where $P_{\text{in}}(f)$ is the measured input frequency response and $P_{\text{out}}(f)$ is the measured output frequency response [23–25].

The -3 dB frequency, $f_{3\text{dB}}$, shall be determined as the lowest frequency at

$$|H(f)| = 0.5 \tag{5.42}$$

Typical fiber transfer function is shown in Figure 5.20 along with $f_{3\text{dB}}$, which is the conventional bandwidth in MMFs. By convention, the baseband response is linearly referred to 1 km.

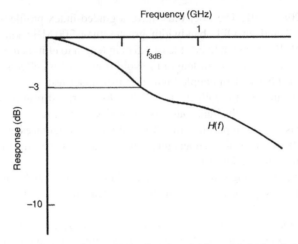

FIGURE 5.20 Typical fiber transfer function, $H(f)$, and 3 dB bandwidth, f_{3dB}.

Bandwidth is one of the most important attributes of GI-MMFs [26] and therefore its measurement and characterization technique has evolved along with development of MMF design and light sources. Generally speaking, OFL measurement system is relevant to GI-MMF used with LED of bandwidth ~300 MHz km, while RML is for GI-MMF with VCSEL light source of bandwidth ~1 GHz km. These two methods share the common equipment except the launching scheme. Recent GI-MMF used with high speed, small beam diameter laser diodes, whose bandwidth exceeds ~10 GHz km, furthermore requires another important attribute, DMD, which is based on position selective excitation of modes and their delay measurement.

In fact, DMD has been used to characterize and optimize the refractive index profiles [27] internally by the fiber manufacturers. Because the DMD was not a commercially specifiable fiber property, such as the spectral attenuation or the bandwidth, there were neither internationally standardized test procedures for DMD nor any conventions for specifying it until recently. With the advent of 850 nm laser-optimized 50 μm multimode fiber for the high speed Ethernet applications, fiber manufacturers not only needed to measure DMD in a more consistent manner but were also required to measure it more accurately with a higher resolution to ensure the 10 Gbps performance. To achieve these goals, the TIA and IEC working groups drafted test procedures for DMD measurement, TIA FOTP-220 [28] and IEC 60793-1-49 [29].

Schematic diagram for DMD measurement is given in Figure 5.21a, where the probe laser pulse with a small beam diameter is scanned across the radial position within the core of test GI-MMF.

For each radial position with a certain step between the spots, the probe laser pulse excites local modes with certain delays and their output pulses are recorded using a high speed detector and sampling scope. The traces of output pulse are then overlaid for radial positions in a two-dimensional plot as in Figure 5.21b. In this plot, the difference in relative delays of the fastest and slowest pulses after correction of input pulse width is the DMD of the test fiber.

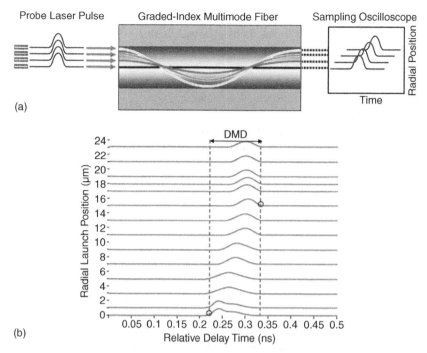

FIGURE 5.21   Principles of DMD measurement and characterization: (a) schematic DMD measurement setup and (b) an example of delay scan and characterization of DMD.

In measurements of DMD, the most critical aspects are the characteristics of the pulsed probe laser and the scanning of the laser spot during measurement. The light source specification is made very stringent in order to guarantee high speed transmission over 10 Gbps.

In DMD measurements, a probe laser should have

- a narrow spectral width, within ±10 nm of the nominal wavelength, to minimize errors caused by source chromatic dispersion;
- a temporal width that is less than 110% of the DMD to be measured; and
- a high-power pulse to maximize signal-to-noise ratio overcoming the typical attenuation (2.5 dB/km) of 50 μm fibers at 850 nm.

These requirements limit the choice of sources to mode lock Ti: sapphire lasers and distributed feedback semiconductor lasers.

These laser pulses are carried by a probe fiber that is placed between the light source and the test GI-MMF. The probe fiber should have

- a single-mode guidance at the probe laser wavelength (850 or 1310 nm);
- a Gaussian spot with a mode field diameter (MFD) given by $(8.7\lambda - 2.39)$ ±0.5 μm, where λ is the probe laser center wavelength given in micrometers, in

order to limit excited modes and maximize DMD resolution. Typical MFD is about 5 and 9 μm at 850 and 1310 nm wavelength.

The probe fiber, which is carrying the laser pulses, is aligned and scanned across the core of the test GI-MMF using a scanning system. The scanning system should have

- an output beam perpendicular to the end face of the test fiber less than 1.0°;
- a gap between the probe fiber and the end face of the test fiber less than 10 μm;
- the spot of the probe fiber centered within ± 1.0 μm from the center of test fiber; and
- consistent launch spot increment in steps <2 μm with an accuracy <0.5 μm.

The above requirements for the probe fiber scanning system are schematically shown in Figure 5.22.

At each radial position (r), the input spot excites a different subset of mode groups to result in variation of time delays and pulse widths of the output pulses. By measuring the output pulse $T(r)$ versus position r, we can deduce mode delays and refractive index errors within the core. The input pulse from the probe fiber is first measured to define T_{ref} as in Figure 5.23a. For the measured $T(r)$, T_{fast} and T_{slow} are then chosen as in Figure 5.23b. Note that the leading and trailing edge times are referenced to 25% of the pulse peak. DMD of the test fiber is given by

$$DMD = (T_{fast} - T_{slow}) - T_{ref} \qquad (5.43)$$

Qualitative comparison of DMD measurements is shown in Figure 5.24 for two contrasting cases. In "bad" DMD, the pulse delay fluctuates over the core because the refractive index profile contains local deviations from the optimal α-profile in Section 5.3. In contrast, "good" DMD case shows a narrow distribution of time delays with a smaller DMD value.

In order to quantify DMD values in a consistent manner and guarantee an effective modal bandwidth for GI-MMF used with laser light sources, six DMD templates have been established in TIA/EIA-492AAAC-A, "Detail specification for 850-nm

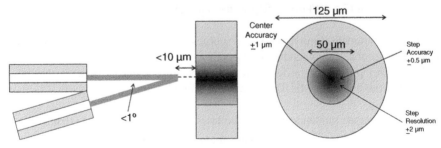

FIGURE 5.22 Requirements for probe fiber scanning in DMD measurements.

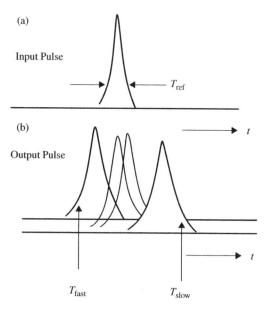

FIGURE 5.23 Temporal response of the input pulse from the probe fiber (a) and the output pulse scanned over the test fiber core (b).

laser-optimized, 50 μm core diameter/125-μm cladding diameter class Ia graded-index multimode optical fibers," as shown in Table 5.2 and Figure 5.25.

The allowable DMD range is plotted over the vertical axes versus the radial offset position of the single-mode fiber probe, which is measured by TIA FOTP-220 or equivalently IEC 60793-1-49 [28,29]. These DMD templates can be used to ensure a minimum effective modal bandwidth–length product of 2000 MHz km over 50 μm fiber, when using sources meeting the requirements described above. The simplest

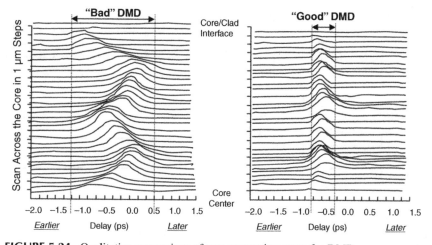

FIGURE 5.24 Qualitative comparison of two contrasting cases for DMD measurements.

TABLE 5.2 DMD Templates for Laser-Optimized 50 μm Core GI-MMF

Template Number	Inner Mask DMD (ps/m) for $R_{INNER} = 5\,\mu m$ to $R_{OUTER} = 18\,\mu m$	Outer Mask DMD (ps/m) for $R_{INNER} = 0\,\mu m$ to $R_{OUTER} = 23\,\mu m$
1	≤ 0.23	≤ 0.70
2	≤ 0.24	≤ 0.60
3	≤ 0.25	≤ 0.50
4	≤ 0.26	≤ 0.40
5	≤ 0.27	≤ 0.35
6	≤ 0.33	≤ 0.33

way to ensure a 2000 MHz km minimum modal bandwidth–length product would be to specify a maximum DMD value of approximately 0.2 ps/m FWHM for all probe positions. Such an ideal fiber would exhibit $\geq$ 2000 MHz km modal bandwidth–length product with an arbitrary excitation power mode distribution, and require no limitation on the laser transmitters. While such an ideal fiber is physically possible, it is practically difficult to manufacture GI-MMFs in massive scale to such tight tolerances. When the launch condition requirements on the transmitters are coupled to the DMD requirements on the fiber, a cost-effective balance can be achieved between

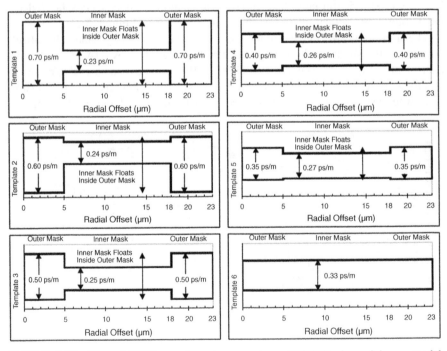

FIGURE 5.25 Allowed DMD templates to guarantee a 2000 MHz km minimum modal bandwidth–length product [30].

fiber tolerances and transmitter tolerances. A careful study, using fibers contributed by several different fiber manufacturers and laser transmitters from several different source manufacturers, and including extensive and detailed simulations, shows that the above coupled specifications on fiber and sources will yield a minimum effective modal bandwidth–length product of 2000 MHz km.

The attenuation, bandwidth, and gigabit Ethernet (GbE) link distance characteristics for GI-MMFs in present commercial markets are listed in Table 5.3.

5.5 MULTIMODE OPTICAL FIBER STANDARDS

There are four major standards for MMFs that specify the geometrical, mechanical, and optical characteristics: ITU-T (G.651) [26], IEC (60793-2-10, 60793-2-20, 60793-2-30, 60793-2-40) [31–34], TIA (492AAAA-B, 492AAAB-A, 492AAAC-A, 492AAAC-B) [35–38], and ISO/IEC (11801) [11]. Among these standards, ITU-T G.651 only specifies the GI-MMF with 50 μm core and 125 μm cladding (abbreviated as 50/125 μm hereafter). The hierarchy structures of the standards are schematically shown for short-haul optical communications in Figure 5.26. ISO/IEC standards are recognized as international but TIA standards are valid within North America. TIA TR-42 [39] and corresponding IEC TC86 [40] committees cover basic specifications and their test methods and they form the foundation of standards for optical fibers in short-haul optical communication applications. Their subordinate bodies are IEC 60793-1 and 2 and TIA 455 and 492 series, which give detailed descriptions of performance requirements and test procedures for manufacturers. TIA 568B [41] and ISO/IEC 11801 [11] deal with specifications for end users to install fiber optic cables, to form the top of food chain.

IEC covers wide ranges of MMFs including step-index profiles and plastic material for both core and cladding. Therefore, we will mainly focus on IEC standards and discuss their cross-references with corresponding TIA and ISO/IEC standards.

MMF index profile given in equation 5.1 can be further normalized as

$$\delta(x) = 1 - x^\alpha \tag{5.44}$$

$$\delta(\chi) = \frac{n(x) - n(1)}{n(0) - n(1)} \tag{5.45}$$

$$x = \frac{r}{a} \quad (0 \le r \le a) \tag{5.46}$$

Here a is the core radius and $n(x)$ is the refractive index at normalized position x within the core.

IEC standards categorize MMF by the types of material at the core and cladding, and then the refractive index profile parameter a. Table 5.4 summarizes the IEC 60793 multimode optical fiber categories [31–34].

A1 and A2 specify attributes for GI and SI glass MMFs, respectively. A3 fiber is in general called hard plastic clad fiber (HPCF). A4 covers plastic optical fibers (POFs) whose core and cladding are composed of plastic material, mainly PMMA.

TABLE 5.3 GI-MMF Characteristics Available in Present Commercial Market [9,10]

Fiber Type	Maximum Attenuation 850/1300 nm (dB/km)	Minimum OFL BW 850/1300 nm (MHz km)	Minimum EMB 850 nm (MHz km)	1 GbE Distance 850/1300 nm (m)	10 GbE Distance 850 nm (m)
62.5/125 μm standard	3.5/1.0	200/500	>220	300/550[a]	33
62.5/125 μm enhanced	3.5/1.0	200/500	>385	500/1000[a]	33
50/125 μm standard	3.5/1.5	500/500	>510	600/600[a]	82
50/125 μm laser-optimized standard	3.5/1.5	700/500	>850	750/600[a]	150[b]
50/125 μm laser-optimized enhanced	3.0/1.5	1500/500	>2000	1000/600[a]	300[b]
50/125 μm laser-optimized premium	3.0/1.5	–	>4700	1000/600[a]	550[c]

OFL, overfilled launch; EMB, effective modal bandwidth.

[a] Guaranteed distance for gigabit Ethernet–compliant systems.

[b] Guaranteed distance for 10Gb Ethernet–compliant systems.

[c] The 550 m distance is equivalent to a 4700 EMB system with standards-compliant transceiver and fiber characteristics, 3.0 dB/km cable attenuation, and 1.0 dB total connector loss.

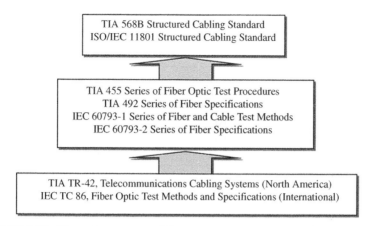

FIGURE 5.26 Hierarchy structure of fiber optic standards for short-haul applications.

A1 fibers are further categorized into four subsets, A1a.1, A1a.2, A1b, and A1d. The main difference among A1a, A1b, and A1d is the core diameter 50, 62.5, and 100 μm for A1a, A1b, and A1d, respectively. A1a is then further divided into A1a.1 and A1a.2 depending on the light source type and launching conditions. A1a.2 is the most recent standard that reflects the latest technical development for large-bandwidth short-haul optical communication, which is conventionally called an 850 nm laser-optimized GI-MMF. In addition to these fibers, A1d applies to 100/140 μm graded-index fiber. Table 5.5 summarizes GI-MMF categories under IEC 60793-2-10 A1 category along with main light sources [31].

ISO/IEC 11810 specifies multimode optical fiber types in an application-oriented viewpoint with transmission length and bandwidth range into the categories OM1, OM2, and OM3 [11]. The specifications for the data rates and link distances are

TABLE 5.4   Categories of Multimode Fibers in IEC Standards

Category	Material	Type	α Range	IEC Standards
A1	Glass core/glass cladding	Graded-index fiber	$1 \le \alpha < 3$	60793-2-10
A2	Glass core/glass cladding	Step- and quasi-step-index fiber	$3 \le \alpha < \infty$	60793-2-20
A3	Glass core/plastic cladding	Step-index fiber	$10 \le \alpha < \infty$	60793-2-30
A4	Plastic core/plastic cladding	Step- or graded-index fiber	$1 \le \alpha < \infty$	60793-2-40

TABLE 5.5   Categories of A1 Graded-Index Multimode Fibers in IEC 60793-2-10

Category	Core Diameter (μm)	Cladding Diameter (μm)	Main Light Source
A1a.1	50	125	LED/LD
A1a.2	50	125	LD/VCSEL
A1b	62.5	125	LED
A1d	100	140	LED

TABLE 5.6   Fiber Types for Various Transmission Capacity Ranges in ISO/IEC 11810 [42]

Data Rate (Mbps)	Distance (m)		
	300	500	2000
10 (LED)	OM1	OM1	OM1
100 (LED)	OM1	OM1	OM1
1,000 (laser)	OM1	OM2	OS1
10,000 (laser)	OM3	OS1	OS1

closely related with recent gigabit and 10 Gb Ethernet developments supported by IEEE 802.3 standards [42]. The transmission ranges that cannot be covered by multimode fibers are recommended to use single-mode fibers specified as OS1. OS1 is more detailed in ITU-T G.652 [43] and IEC 60793-2-50 [44], which were described in Chapters 2 and 3.

The ranges of MMF types that cover the transmission capacity given by the link distance and data rate are summarized in Table 5.6.

These requirements for OM1 and OM2 are a subset of the more general IEC requirements for A1a.1 and A1b, which correspond to TIA 492AAAB-A and 492AAAC-A and -B, respectively. The requirements for fiber type A1a.2 contained in IEC 60793-2-10 and the ISO/IEC 11801 requirements for fiber type OM3 are identical. OM3 and A1a.2 correspond to TIA 492AAAC-A and -B.

In Table 5.7 the standards in TIA, IEC, and ISO are summarized. The related documents in different standard bodies are grouped together using a superscript index. For example, TIA 492AAAA is related with IEC 60793-2-10 A1b fiber specifications and ISO 11801 OM1 fiber specifications, which are all labeled with superscript (1) in Table 5.7.

Cross-reference among three standards, IEC, TIA, and ISO/IEC, is summarized in Table 5.8. Note that the effective modal bandwidth (EMB) that is assured by the six

TABLE 5.7   MMF Types in Standard Organizations

Organization	Documents	Performance
TIA	492AAAA[1]	62.5 μm fibers with 160/500 MHz km OFL BW
	492AAAB[2]	50 μm fibers with 500/500 MHz km OFL BW
	492AAAC[3]	Laser-optimized 50 μm fibers with 2000 MHz km EMB at 850 nm
IEC	60793-2-10	A1a.1[2] 50 μm fibers with specific OFL BWs
		A1a.2[3] Laser-optimized 50 μm fibers with 2000 MHz km EMB at 850 nm
		A1b[1] 62.5 μm fibers with specific OFL BWs
ISO	11801	OM1[1] 200/500 MHz km OFL BW
		OM2[2] 500/500 MHz km OFL BW
		OM3[3] Laser-optimized 50 μm fibers with 2000 MHz km EMB at 850 nm

TABLE 5.8 Cross-Reference Among IEC, TIA, and ISO/IEC Standards

Attribute	IEC 60793-2-10			ISO/IEC 11801		
Fiber type	*A1a.1*	*A1a.2*	*A1b*	*OM1*	*OM2*	*OM3*
Core diameter (μm)	50	50	62.5	50 62.5	50 62.5	50
IEC fiber type cross-reference	–	–	–	A1a.1 A1b	A1a.1 A1b	A1a.2
Minimum BW OFL 850 nm (MHz km)	200–800	1500	100–800	200	500	1500
Minimum BW OFL 1300 nm (MHz km)	200–1200	500	200–1000	500	500	500
Minimum EMB 850 nm (MHz km)	Not specified	2000	Not specified	Not specified	Not specified	2000
TIA 492AAA*-cross-reference	B-A	C-A, B	A-B	A-B	A-B	C-A, B

BW, bandwidth; OFL, overfilled launch; EMB, effective modal bandwidth.

TABLE 5.9 Optical Characteristics of MMF in IEC 60793-2-10 Standard

Attributes	Unit	Limit		
		A1a.1	A1a.2	A1b
Maximum attenuation coefficient at 850 nm	dB/km	2.4 – 3.5	2.5	2.8 – 3.5
Maximum attenuation coefficient at 1300 nm	dB/km	0.7 – 1.5	0.8	0.7 – 1.5
Minimum BW OFL at 850 nm	MHz·km	200 – 800	1500	100 – 800
Minimum BW OFL at 1300 nm	MHz·km	200 – 1200	500	200 – 1000
Minimum EMB 850 nm	MHz·km	Not specified	2000	Not specified
Numerical aperture	Unit less	0.20 ± 0.02 or 0.23 ± 0.02	0.20 ± 0.015	0.275 ± 0.015
Maximum Macrobending loss[a]	dB	0.5	0.5	0.5
Zero dispersion wavelength, λ_0	nm	$1295 \leq \lambda_0 \leq 1320$		$1320 \leq \lambda_0 \leq 1365$
Zero dispersion slope, S_0	ps/nm²·km	$\leq 0.001\,(\lambda_0\text{-}1190)$ for 1295 nm $\leq \lambda_0 \leq$ 1300 nm ≤ 0.11 for 1300 nm $\leq \lambda_0 \leq$ 1320 nm		≤ 0.11 for 1320 nm $\leq \lambda_0 \leq$ 1348 nm $\leq 0.001(1458 - \lambda_0)$ for 1348 nm $\leq \lambda_0 \leq$ 1365 nm

[a] 100 turns on mandrel diameter of 75 mm at wavelength of 850 and 1310 nm.

DMD templates is defined only for IEC A1a.2, TIA 492 AAAAC-A and -B, and ISO OM3 fibers, for laser light sources.

More detailed optical characteristics for MMFs under IEC specifications are listed in Table 5.9.

Specifications of MMFs are rapidly adapting to recent developments of short-haul applications such as Ethernet, fiber channel, and VSR optical communications. Recent high speed applications are mainly using 850 nm VCSEL light sources and their data rate exceeding 10 and 40 Gbps demonstrations is being reported worldwide as in Figure 5.27.

Most popular applications in short-haul optical communications are Ethernet, fiber channel, and very-short-reach communications, whose standards are published in IEEE 802.3 series, ANSI-INCITS standards series, and ITU/T, IOF specifications, respectively. For these applications, the light source, optical fibers, and link distances are predefined for various communication capacities. In Table 5.10, the link distances (in meter) are shown in shaded region for various data rate, light sources, and GI-MMFs.

Laser-optimized multimode fiber was introduced in 1999. It supports 300 m link lengths for 10 Gb/s applications and is tested to ensure a 2000 MHz km EMB. Its industry standard 50 µm core size couples sufficient power from LED sources to support legacy applications such as Ethernet, token ring, FDDI, and Fast Ethernet for virtually all in-building networks and most campus networks. The 50 µm core size is also directly compatible with laser-based applications such as gigabit Ethernet, fiber channel, and so on. Furthermore, it is the recommended multimode fiber type in ANSI/EIA/TIA-942, Telecommunications Infrastructure Standard for Data Centers.

OM3 fiber is a logical choice for short-range applications that need to support 1 Gbps or multi-gigabit speeds. Today 1 Gbps-ready backbone solutions are the norm. OM3 fiber has a significantly higher bandwidth advantage for longer reach 1 and 10 Gbps data rates to make OM3 the multimode fiber of choice in LANs, SANs, data center interconnects, and recent access applications.

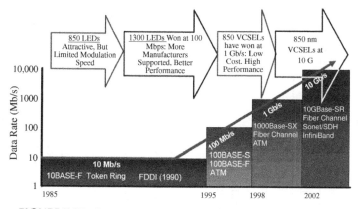

FIGURE 5.27   Recent trends in short-haul optical communications.

TABLE 5.10 Summary of GI-MMFs Applications

Applications			50μm Core MMF			62.5μm Core MMF			Bandwidth (MHz km)
			850 nm		1300 nm	850 nm		1300 nm	
Standards	Data Rate	Light Source	500	2000	500	160	200	500	Fiber Types
Ethernet IEEE 802.3									
1000BASE-SX	1 Gbps	850 nm VCSEL	550	860		220	275		OM1, OM2
1000BASE-LX	1 Gbps	1310 nm FP-LD			550			550	OM1, OM2
10GBASE-SR	10 Gbps	850 nm VCSEL	82	300		26	33		OM1, OM2, OM3
10GBASE-LX4	10 Gbps	1300 nm DFB-LD			300			300	OM1, OM2
Fiber Channel ANSI-INCITS									
100-M5/6-SN-I	1 Gbps	850 nm VCSEL	500	860		250	300		OM1, OM2,
200-M5/6-SN-I	2 Gbps	850 nm VCSEL	300	500		120	150		OM1, OM2
400-M5/6-SN-I	4 Gbps	850 nm VCSEL	150	270		55	70		OM1, OM2
1200-M5/6E-SNS	10 Gbps	850 nm VCSEL	82	300		26	33		OM1, OM2, OM3
1200-M5/6-SN4S	10 Gbps	850 nm VCSEL	290	550			118		OM1, OM2, OM3
1200-M5/6-LC4S	10 Gbps	1300 nm DFB-LD			290			290	OM2
VSR ITU/T-SONET/SDH IOF									
OC-192 VSR4-01	10 Gbps	850 nm VCSEL	550	850		220	275		OM3
OC-192 VSR4-02	10 Gbps	1300 nm FP,DFB-LD							OM3
OC-192 VSR4-03	10 Gbps	850 nm VCSEL	300	620					OM3
OC-192 VSR4-04	10 Gbps	850 nm VCSEL	85	300		25	33		OM3
OC-768 VSR5-01	40 Gbps	850 nm VCSEL	100	300					OM3

Future direction of research and development in GI-MMF would be

(1) Enhance scalability
   - Extended link lengths at 10 Gb/s
   - 40 Gb/s data rates via CWDM
(2) Adapt parallel optics
   - Ribbon cables
   - Array VCSELs

Fiber manufacturers are developing next-generation 50 μm GI-MMFs with extended bandwidth (preliminarily being referred to as "OM4" fiber) that potentially could be used to extend the system cost benefits enabled by 850 nm VCSEL technology to ultralong building backbones and medium length campus backbones. It would support 10 Gb/s Ethernet, fiber channel, and OIF applications to distances as much as 550 m (subject to some conditions) by utilizing an EMB of 4700 MHz km, more than double the IEEE requirement for 10 Gb/s 300 m support.

Parallel optics components are also being developed to increase the capacity by using multiple parallel strands of MMFs in the form of ribbon cables. Light sources should conform to the parallel optics configurations to pursue VCSEL arrays.

REFERENCES

[1] A. Shah and G. Ramakrishnan, "FDDI: a high speed network," Prentice Hall, 1994.

[2] C. E. Spurgeon, "Ethernet: the definitive guide," O'Reilly Media, 2000.

[3] U. Troppens, R. Erkens, and W. Müller-Friedt, "Storage networks explained: basics and applications of fibre channel SAN, NAS, iSCSI, Infiniband, and FCoE," Wiley, 2004.

[4] R. Ramswami, K. N. Sivarajan, and G. H. Sasaki, "Optical network: a practical perspective," 3rd edition, Morgan Kaufmann Publisher, 2010.

[5] K. J. Dreyer, D. S. Hirchorn, J. H. Thrall, and A. Metha, "PACS: a guide to the digital revolution," 2nd edition, Springer, 2006.

[6] http://www.mostcooperation.com/home/index.html.

[7] D. Gloge and E. A. J. Marcatili, "Multimode theory of graded core fibers," The Bell System Technical Journal, vol. 52, no. 9, pp. 1563–1578, 1973.

[8] P. F. Kolesar and D. J. Mazzarese, "Understanding multimode bandwidth and differential mode delay measurements and their applications," Proceedings of the 51st International Wire and Cable Symposium (IWCS), Lake Buena Vista, FL, pp. 453–460, 2002.

[9] LaserWave® multimode optical fiber, OFS.

[10] InfiniCor®, multimode optical fiber, Corning.

[11] ISO/IEC 11801, "Information technology-generic cabling for customer premises," 2010.

[12] IEEE 802.3u-1995, "IEEE standard for local and metropolitan area networks—supplement—media access control (MAC) parameters, physical layer, medium attachment units and repeater for 100 mb/s operation, type 100BASE-T" (Clauses 21–30).

[13] IEEE 802.3z-1998, "Supplement to carrier sense multiple access with collision detection (CSMA/CD) access method and physical layer specifications: media access control (MAC) parameters, physical layer, repeater and management parameters for 1000 Mb/s operation."

[14] IEEE 802.3ae-2002, "IEEE standards for local area networks: carrier sense multiple access with collision detection (CSMA/CD) access method and physical layer specifications: media access control (MAC) parameters, physical layer, repeater and management parameters for 10 Gb/s operation."

[15] R. Olshansky and D. B. Keck, "Pulse broadening in graded index optical fiber," Applied Optics, vol. 15, no. 2, pp. 483–491, 1976.

[16] I. P. Kaminow, D. Marcuse, and H. M. Presby, "Multimode fiber bandwidth: theory and practice," Proceedings of the IEEE, vol. 68, no. 10, pp. 1209–1213, 1980.

[17] N. Shibata and T. Edahiro, "Refractive index dispersion for GeO_2-, P_2O_5-, B_2O_3 doped silica glasses in optical fibers," Transactions IECE Journal, vol. E65, no. 3, pp. 162–172, 1982.

[18] K. Iwasaki, H. Yokota, S. Horiguchi, and K. Kitayama, "Characteristics of graded index type fiber in the short and long wavelength region," National Convention IEICE Japan, vol. 882, pp. 4–142, 1980.

[19] D. Marcuse, "Calculation of bandwidth from index profiles of optical fibers. 1: theory," Applied Optics, vol. 18, no. 12, pp. 2073–2080, 1979.

[20] IEC 60793-1-41, "Optical fibres—part 1-41: measurement methods and test procedure-bandwidth," 2010.

[21] M. Horiguchi, Y. Ohmori, and H. Takata, "Profile dispersion characteristics in high-bandwidth graded-index fibers," Applied Optics, vol. 19, no. 18, pp. 3159–3167, 1980.

[22] W. F. Love, "Novel mode scrambler for use in optical-fiber bandwidth measurements," Topical Meeting on Optical Fiber Communications, Technical Digest, Washington, D.C., p. 118, March 6–8, 1979, paper ThG2.

[23] I. Kobayashi, "Bandwidth measurement in multimode optical fibers," Symposium on Optical Fiber Measurements, Technical Digest, National Bureau of Standards (U.S.) Special Publication 597, pp. 49–54, 1980.

[24] T. Tanifuji, T. Horiguchi, and T. Masamitsu, "Baseband-frequency-response measurement of graded-index fiber using step-index fiber as an exciter," Electronics Letters, vol. 15, no. 7, pp. 204–205, 1979.

[25] D. L. Franzen and G. W. Day, "Measurement of optical fiber bandwidth in the time domain," National Bureau of Standards (U.S.), Technical Note 1019, 1980.

[26] ITU-T G.651, "Characteristics of a 50/125 μm multimode graded index optical fiber cable," 1998.

[27] D. Marcuse, "Principles of optical fiber measurement," Chapter 6, Academic Press, 1981.

[28] TIA-455-220-A FOTP-220, "Differential mode delay measurement of multimode fiber in the time domain," 2003.

[29] IEC 60793-1-49, "Optical fibres—part 1-49: measurement methods and test procedures—differential mode delay," 2.0 b edition, 2006.

[30] TIA/EIA-492AAAC-A, "Detail specification for 850-nm laser-optimized, 50 μm core diameter/125-μm cladding diameter class Ia graded-index multimode optical fibers."

[31] IEC 60793-2-10, "Optical fibres—part 2-10: product specifications-sectional specification for category A1 multimode fibres," 3.0 b edition, 2007.

[32] IEC 60793-2-20, "Optical fibres—part 2-10: product specifications—sectional specification for category A2 multimode fibres," 2.0 b edition, 2007.

[33] IEC 60793-2-30, "Optical fibres—part 2-10: product specifications—sectional specification for category A3 multimode fibres," 2.0 b edition, 2007.

[34] IEC 60793-2-40, "Optical fibres—part 2-10: product specifications—sectional specification for category A4 multimode fibres," 3.0 b edition, 2009.

[35] TIA-492AAAA-B, "Detail specification for 62.5-μm core diameter/125-μm cladding diameter class Ia graded-index multimode optical fibers," 2009.

[36] TIA-492AAAB-A, "Detail specification for 50-μm core diameter/125-μm cladding diameter class Ia graded-index multimode optical fibers," 2009.

[37] TIA-492AAAC-A, "Detail specification for 850-nm laser-optimized, 50-μm core diameter/125-μm cladding diameter class Ia graded-index multimode optical fibers," 2003.

[38] TIA TIA-492AAAC-B, "Detail specification for 850-nm laser-optimized, 50-μm core diameter/125-μm cladding diameter class Ia graded-index multimode optical fibers," 2009.

[39] http://www.tiaonline.org/standards/committees/committee.cfm?comm=tr-42.

[40] http://tc86.iec.ch/index-tc86.html.

[41] TIA-568B, "Commercial building telecommunications cabling standard—part 1: general requirements," 2001.

[42] IEEE 802.3, "IEEE standard for information technology—telecommunications and information exchange between systems—local and metropolitan area networks—specific requirements. Part 3: carrier sense multiple access with collision detection (CSMA/CD) access method and physical layer specifications," 2008.

[43] ITU-T G.652, "Characteristics of a single-mode optical fibre and cable," 2009.

[44] IEC 60793-2-50, "Optical fibres—part 2-50: product specifications—sectional specification for class B single-mode fibres," 2008.

CHAPTER SIX

Optical Nonlinearity Control in Optical Fibers

6.1 HISTORICAL REVIEW OF OPTICAL NONLINEARITY IN OPTICAL FIBERS

Light as a bosonic particle has been treated as noninteracting until 1950s. Spectrally bright and coherent lasers, however, quickly disclosed that light does interact with optical materials when its intensity exceeds a certain level. Nonlinear optics in bulk media has flourished with the advent of laser throughout 1960s to 1990s and discovered the following facts, which cannot be explained in terms of prior linear optics:

(1) The refractive index of an optical medium changes with the propagating light's intensity, which subsequently modifies the speed of light in the medium.

(2) The frequency of light can change as the light interacts with a nonlinear optical medium.

(3) The physical quantities of light, such as amplitude, phase, and polarization, can be controlled by other lights through a nonlinear medium.

One of the most evident applications of nonlinear optics in present technologies would be second harmonic generation (SHG), where the frequency of incident light, ω, is doubled to 2ω in a nonlinear crystal. SHG is being widely used in generation of green light out of IR lasers for display, biomedical applications, and commercial laser pointers.

Silica Optical Fiber Technology for Devices and Components: Design, Fabrication, and International Standards, First Edition. By Kyunghwan Oh and Un-Chul Paek.
© 2012 John Wiley & Sons, Inc. Published 2012 by John Wiley & Sons, Inc.

The main constituent material of single-mode optical fibers is vitreous silica and in its bulk form silica glass shows a significantly low nonlinear optical property in comparison to other optical glasses or crystals. Despite the inherent low nonlinearity of silica, silica optical fibers have shown a rich collection of nonlinear effects due to their unique optical guidance: a tight confinement of light over a long distance with a low attenuation, which is not attainable in bulk nonlinear media. Optical fibers have played the main role as passive transmission media and these nonlinear effects have been traditionally regarded as seriously detrimental effects in preserving signal integrity especially in telecommunication applications. When erbium-doped fiber amplifiers (EDFAs) were introduced to initiate wavelength division multiplexing (WDM) systems to carry multitudes of optical signal over a single strand of optical fiber, nonlinear fiber optics became even more a critical issue [1–5].

Nonlinearities in optical fibers can be generally categorized into two classes:

(1) Scattering effects, the frequency-dependent optical power transfer, such as stimulated Brillouin scattering (SBS) and stimulated Raman scattering (SRS)
(2) Kerr effect, the intensity-dependent refractive index–related phenomena, such as self-phase modulation (SPM), cross-phase modulation (XPM), modulation instability (MI), and four-wave mixing (FWM).

In order to cope with these nonlinearities in optical telecommunication applications, international standards have been recently issued for consistent measurements, clear specifications, and detailed descriptions on system impairments [6,7]. In these standards, most of nonlinear effects other than Raman scattering in optical amplifier applications have been regarded as optical phenomena to be suppressed in order to secure the signal integrity of the communication systems.

In contrast to this conventional viewpoint to suppress or get rid of the nonlinearities for telecom applications, recently totally opposite viewpoint is rapidly growing to enhance nonlinearities in optical fibers [8]. In this relatively new viewpoint, optical nonlinearities are being actively exploited [9]: (1) to change the spectral components, amplify amplitude, and manipulate the phases of optical signals; (2) to provide an all-fiber, all-optical means to process the optical signals [10–13]. Some of the latest applications of optical fiber nonlinearities include the supercontinuum generation [14] and slow light and fast light observation [15] in optical fibers.

A variety of parameters influence nonlinear optical effects in optical fibers, including

(1) fiber core material,
(2) fiber chromatic dispersion characteristics,
(3) effective area of the fiber,
(4) degree of longitudinal uniformity of the fiber characteristics,
(5) fiber length and attenuation,
(6) number and spacing of WDM signal channels, and
(7) signal intensity and light source linewidth

In this chapter, we will discuss and review fiber waveguide parameters, namely, from (1) to (5) in the above list, so that the readers would understand the main concepts in optical fiber design and direction of fabrication process to manipulate optical nonlinearities in optical fibers.

Among nonlinearities, we will focus on stimulated Raman and Brillouin scatterings in silica fibers for the sake of readers' understanding because these nonlinearities are highly dependent on fiber design parameters as listed above. Optical Kerr effect–related nonlinearities such as self-phase modulation, cross-phase modulation, modulation instability, and four-wave mixing have been discussed in prior books and review papers [1–5,8], so we will not further discuss these issues in this book and readers are recommended to consult the references.

In this chapter, we will furthermore focus our discussion on silica fibers with index controlling dopants, GeO_2, P_2O_5, and F, as in the cases of ITU-T G652, 653, and 655 fibers in Chapter 2. There are intensive studies for optical nonlinearity in microstructured fibers or photonic crystal fibers that requires further explanation on their unique guiding properties. We will discuss nonlinearity in photonic crystal fibers separately in Chapter 8.

6.2 ORIGIN OF OPTICAL NONLINEARITIES IN OPTICAL FIBERS

In the case of general linear dielectric medium, its electromagnetic properties are represented by the relation between the polarization vector in the medium and incident electric field as described in Chapter 2:

$$\vec{P} = \varepsilon_0 \chi_e \vec{E} = (\varepsilon - \varepsilon_0)\vec{E} \tag{6.1}$$

If we assume that the medium is homogeneous, isotropic, and nondispersive, then the above equation holds with a scalar susceptibility, χ_e, which will subsequently determine the refractive index of the medium by

$$n^2 = 1 + \chi_e \tag{6.2}$$

In this introductory section, we will also maintain these assumptions for a nonlinear medium but the relation between polarization and electric field amplitudes is expanded to a type of Taylor series as follows:

$$P \approx \varepsilon_0 \chi E + 2dE^2 + 4\chi^{(3)} E^3 \tag{6.3}$$

where d and $\chi^{(3)}$ are coefficients describing the second- and third-order nonlinear effects [16].

In silica glass where the isotropy of optical characteristics is inherently enforced, the second nonlinearity d is zero and therefore the nonlinearity in silica glass optical fibers is determined by the third-order nonlinearity $\chi^{(3)}$, whose value ranges from

10^{-34} to 10^{-29} in MKS unit for conventional bulk optical materials. The polarization can be decomposed into linear term and nonlinear term as follows:

$$P = P_L + P_{NL}$$

$$P_L = \varepsilon_0 \chi E, \qquad P_{NL} = 4\chi^{(3)} E^3 \tag{6.4}$$

Assuming monochromatic wave with harmonic time dependence, the amplitude of the incident electric field can be expressed in the following equation:

$$E(t) = \mathrm{Re}\{E(\omega)e^{j\omega t}\} = \frac{1}{2}\left[E(\omega)e^{j\omega t} + E^*(\omega)e^{-j\omega t}\right] \tag{6.5}$$

Substituting the above expression for the electric field into P_{NL}, we have two nonlinear polarization terms, $P_{NL}(3\omega)$ and $P_{NL}(\omega)$:

$$P_{NL} = 4\chi^{(3)} E^3 = 4\chi^{(3)} \frac{1}{8}\left[E(\omega)e^{j\omega t} + E^*(\omega)e^{-j\omega t}\right]^3$$

$$= \frac{\chi^{(3)}}{2}\left[EEE\,e^{j3\omega t} + E^*E^*E^*\,e^{-j3\omega t} + 3EEE^*\,e^{j\omega t} + 3E^*E^*E\,e^{-j\omega t}\right]$$

$$= \frac{\chi^{(3)}}{2}\left[2\mathrm{Re}\left\{E^3(\omega)e^{j3\omega t}\right\} + 6|E(\omega)|^2\mathrm{Re}\left\{E(\omega)e^{j\omega t}\right\}\right]$$

$$= P_{NL}(3\omega) + P_{NL}(\omega)$$

$$P_{NL}(3\omega) = \chi^{(3)} E^3(\omega)e^{j3\omega t} \tag{6.6}$$

$$P_{NL}(\omega) = 3\chi^{(3)}|E(\omega)|^2 E(\omega)e^{j\omega t} \tag{6.7}$$

Due to the presence of nonzero third-order nonlinearity, $\chi^{(3)}$, the polarization has a 3ω component $P_{NL}(3\omega)$, which corresponds to third harmonic generation. Third harmonic generation efficiency in optical fiber is very weak because it requires three photons at the same time at the same location.

In the following discussions, we will focus on $P_{NL}(\omega)$, which is related to the incremental change in the electric susceptibility, $\Delta\chi$:

$$P_{NL}(\omega) = \varepsilon_0 \Delta\chi\, E(\omega) = 3\chi^{(3)}|E(\omega)|^2 E(\omega) \tag{6.8}$$

From the relation between the optical intensity $I(\omega)$ and electric field amplitude $E(\omega)$, $I(\omega) = (1/2\eta)|E(\omega)|^2$, and equation 6.8, we can express $\Delta\chi$ as follows:

$$\Delta\chi = \frac{6\eta\chi^{(3)}}{\varepsilon_0}I(\omega) \tag{6.9}$$

Here η is the impedance of the electromagnetic wave in the medium, given by $\eta = (\eta_0/n) = (1/n)\sqrt{\mu_0/\varepsilon_0}$.

The refractive index n can now be generalized in a nonlinear medium using equations 6.2 and 6.9 along with the following relations:

$$n^2 = 1 + \chi, \qquad 2n\,\Delta n = \Delta\chi$$

$$\Delta n = \frac{\Delta\chi}{2n} = \frac{3\eta_0\chi^{(3)}}{\varepsilon_0 n^2} I(\omega)$$

$$n(I) \approx n + \Delta n = n_0 + n_2 I \qquad (6.10)$$

$$n_2 = \frac{3\eta_0\chi^{(3)}}{\varepsilon_0 n_0^2} \qquad (6.11)$$

As a result of the third-order nonlinearity $\chi^{(3)}$ in a glass medium, we found the refractive index n is linearly dependent on the intensity of the incident light $I(\omega)$ with a proportional constant n_2, the nonlinear index as in equation 6.10. Here we denoted the linear refractive index as n_0 in order to make a clear distinction with the nonlinear index n_2. This phenomenon is often referred as optical Kerr effect.

The nonlinear refractive index n_2 has the physical unit of (m^2/W) and its value in silica bulk glass is 2.2–2.6×10^{-20} m^2/W.

For a wide range of material choices, an empirical rule called Miller's rule has been applied to correlate the linear refractive index n_0 and nonlinear index n_2 (Figure 6.1) as follows [17,18]:

$$\chi^{(3)} = (\chi^{(1)})^4 \times 10^{-10} \quad [\text{esu}] \qquad (6.12)$$

or equivalently as

$$n_2^2 = \frac{5k}{32\pi^2 c} \frac{(n_0^2 - 1)^4}{n_0^2} \times 10^{-10} \qquad (6.13)$$

Another empirical correlation between the linear refractive indices measured at various wavelengths and the nonlinear index has been reported for doped bulk silica glass, which is being widely accepted among optical fiber manufacturers [19,20]:

$$n_2 = 2.867 \frac{68(n_D - 1)(n_D^2 + 2)^2}{v_D[1.517 + v_D[((n_D + 1)(n_D^2 + 2)^2)/]6n_D]^{1/2}} \quad [\times 10^{-20}\ m^2/W] \qquad (6.14)$$

Here $v_D = (n_D - 1)/(n_F - n_C)$ is Abbe number defined for the linear refractive indices n_F, n_D, and n_C measured at the wavelength of 486.13, 587.56, and 565.27 nm, respectively.

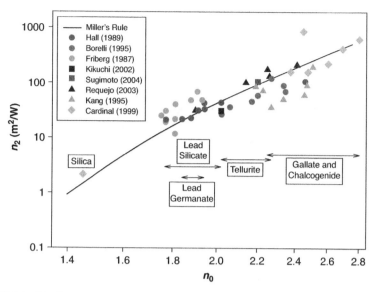

FIGURE 6.1 Nonlinearity index, n_2, versus linear refractive index, n_0, for various bulk optical glasses. Here the vertical axis is in the unit of 10^{-20} m^2/W. The solid curve is fitted to the Miller's rule [18].

In the case of optical fibers, the nonlinear properties of constituent material can be furthermore varied by the propagating light only within the effective area of the guided mode, A_{eff}, which is highly dependent on waveguide design parameters. It is therefore appropriate to introduce a new parameter, the effective nonlinearity γ, in optical fibers and waveguides as follows [8]:

$$\gamma = \frac{2\pi}{\lambda}\frac{n_2}{A_{eff}} \quad \left[(\text{W km})^{-1}\right] \tag{6.15}$$

As discussed in Chapter 2, A_{eff} is expressed in terms of mode field radius w of the fundamental mode:

$$A_{eff} = k\pi w^2 \tag{6.16}$$

Here k is a correction factor that depends on waveguide structures, which will be discussed in the next sections.

For example, standard single-mode fiber has $\gamma \approx 1$–2 (W km)$^{-1}$. It is noteworthy that the effective nonlinearity of an optical fiber can be controlled by modifying the fiber's effective mode area, using different host materials, or both.

In optical fibers, the nonlinear refractive index, n_2, has been systematically controlled by doping GeO$_2$ in silica for the core and doping fluorine in the cladding as shown in the schematic refractive index profiles in Figure 6.2.

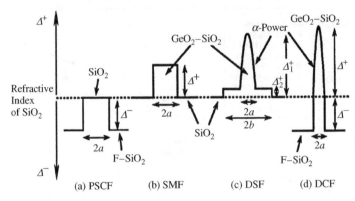

FIGURE 6.2 Various refractive index profiles to control n_2 in optical fibers. PSCF, pure silica core fiber; SMF, single-mode fiber; DSF, dispersion-shifted fiber; DCF, dispersion-compensating fiber [21].

For these conventional transmission optical fibers, the nonlinear effective indices were measured and plotted as a function of the refractive index difference between the core and cladding in Figure 6.3. It is noted that the n_2 values increase with the refractive index difference.

The correlation between n_2 and the relative refractive index difference Δ for GeO$_2$- or F-doped core silica fibers has been experimentally investigated [30] to obtain empirical linear relationships given by

$$n_2(\text{GeO}_2) \simeq 2.507 + 0.505\Delta \quad [\times 10^{-20}\ \text{m}^2/\text{W}]$$
$$n_2(\text{F}) \simeq 2.507 - 0.220\Delta \quad [\times 10^{-20}\ \text{m}^2/\text{W}] \tag{6.17}$$

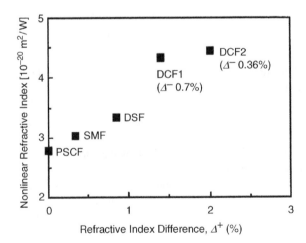

FIGURE 6.3 Nonlinear index n_2 for various transmission fibers whose refractive index profiles are shown in Figure 6.2 [21].

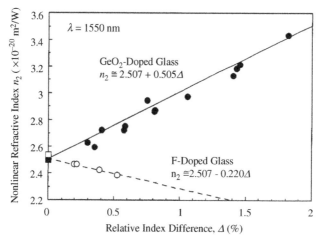

FIGURE 6.4 Relationship between the relative index difference D and the nonlinear refractive index n_2 of GeO$_2$- and F-doped glass in optical fibers [20].

The experimental graphs overlaid with these empirical relations are shown in Figure 6.4.

As defined in equation 6.15, the effective optical nonlinearity, γ, can be controlled by the choice of glass types as summarized in Figure 6.4 with the dopant concentration–dependent n_2. And it should be also emphasized that A_{eff} is indeed another important degree of freedom in fiber waveguide design to tailor the nonlinearity. Hansryd et al. reported measurements of γ in various silica optical fibers based on optical parametric amplification experiments [22]. The values are listed in Table 6.1.

In the following sections, we will describe nonlinear optical phenomena in terms of both optical fiber material and waveguide design parameters.

6.3 SPECIFICATIONS OF NONLINEAR OPTICAL PROCESSES IN OPTICAL FIBERS

In the last section, we found that the nonlinear refractive index n_2 and the third-order nonlinearity $\chi^{(3)}$ are related by the following equations:

$$n(I) \approx n + \Delta n = n_0 + n_2 I \tag{6.10}$$

$$n_2 = \frac{3\eta_0 \chi^{(3)}}{\varepsilon_0 n_0^2} \tag{6.11}$$

Now let us consider the phase shift of an electromagnetic wave incurred by the light intensity–dependent refractive index $n(I)$ over length L:

$$\phi = \frac{2\pi L}{\lambda_0} n(I) = \frac{2\pi L}{\lambda_0} n_0 + \frac{2\pi L}{\lambda_0} n_2 I = \phi_0 + \Delta\phi \tag{6.18}$$

TABLE 6.1   Characteristics of Optical Fibers and Their Effective Nonlinearity [22]

Fiber	SMF	DSF	DCF	HNLF
Attenuation (dB/km)	0.2	0.2	0.45	0.7
Effective area, A_{eff} (μm^2)	85	50	19	12
Effective optical nonlinearity, γ ((W km)$^{-1}$)	1.8	2.7	5.0	15

Note that all the values are measured at 1550 nm. SMF, single-mode fiber; DSF, dispersion-shifted fiber; DCF, dispersion-compensating fiber; HNLF, highly nonlinear fiber.

Here ϕ_0 and $\Delta\phi$ are linear and nonlinear phase shifts, respectively.

In general, the third-order nonlinearity $\chi^{(3)}$ can be a complex number with its real and imaginary parts, as follows:

$$\chi^{(3)} = \chi_R^{(3)} + j\chi_I^{(3)} \tag{6.19}$$

Substituting the complex $\chi^{(3)}$ to the nonlinear phase shift, $\Delta\phi$, we have a complex nonlinear phase shift:

$$\Delta\phi = \frac{2\pi L}{\lambda_0} n_2 I = \frac{2\pi L}{\lambda_0} \frac{3\eta_0 \chi^{(3)}}{\varepsilon_0 n_0^2} I = \frac{6\pi\eta_0 L}{\lambda_0 \varepsilon_0 n_0^2} (\chi_R^{(3)} + j\chi_I^{(3)}) I$$

$$\Delta\phi = \Delta\phi_R + j\Delta\phi_I \tag{6.20}$$

For these phase shifts, we will have the following phasor term for a monochromatic wave:

$$e^{j(\omega t - \phi)} = \exp(j\omega t) \cdot \exp(-j(\phi_0 + \Delta\phi_R)) \cdot \exp(\Delta\phi_I) \tag{6.21}$$

It is noteworthy that the linear phase shift, ϕ_0, and the real part of nonlinear phase shift, $\Delta\phi_R$, together produce phase modulations. In contrast to these components, the imaginary part of nonlinear phase shift, $\Delta\phi_I$, contributes to optical gain or loss:

$$\Delta\phi_R = \frac{6\pi\eta_0}{\lambda_0 \varepsilon_0 n_0^2} \chi_R^{(3)} IL \rightarrow \exp(-j\Delta\phi_R) \quad \text{(Phase modulation)}$$

$$\Delta\phi_I = \frac{6\pi\eta_0}{\lambda_0 \varepsilon_0 n_0^2} \chi_I^{(3)} IL \rightarrow \exp(\Delta\phi_I) \quad \text{(Gain or loss)} \tag{6.22}$$

From equation 6.22, we can classify two fundamental nonlinear optical processes [8,16]:

(1) Optical process by imaginary part of third-order nonlinearity $\chi_I^{(3)}$ to cause gains or losses: stimulated scattering process—Raman scattering and Brillouin scattering

(2) Optical process by real part of third-order nonlinearity $\chi_R^{(3)}$ to cause phase modulations: optically induced phase modulation—self-phase modulation, cross-phase modulation, modulation instability, and four-wave mixing

The third-order nonlinearity $\chi^{(3)}$ indeed provides rich nonlinear optical process in silica optical fibers. We will focus on the stimulated scattering process in this book.

In the following sections, we will discuss fiber design parameters to control the stimulated scattering processes, whose gain spectra are highly dependent on the core material and waveguide design parameters.

Before we move on to specific nonlinear optical processes, we will define and explain fundamental physical properties, namely, the effective area and the effective length.

Effective area (A_{eff}) is a parameter that is closely related to optical fiber nonlinearities. A_{eff} is defined as follows:

$$A_{eff} = \frac{2\pi \left[\int_0^\infty I(r) r \, dr \right]^2}{\int_0^\infty I(r)^2 r \, dr} \qquad (6.23)$$

where $I(r)$ is the field intensity distribution of the fundamental mode at the radial position r.

Note that the integration in the equation is carried out over the entire cross-sectional area of the fiber. The physical interpretation of the effective area is the area where the optical intensity I is uniform and constant there within, and zero outside. In other words, highly complicated intensity distribution over the core and cladding is represented by a stepwise uniform intensity over the effective area.

A_{eff} varies significantly depending on the refractive index profiles and the operating wavelength. The nominal values for conventional single-mode fibers (SMFs), dispersion-shifted fibers (DSFs), and dispersion-compensating fibers (DCFs) are ~80, 50, and 20 μm^2 at the operating wavelength of 1550 nm.

Namihira [23] first reported the relationship between the effective area, A_{eff}, and the mode field diameter, $2w$, in optical fibers assuming a proportional constant k as follows:

$$A_{eff} = k\pi w^2 \qquad (6.16)$$

Recently in the international standard, ITU-T G650.2 [6], the constant k has been tabulated for various standard telecom fibers and the typical values are listed in Table 6.2. Here CSF and NZ-DSF are cutoff shifted fiber and nonzero dispersion-shifted fiber. The fiber types and corresponding international standards are described in Chapter 2.

A_{eff} renormalizes the transverse optical intensity profile over the fiber cross-section into a uniform cross-section. Similarly L_{eff} renormalizes the longitudinal optical

TABLE 6.2   The Correlation Factor k Between A_{eff} and Mode Field Diameter $2w$[6]

Fiber Type		Operating Wavelength (nm)	
Acronyms	Standards	~1310	~1550
SMF	ITU-T G652, IEC B1	0.970–0.980	0.960–0.970
CSF	ITU-T G654, IEC B1.2	–	0.975–0.985
DSF	ITU-T G653, IEC B2	0.940–0.950	0.950–0.960
NZ-DSF	ITU-T G655, IEC B2	–	1.02–1.16

intensity profile along the fiber axial direction into a uniform intensity propagation length. The effective length is defined as follows:

$$L_{eff} = \int_0^L e^{-\alpha(\lambda)z} \, dz = \frac{1 - e^{-\alpha(\lambda)L}}{\alpha(\lambda)} \tag{6.24}$$

Here $\alpha(\lambda)$ is the attenuation coefficient given in m^{-1}. For a typical SMF the loss at 1550 nm is about 0.2 dB/km, which corresponds to $\alpha(1550 \text{ nm}) \approx 5 \times 10^{-5} \text{ m}^{-1}$, and L_{eff} is about 21 km.

In order to fully investigate the nonlinear effects in optical fibers, the effective length of an optical fiber should be compared with other nominal lengths such as nonlinear length (L_{NL}), dispersion length (L_D), coherence length (L_C), or polarization beat length (L_B), which are to be discussed in the following chapters. General rule is that the shortest lengths among these nominal lengths will determine the limitation of nonlinear processes [8].

6.4   COMPARISON OF RAMAN AND BRILLOUIN SCATTERING IN SINGLE-MODE OPTICAL FIBERS

When a light signal is injected into an optical fiber and guided in the fundamental mode, the light will be scattered, as it propagates, by the optical fiber medium in two different physical origins, Raman and Brillouin scatterings. These nonlinear scattering processes convert the incident pump light of the frequency ν_P into Stokes and anti-Stokes components that are separated by characteristics frequency shifts, ν_R and ν_B, from the pump. As discussed in the previous section, these Stokes and anti-Stokes components will experience an optical gain by the energy transfer from the pump photon in these nonlinear scattering processes. Schematic diagram for Raman and Brillouin scatterings is shown in Figure 6.5.

Raman scattering is attributed to the light interaction with optical phonon that is directly related to the vibrations of the constituent molecule, SiO_2, in silica glass optical fibers, while the origin of Brillouin scattering is related to acoustic phonon, which is acoustic wave along the fiber generated by electrorestriction force as the pump light propagates [1,5,8].

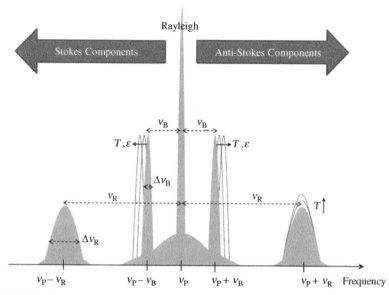

FIGURE 6.5 Schematic diagram of Raman and Brillouin scatterings in optical fiber. ν_R and ν_B correspond to the frequency shift; $\Delta\nu_R$ and $\Delta\nu_B$ are the spectral width of the Raman and Brillouin scatterings, respectively. In Brillouin scattering ν_B shifts by temperature (T) and strain (ε) changes in both Stokes and anti-Stokes components. In Raman scattering only the peak of anti-Stokes components changes with temperature.

Despite the similar physical origin–light interaction with phonons in optical fiber, Raman and Brillouin scatterings show very different characteristics and summary of comparison is listed in Table 6.3.

Raman scattering in optical fiber could be in both backward and forward directions for the forward-propagating pump light. However, Brillouin scattering is confined only to the backward direction in order to satisfy both energy and momentum conservation in conventional optical fibers.

The amount of spectral separation from the pump frequency, ν_P, is designated as the frequency shift and the Raman frequency shift, ν_R, is $\sim$13 THz and Brillouin frequency shift, ν_B, is $\sim$11 GHz in Brillouin scattering. Note that Raman scattering is separated further away from the pump frequency by a factor of $\sim$1000 in comparison to that of Brillouin scattering [8].

It is noteworthy in terms of fiber design that the Brillouin frequency shift ν_B shows a linear relationship with the concentration of conventional dopants such as GeO_2, P_2O_5, and F in silica glass, which will be discussed in the following sections. In contrast, ν_R does not show strong correlation with dopant concentrations.

When a light whose frequency corresponds to Stokes components is copropagating with the pump light, the energy of pump light is effectively transferred to the Stokes and anti-Stokes components, which exponentially grow by the optical gain provided by stimulated scattering processes [24,25].

TABLE 6.3   Comparison of Raman and Brillouin Scattering

	Raman	Brillouin
Physical origin	Light interaction with molecular vibration (SiO_2)	Light interaction with acoustic wave generated by electrorestriction
Scattering direction	Forward and backward	Backward
Frequency shift at 1.5 μm	$\nu_R \approx 13$ THz, more or less fixed	$\nu_B = 2n_{eff}V_a/\lambda \approx 11$ GHz, increases linearly with dopant concentration
Bandwidth	$\Delta\nu_R \geq 10$ THz	$\Delta\nu_B \leq 50$ MHz, Lorentzian shape
Peak gain	$g_R \approx 1 \times 10^{-13}$ m/W, increases with GeO_2 concentration	$g_B \approx 5\nu \times 10^{-11}$ m/W, almost constant
Threshold power	$P_R^{th} \simeq 16\frac{A_{eff}}{g_R L_{eff}} \approx 600$ mW	$P_B^{th} \simeq 21\frac{A_{eff}}{g_B L_{eff}} \approx 1$ mW
Temperature sensitivity	Peak increases with T, only at anti-Stokes band	ν_B linearly increases with T, at both Stokes and anti-Stokes bands
Strain sensitivity	None	ν_B linearly increases with ε, at both Stokes and anti-Stokes bands

The bandwidth in the gain spectrum is also contrasting between Raman and Brillouin scatterings. The 3 dB bandwidth of Raman gain, $\Delta\nu_R$, exceeds 10 THz, while that of Brillouin scattering, $\Delta\nu_B$, is less than 50 MHz in conventional silica single-mode fibers. Brillouin gain spectra are well fitted by Lorentzian distribution with its width parameter $\Delta\nu_B$. Raman gain spectrum is not fitted by a single function and highly depends on the composition of the core material that determines molecular vibrations.

In terms of the peak gain, Brillouin scattering offers higher value $g_B \approx 5 \times 10^{-11}$ m/W than that of Raman scattering, $g_R \approx 1 \times 10^{-13}$ m/W. Due to this high peak gain, Brillouin scattering provides a lower threshold power $P_B^{th} \simeq 21(A_{eff}/(g_B L_{eff})) \approx 1$ mW than Raman scattering $P_R^{th} \simeq 16(A_{eff}/(g_R L_{eff})) \approx 600$ mW [26].

The Brillouin frequency shift, ν_B, increases with both temperature and strain with a high linearity, which has been intensively exploited for distributed fiber optic sensor systems by monitoring either Stokes or anti-Stokes band. In Raman scattering only the magnitude of the Stokes band can be changed with respect to temperature, which is widely used for distributed temperature sensing [27–30].

6.5   CONTROL OF RAMAN SCATTERING IN SILICA OPTICAL FIBERS

Raman scattering is named after Sir Chanrasekhara Venkata Raman [31] who observed the Stokes-shifted radiation in liquids to win the Nobel Prize in physics in 1930. When light is scattered from a molecule, most photons are elastically scattered to produce Rayleigh scattering components that have the same frequency and wavelength as the incident photons. However, a small fraction of the scattered light has a frequency different from, and usually lower than, the frequency of the incident photons. The difference between this inelastic scattering components and the incident photon can be described by annihilation and creation of phonon, the quanta for vibrational motion of the molecules.

For arbitrary intensity of incident pump light, the Stokes-shifted bands can be observed with their characteristic frequency shift and spectral distribution determined by the molecular structure. This is designated as spontaneous Raman scattering. When frequency difference between pump photon and signal photon matches the Raman frequency shift, ν_R, the pump photon energy is efficiently transferred to the Stokes-shifted signal to provide the optical gain in the SRS process. We will not discuss anti-Stokes components in the following sections because their intensity is negligible in comparison to Stokes components at room temperature. The same principle is applied to optical fibers, which are made of silica glass [32]. The SRS in an optical fiber is schematically shown in Figure 6.6.

The Raman shift frequency, ν_R, is related to vibrational modes of silica molecules, bending motion of the Si–O–Si bond, as shown in the circular inset in Figure 6.6. When the pump photon at the frequency of ν_P and signal photon $\nu_S = \nu_P - \nu_R$ are propagating together along the optical fiber, pump photon energy is transferred to the signal by emitting optical phonons.

In the energy-level diagram in Figure 6.6, a pump photon is converted into a signal photon that is an exact replica of the initial one, and the remaining energy produces an optical phonon. The initial signal photon, therefore, has been amplified by SRS. In this section, we will focus on silica optical fibers and discuss the effects of dopants, mainly GeO_2 and P_2O_5, and the waveguide structure over the SRS spectra. Stolen et al. [33] reported Raman gain spectra for silica core single-mode fiber as shown in Figure 6.7.

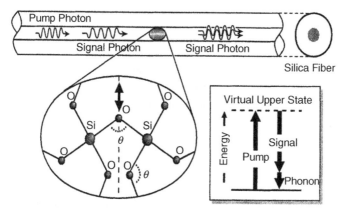

FIGURE 6.6 Schematic diagram for stimulated Raman scattering in a silica optical fiber [32]. Here the pump and signal photons' energy is $h\nu_P$ and $h\nu_S$ and they satisfy $h\nu_P = h\nu_S + h\nu_R$.

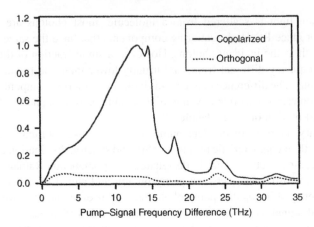

FIGURE 6.7 Raman gain spectra in silica core fiber. The solid line is for the case where the pump and the signal polarization is parallel, and the dotted line for the orthogonal case [33].

The curve in Figure 6.7 is normalized to 1.0 at the 13.2 THz peak corresponding to 440 cm^{-1}. The peak gain for a pump wavelength of 532 nm is 1.86×10^{-11} cm/W and varies with pump wavelength as $1/\lambda_p$. According to Lines [34] the most accurate estimate for peak Raman gain in SiO$_2$ glass is given as

$$g_R^{SiO_2} \approx \frac{1.2 \times 10^{-11}}{\lambda_p} \quad [cm/W] \tag{6.25}$$

where the pump wavelength is measured in micrometers.

The spectral shape and the spectral location of Raman gain in optical fiber vary significantly depending on the doping material in silica. Shibata et al. [35] reported variation of Raman scattering spectra for vitreous GeO$_2$, P$_2$O$_5$, and B$_2$O$_3$ glasses, which are most common dopants in silica optical fibers.

The Raman spectra of those oxide glasses are shown in Figure 6.8 in comparison to SiO$_2$ glass. The intensity unit has been chosen so that the 440 cm^{-1} peak in the spectrum of SiO$_2$ glass has an intensity of 1.0. The 420 cm^{-1} peak in GeO$_2$ has a relative intensity of 7.4. The 808 cm^{-1} peak in B$_2$O$_3$ glass has an intensity of 4.6. Vitreous P$_2$O$_5$ shows two prominent peaks: the one at 640 cm^{-1} has an intensity of 4.9 and the one at 1390 cm^{-1} has a peak intensity of 3.0.

Relative Raman scattering cross-sections can be obtained by multiplying correction factor:

$$M = \frac{n^2}{(1 - R)^2} \tag{6.26}$$

where n is the refractive index of the glass and R is its reflectivity [36]. The Raman scattering cross-sections of SiO$_2$, GeO$_2$, B$_2$O$_3$, and P$_2$O$_5$ glasses have relative strengths of 1.0 (at 444 cm^{-1}), 9.2 (at 420 cm^{-1}), 4.7 (at 808 cm^{-1}), and 5.7 and

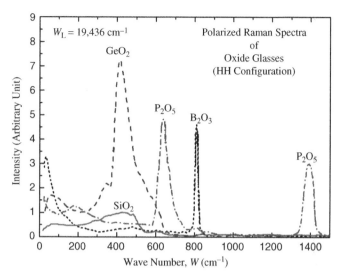

FIGURE 6.8 Relative Raman spectra of vitreous SiO_2, GeO_2, P_2O_5, and B_2O_5 glasses [35].

3.5 (at 640 and 1390 cm^{-1}), respectively. Comparison of Raman peak position, relative intensity and relative cross-sections is summarized in Table 6.4.

In optical fiber design, the most important glass composition would be $(SiO_2)_{1-x}$ $(GeO_2)_x$ glass, where x is the GeO_2 concentration (mol%). In most of telecom fibers, GeO_2 is the primary dopant to tailor the refractive index profile due to its low loss and the highly linear correlation with refractive index and its concentration, as described in Chapter 2.

Davey et al. have reported the relative peak intensity and spectral shape of Raman scattering as a function of GeO_2 doping concentration [37]. Linear correlation between the relative peak Raman intensity and the doping concentration was experimentally obtained as shown in Figure 6.9a. The data were interpolated using a linear least square fit, which gave a relationship of the following form:

$$\sigma_0(SiO_2, v)\sigma_0(x_{GeO_2}, v) = [1 + C(v) \cdot x_{GeO_2}] \cdot \sigma_0(SiO_2, v) \qquad (6.27)$$

TABLE 6.4 Comparison of Raman Peak, Relative Intensity, and Cross-Section

Glass	Refractive Index	Peak Position (cm^{-1})	Peak Position (THz)	Relative Intensity	Relative Cross-Section
SiO_2	1.46	440	13.2	1	1
GeO_2	1.60	420	12.6	7.4	9.2
B_2O_3	1.48	808	24.2	4.6	4.7
P_2O_5	1.55	640	19.2	4.9	5.7
P_2O_5	1.55	1390	41.7	3.0	3.5

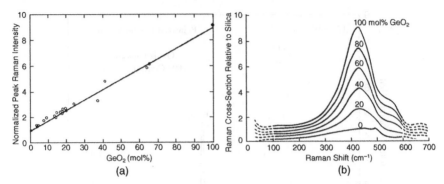

FIGURE 6.9 (a) Peak Raman intensity as a function of GeO_2 mol%, normalized to silica in germanosilicate glass; (b) Raman cross-section relative to silica in germanosilicate glass [37].

Here $C(v)$ is a normalized Raman cross-section spectrum of GeO_2 glass, whose peak value was set to 0.08 [38]. $\sigma_0(SiO_2, v)$ is the Raman cross-section spectrum of SiO_2 glass as shown in Figure 6.7. Using the above equation, the Raman cross-section of $(SiO_2)_{1-x}(GeO_2)_x$ is plotted in Figure 6.9b. As the GeO_2 concentration increases, the peak shifts slightly to lower energy from 440 to 420 cm^{-1}, which is consistent with Table 6.4. The spectrum broadens on its high-energy side and sharpens on its low-energy side.

These data are helpful to estimate the peak value and spectrum of Raman gain only in a step-index germanosilica optical fiber. However, the index profiles for modern telecom fibers are far from simple step index as we discussed in Chapter 2 and we need a consistent method to estimate the spectral shape of Raman gain of arbitrary refractive index profile fibers, whose indices are raised by adding GeO_2.

Bromage et al. [38] have reported a method to predict the Raman gain spectra of germanosilicate optical fiber with various refractive index profiles. In the analysis the main feature of Raman gain spectrum g_R was assumed to result from bond-bending motion of the bridging oxygen in the M–O–M′ bond (M = Si or Ge). The relative distribution of the Si–O–Si:Ge–O–Si:Ge–O–Ge bonds is known to be $(1-x)^2:2x(1-x):x^2$ in $(SiO_2)_{1-x}(GeO_2)_x$ glass [39]. Since common optical fibers have low GeO_2 concentration such that $x \ll 1$, the dominating bridges are Si–O–Si and Ge–O–Si with a relative distribution $(1-2x):2x$. Therefore, Raman gain spectra can be decomposed into two contributions, one from pure silica (Si–O–Si) and the other from binary germania and silica (Ge–O–Si). Each contribution is calculated by integrating the overlap of the radial profiles of pump-and-signal intensity profiles, weighted by the fractional radial distribution of the Si–O–Si and Ge–O–Si bridges:

$$g_R(v) = C_{SiSi}(v) \int_0^\infty (1 - 2x(r))I_P(r)I_S(r)r\,dr + C_{GeSi}(v) \int_0^\infty 2x(r)I_P(r)I_S(r)r\,dr$$

$$(6.28)$$

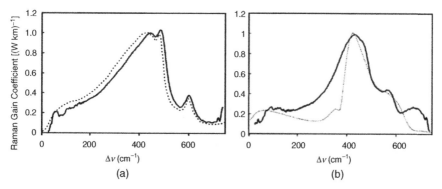

FIGURE 6.10   Amplitude normalized spectra for (a) C_{SiSi} and (b) C_{GeSi} as a function of the frequency separation between pump and signals in cm^{-1}. The dashed curves are measurements made on vitreous silica and germania [38].

Here $I_P(r)$ and $I_S(r)$ are the normalized intensity profiles of pump and signal photons in the given optical fiber waveguide structure, respectively [38].

The refractive index profile of a given optical fiber will give information on $x(r)$, GeO$_2$ doping profile, and the intensity profiles $I_P(r)$ and $I_S(r)$ can be found from the guided modes at the pump and signal wavelengths, respectively.

Bromage et al. [38] reported $C_{SiSi}(\nu)$ and $C_{GeSi}(\nu)$ by comparing germanosilicate optical fibers with various refractive index profiles, and their spectra are shown in Figure 6.10.

The peak values are 4.20 m^2/(W km) for C_{SiSi} at 445 cm^{-1} and 14.84 m^2/(W km) for C_{GeSi} at 430 cm^{-1}. As they are proportional to the Raman cross-section, they also scale with pump wavelength as $1/\lambda_P$.

For a given fiber index profile and pump wavelength, the Raman gain spectra can be calculated for any germanosilicate fiber with arbitrary refractive index profile. The simulation results for conventional telecom fibers are shown in Figure 6.11.

Note that the peak Raman gain coefficient ranges from 0.4 to 6.1 (W km)$^{-1}$, which is mainly attributed to GeO$_2$ concentration and mode profile. The recently developed Raman fiber with a high GeO$_2$ concentration and good overlap between the pump and signal modes shows the peak gain over 6 (W km)$^{-1}$, which is about 15 times larger than conventional SMFs. The simulations shown in solid lines are in very good agreement with the experimental results.[38].

The correlation between the core diameter and the GeO$_2$ concentration in an α-profile optical fiber has also been proposed using the formula in equation 6.28, and the results are shown in Figure 6.12.

For a simple graded-index α-profile, the relative index difference is given by

$$\Delta(r) = \Delta(0)\left[1 - \left(\frac{2r}{d}\right)^{\alpha}\right] \tag{6.29}$$

Calculated peak Raman gain coefficients for a range of peak GeO$_2$ concentration, corresponding to $\Delta(0)$, and core diameters (d) are shown as the numbers on the

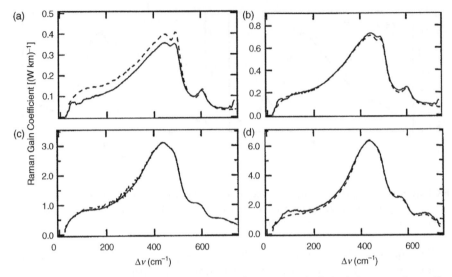

FIGURE 6.11 Comparison of predicted (solid) and measured (dashed) Raman gain coefficients for (a) pure silica core fiber, (b) TrueWave reduced slope transmission fiber, (c) dispersion-compensating fiber, and (d) Raman gain fiber [38].

corresponding contours in Figure 6.12. Here $\alpha = 4$ and $\lambda_P = 1455$ nm were assumed. As the peak GeO_2 concentration increases, the gain coefficient increases, which is consistent with Figure 6.9b. It is noteworthy that for a given GeO_2 concentration, there is a core diameter that gives the maximum gain. This core diameter corresponds to the index profile that has the largest value of the overlap integral between pump and signal intensities.

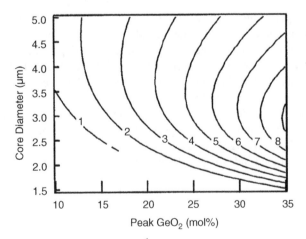

FIGURE 6.12 Peak Raman gain $((W\ km)^{-1})$ as a function of core diameter and peak GeO_2 concentration [32].

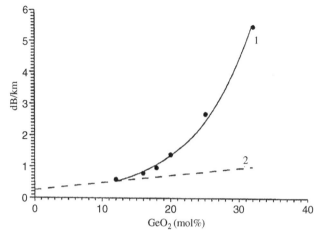

FIGURE 6.13 Optical loss in high Ge-doped fibers at 1.55 μm. Dashed line represents Rayleigh scattering loss estimation and solid circles are experimental data [40].

In actual optical fiber fabrication, however, high GeO_2 concentration in silica optical fiber induces an excess loss, which is much higher than Rayleigh scattering loss. Dianov [40] reported the experimental measurements of high GeO_2 fibers as a limiting factor to design Raman fibers.

Figure 6.13 shows an experimental dependence of optical losses of germano-silicate fibers fabricated by a standard modified chemical vapor deposition (MCVD) technique, on GeO_2 (mol%) at the measurement wavelength of 1.55 μm (curve 1). The dotted line (curve 2) is the loss estimated by Rayleigh scattering. For GeO_2 content over 15 mol% the loss increases well over the Rayleigh scattering estimates. There are several sources of excessive optical losses in high GeO_2-doped fibers: generation of various defects with high optical loss such as oxygen vacant Ge defects and waveguide imperfections such as central dip in the core [41].

Flow control in the final collapse phase of MCVD can avoid the central dips. For example, the final layers can be chemically etched by flowing SiF_4 or fluorine-containing gases as described in Chapter 3. Additional flow of $GeCl_4$ during the final collapse pass and sealing pass could also reduce the central dip. Graded-index technique would be also helpful to reduce the mismatch between the silica cladding and GeO_2-doped core layers.

Typical attenuation spectrum of a low-loss high GeO_2 concentration silica optical fiber fabricated by MCVD is shown in Figure 6.14, which has been used for 1.3 μm Raman amplification experiments [42]. The fiber had 20 mol% GeO_2 in the core, LP_{11} mode cutoff was 0.97 μm, and the loss at 1.31 μm was 0.90 dB/km.

Alternative to GeO_2, P_2O_5 has been widely used in silica optical fiber fabrication since it has similar refractive index–raising efficiency as GeO_2 with a relative low loss. Furthermore, P_2O_5 can significantly reduce the glass processing temperature, which has been widely used in multimode fiber fabrication using MCVD.

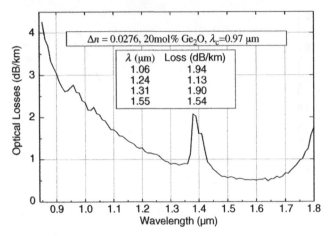

FIGURE 6.14   Attenuation spectrum of high GeO_2 silica fiber fabricated by MCVD [41].

As reported by Galeener et al. [36], P_2O_5 glass shows unique potential in Raman scattering as shown in Figure 6.8 and Table 6.4. P_2O_5 glass shows prominent peaks at $640\,cm^{-1}$ (19.2 THz) and $1390\,cm^{-1}$ (41.7 THz), whose intensities are 5.7 and 3.5 times larger than SiO_2 peak at $420\,cm^{-1}$ (13.2 THz). Raman characteristics of P_2O_5–SiO_2 fibers have been reported by Suzuki et al. [43,44], where the unique Raman gain peak near $1320\,cm^{-1}$ (40 THz) has been experimentally confirmed [45].

Similar to GeO_2–SiO_2 glass shown in Figure 6.9a, the peak Raman gain in P_2O_5–SiO_2 glass optical fiber showed a linear increase with P_2O_5 concentration as in Figure 6.15a [44]. Normalized Raman gain spectra of GeO_2–SiO_2 and P_2O_5–SiO_2 optical fibers are shown in Figure 6.15b [46]. GeO_2–SiO_2 glass optical fiber such as DCF exhibits a single spectral structure with the major peak near 13.2 THz with the peak value ~ 3.0 $(W\,km)^{-1}$ and the attenuation of 0.32 dB/km at 1550 nm.

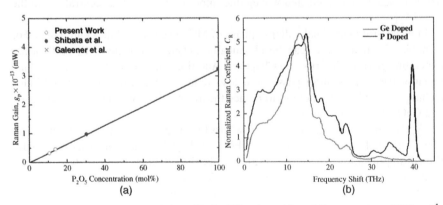

FIGURE 6.15   Raman characteristics of P_2O_5–SiO_2 glass: (a) peak Raman gain at $1320\,cm^{-1}$ (40 THz) versus P_2O_5 concentration in silica glass [35,36,44]. (b) Normalized Raman gain spectra for GeO_2–SiO_2 and P_2O_5–SiO_2 optical fibers [46].

To date P_2O_5–SiO_2 glass optical fibers have shown higher losses, about $\sim$1 dB/km at 1550 nm, than GeO_2-doped fibers, but they have a more useful spectrum, with a double spectral peak structure: a broad one covering 13–14 THz and the other peak at 40 THz. For $\sim$10 mol% P_2O_5-doped silica fiber, the peak Raman gains for the broadband and the peak are $\sim$4.5 and $\sim$3.5 (W km)$^{-1}$, respectively [41]. Especially the Raman peak at 40 THz, the longest Raman frequency shift, ν_R, obtainable in silica-doped optical fiber, has drawn intense attention because it enables the use of wide separation between the pump ν_P and signal ν_S ($\nu_P - \nu_S = \nu_R$). Especially fiber lasers can take advantage using fewer Stokes shifts. For example, a 1480 nm Raman fiber laser can be made of a GeO_2-doped fiber and pumped at 1117 nm with five Raman shifts [47], yet the same can be made of a P_2O_5-doped fiber and pumped at 1065 nm with two Raman shifts [48]. Large Stokes shift in P_2O_5–SiO_2 optical fiber can significantly relieve the cavity requirements in fiber lasers, which will be discussed in the following section.

6.6   BRIEF REVIEW ON RAMAN AMPLIFIERS AND LASERS

Optical gain from nonlinear scattering such as Raman (and Brillouin) process gives very flexible degree of freedom to choose the optical amplification window. As long as the signal frequency and the pump frequency are separated by the characteristic Raman frequency shift ($\nu_P - \nu_S = \nu_R$), the pump photon energy is efficiently transferred to signal to result in optical gain.

Despite manifest advantages in Raman gain, it is only recent that practical optical fiber amplifiers could be developed for optical telecommunications. The main technological breakthroughs have been made possible by development of high power laser diodes operating near 1400 nm, which provide optical gain in C and L bands of optical communication windows. Low-loss highly nonlinear optical fiber fabrication techniques and fiber grating device technologies also played important roles to revive Raman scattering in fiber optic communications.

Two general categories of Raman amplifiers exist: distributed and discrete Raman amplifiers (DRAs). They improve the noise figure and reduce the nonlinear penalty of the in-line amplifiers, allowing for longer amplifier spans, higher bit rates, closer channel spacing, and operation near the zero-dispersion wavelength. There have been numerous research efforts in Raman amplification in optical fibers and recently published books [49–51] well describe technical issues in amplifier and laser applications. There has been already an international standard issued for fiber Raman amplifiers, ITU-T G.665 (Generic characteristics of Raman amplifiers and Raman amplifier subsystems) [7], in order to consistently characterize their operations in optical communications.

In this chapter, we will focus only on the basic principles of how the Raman gain discussed in the previous sections is applied in optical communication as an amplifier and laser.

Consider the simplest situation where a single pump laser is launched into an optical fiber in order to amplify a signal. Let us assume that the pump and signal

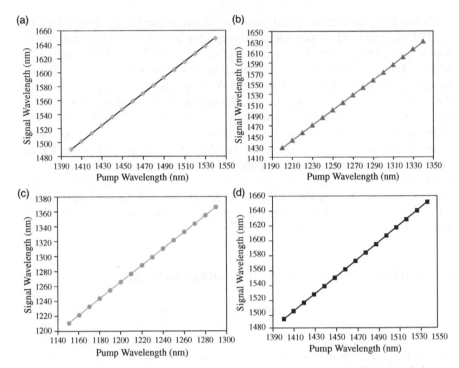

FIGURE 6.16 The pump photon wavelength versus the first Stokes-shifted signal photon wavelength for (a) GeO_2–SiO_2 glass fiber with $\nu_R = 13$ THz, (b) P_2O_5–SiO_2 glass fiber with $\nu_R = 40$ THz, (c) SMF with $\nu_R = 13.2$ THz signal in O band, and (d) SMF with $\nu_R = 13.2$ THz signal in C–L band.

photons are at the wavelength of λ_p and λ_s, respectively. In most of optical communication applications the photons are described in the wavelength scale of either nanometers or micrometers. The relationship between the pump and signal in Raman scattering process, $\nu_P - \nu_S = \nu_R$, should be converted in the wavelength scale in the optical communication windows.

The conversion in wavelength unit is summarized in Figure 6.16 for GeO_2–SiO_2 and P_2O_5–SiO_2 glass fibers, and SMF assuming only the first-order Stokes shift, $\nu_P - \nu_S = \nu_R$. Due to difference in the Raman frequency shifts, $\nu_R = 13$ THz for GeO_2–SiO_2, 40 THz for P_2O_5–SiO_2 glass fiber, and 13.2 THz in SMF, the wavelength ranges for pump–signal are very different depending on the types of fibers. In GeO_2–SiO_2 glass fibers, so-called 14xx nm pumps are required for signal amplification in C and L bands, 1520–1620 nm as indicated in Figure 6.16a. In P_2O_5–SiO_2 glass fibers 12xx nm pumps are required instead (see Figure 6.16b). For the conventional SMFs, the pump wavelength ranges for O band (Figure 6.16c) and C–L band (Figure 6.16d) are 1220–1260 and 1440–1490 nm, respectively. Therefore, composition of fiber, pump wavelength, and signal wavelength should be considered all together in Raman amplifier and laser design.

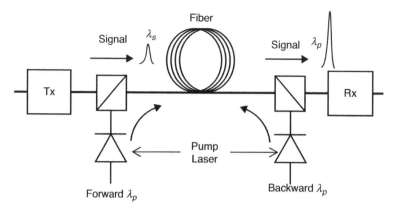

FIGURE 6.17   Schematic diagram of fiber Raman amplifier.

Schematic diagram of most common optical fiber Raman amplifier is shown in Figure 6.17.

Note that two directions of pump can be adopted with respect to the signal propagation direction.

Raman amplification process in optical fiber is governed by the following two sets of two coupled equations. We will assume simple case for the single pump and single signal channel, for basic understanding:

$$\frac{dP_s}{dz} = g_R P_p P_s - \alpha_s P_s \tag{6.30}$$

$$\pm \frac{dP_p}{dz} = -\frac{\lambda_F}{\lambda_p} g_R P_p P_s - \alpha_p P_p \tag{6.31}$$

Here α_s and α_p are fiber attenuation at the first Stokes-shifted signal (λ_s) and pump wavelength (λ_p), respectively. The $\pm$ sign in the second equation indicates the direction of pumping, $+$ sign for forward and $-$ sign for backward. Note that g_R is the Raman gain coefficient in (W km)$^{-1}$ assuming $\nu_P - \nu_S = \nu_R$ as discussed in previous sections.

The first and second terms on the right-hand side of equation 6.30 represent the Raman gain at the first Stokes shift and fiber attenuation at the signal. In equation 6.31, the first term on the right-hand side represents pump depletion by the Raman process. Note that the Raman gain is once again scaled to $1/\lambda_P$ as discussed in the previous sections.

For a first-order approximation we assume that the pump depletion is negligible in the forward pumping case; then we can set $g_R \sim 0$ in equation 6.31 to have solution for $P_p(z)$ as follows:

$$P_p(z) = P_0 \exp(-\alpha_p z) \tag{6.32}$$

where P_0 is the input pump power at $z = 0$.

Equation 6.32 can be substituted into equation 6.30 and then integrated over the fiber length L to obtain the signal power as follows:

$$P_s(L) = P_s(0)\exp(g_R P_0 L_{eff} - \alpha_s L) \tag{6.33}$$

Here L_{eff} is the effective length as defined in equation 6.24, at the pump wavelength λ_P:

$$L_{eff} = \int_0^L e^{-\alpha(\lambda_P)z}\, dz = \frac{1 - e^{-\alpha(\lambda_P)L}}{\alpha(\lambda_P)} \tag{6.34}$$

We can further introduce "on–off gain," the ratio of signal power with and without pump laser, as follows:

$$G_{\text{on–off}} = \frac{P_s(L) \text{ with pump on}}{P_s(L) \text{ with pump off}} = \exp(g_R P_0 L_{eff}) \tag{6.35}$$

The Raman on–off gain has been measured in a good agreement with the above equation as shown in Figure 6.18 [52], where the amplification factor $G_{\text{on–off}}$ increases exponentially with P_0. When the pump power exceeds a certain level, 1 W in Figure 6.18, the gain no longer follows the exponential growth and shows pump depletion. In this case the coupled differential equations 6.30 and 6.31 should be solved numerically.

As long as the frequency difference between the pump and the signal, $\nu_P - \nu_S$, lies within the Raman gain spectrum g_R shown as in Figures 5.20 and 5.24b, the signal will experience the optical gain as it propagates along the optical fiber.

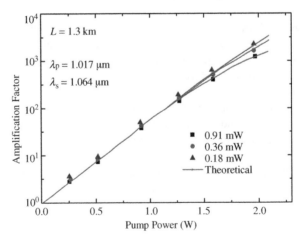

FIGURE 6.18 Variation of Raman on–off gain as a function of pump power. Different symbols correspond to three different levels of input signal power [52].

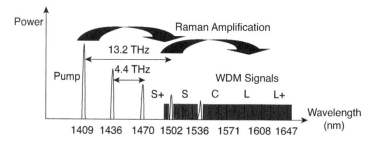

FIGURE 6.19 Schematic diagram for WDM pumping in broadband Raman amplifiers. Here single-mode fiber (SMF) is assumed and the pump and signal wavelength allocations are shown [53].

Raman amplification for single pump–single signal case can be further expanded to so-called WDM pumping for broadband flat optical amplifier for multiple signal channels (Figure 6.19). The idea of "WDM" pumping is to use a set of pumps operating at different wavelengths combined through couplers into a single fiber to realize a composite Raman gain.

In comparison to simple single pump–single signal case, the composite Raman gain created by different wavelengths of pumps requires complicated numerical solution for a set of coupled nonlinear equations for all the pumps and signals, where pump-to-pump, pump-to-signal, and signal-to-signal Raman interactions should be included in addition to pump depletions due to Raman energy transfer, Rayleigh backscattering, fiber loss, spontaneous emission noise, and thermal noise [54,55].

In WDM pumping scheme, we need to determine the number of pumps and their wavelengths by which a desired composite Raman gain can be obtained. For each pump individual Raman gain spectrum shifted by the respective wavelength differences as shown in Figure 6.16 should be added in the logarithmic scale with adequate weighting factors. The weighting factors of individual pumps represent the launched pump power in the transmission optical fiber.

For the given Raman gain spectra g_R of a transmission optical fiber, as in Figures 5.20 and 5.24, the composite gain spectrum of WDM pumped Raman amplifiers can be determined by adjusting the power and spectral position of pumps and their respective gain profiles. The key design parameters and required specification in a WDM pumped Raman amplifier are listed in Figure 6.20 along with simulated composite gain spectrum in C–L bands using four pumps [54].

Flat and broadband optical amplification in various transmission fibers have been experimentally reported, whose bandwidth extends over 100 nm covering both C and L bands [56–58]. Figure 6.21 shows comparison of simulation and experimental results for report of Namiki and Emori [58], which shows a good agreement.

Optical gain provided by stimulated Raman scattering process in optical fiber has been successfully applied to fiber Raman lasers and Raman amplifiers. Fiber Raman lasers (FRL) are usually composed of so-called "cascaded cavity" structure [46] as shown in Figure 6.22. The gain media are either GeO_2–SiO_2 glass fiber [59–61] or

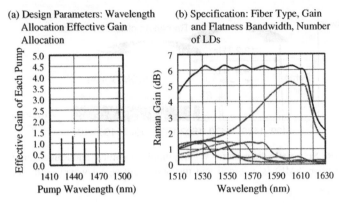

FIGURE 6.20 Key parameters and design specifications for WDM pumped Raman amplifier based on SMF: (a) the backward pump wavelengths and input powers; (b) the total Raman gain of 6 dB over C and L bands using four pumps [54].

P_2O_5–SiO_2 glass fiber [45,48,62,63] and the output wavelengths of FRL are mainly in two bands: one near 1240 nm and the other near 1480 nm. Lasers near 1240 nm have been successfully used to provide a high gain in O band near 1.3 µm [59,64]. The 1480 nm FRL has been applied in preamplifier for C–L-band WDM communications [62]. Raman amplifiers have been also successfully demonstrated in S-band [65].

The cavities are closed by fiber Bragg gratings whose reflections are centered near the first Stokes Raman-shifted frequency in a nested manner. For example, in the case of GeO_2–SiO_2 glass optical fibers the Raman frequency shift ν_R is ~13 THz and the corresponding Stokes-shifted wavelengths in the cascaded Raman fiber laser cavities are summarized in Table 6.5. The Raman frequency shift in P_2O_5–SiO_2 glass fiber is about 40 THz, three times larger than that of GeO_2–SiO_2 glass optical fibers, and the

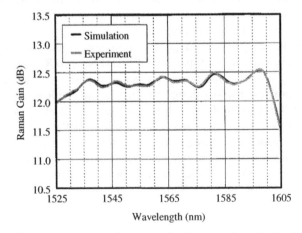

FIGURE 6.21 Comparison between the composite Raman gains calculated with weighting factors and experimentally realized by adjusting pump powers. The fiber used was 25 km DSF [58].

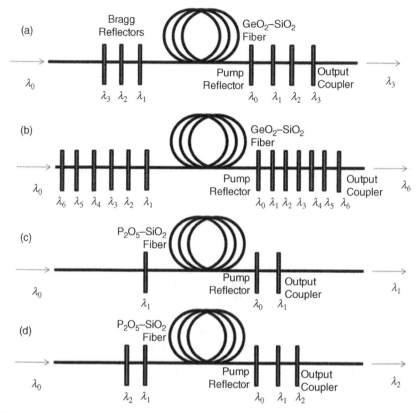

FIGURE 6.22 Schematic diagram for cascaded fiber Raman laser cavity structures. Bragg reflectors are made of fiber Bragg grating inscribed on optical fibers. (a) GeO_2–SiO_2 fiber–based $\sim$1240 nm laser, (b) GeO_2–SiO_2 fiber–based $\sim$1480 nm laser, (c) P_2O_5–SiO_2 fiber–based $\sim$1240 nm laser, and (d) P_2O_5–SiO_2 fiber–based $\sim$1480 nm laser.

TABLE 6.5 nth-Order Stokes-Shifted Wavelength in GeO_2–SiO_2 Fiber Cascaded Raman Laser Cavity

GeO_2–SiO_2 Fiber	Pump λ (nm)	Pump ν (THz)	Raman Shift (THz)	Raman Shift (nm)	Absorption (km^{-1})
0th-Order Stokes λ_0	1064	282	269	1117	0.143
1st-Order Stokes λ_1	1117	269	256	1174	0.118
2nd-Order Stokes λ_2	1174	256	243	1237	0.0969
3rd-Order Stokes λ_3	1237	243	230	1307	0.0852
4th-Order Stokes λ_4	1308	229	216	1387	0.0814
5th-Order Stokes λ_5	1387	216	203	1476	0.194
6th-Order Stokes λ_6	1476	203			0.0436

TABLE 6.6 *n*th-Order Stokes-Shifted Wavelength in P_2O_5–SiO_2 Fiber Cascaded Raman Laser Cavity

P_2O_5–SiO_2 Fiber	Pump λ (nm)	Pump ν (THz)	Raman Shift (THz)	Raman Shift (nm)	Absorption (km^{-1})
0th-Order Stokes λ_0	1064	282	242.0	1240	∼0.31
1st-Order Stokes λ_1	1240	242	201.9	1486	∼0.20
2nd-Order Stokes λ_2	1486	202			∼0.21

corresponding Stokes-shifted wavelengths are summarized in Table 6.6. Except the output coupler, all the Bragg reflectors have high reflection at the designated wavelengths, λ_i. The main principle of the cascaded Raman laser cavity is the laser at λ_i formed by Bragg reflector pair at the wavelength serves as the pump for Stokes-shifted λ_{i+1}.

Typical Stokes-shifted wavelengths λ_i are shown in Tables 6.5 and 6.6 for GeO_2–SiO_2 and P_2O_5–SiO_2 fibers, respectively.

In the schematic of a typical FRL shown in Figure 6.22, the cavity is made of a few tens to hundreds of meters of optical fiber in the middle and it is nested by pairs of fiber Bragg grating (FBG) reflectors. Grubb et al. [60] reported 1480 nm FRL output using 1117 nm pump light. The 1117 nm pump enters the cavity and is Raman converted to the Stokes shift at 1175 nm, selected by a pair of FBG reflectors on both sides of Raman fiber. The 1175 nm radiation is in turn converted into a longer wavelength at the next Stokes order and so on, until the last Stokes order at 1480 nm. Note that the leftover pump is reflected by another high reflectivity at the pump wavelength, pump reflector in Figure 6.22.

The cascaded FRL output spectra are shown for GeO_2–SiO_2 and P_2O_5–SiO_2 fibers in Figure 6.23. It is noted that the final output has intensity higher than the intermediate Stokes shifts by more than 20 dB; the lasing lines at these Stokes shifts do carry a certain portion of energy. Detailed theoretical analyses have been reported

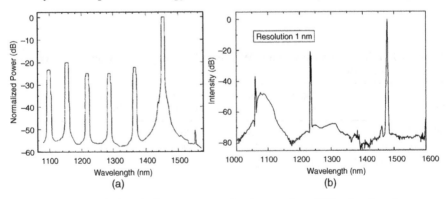

FIGURE 6.23 Typical optical spectrum at the output of an FRL, showing the pump, intermediate Stokes orders, and the final output order: (a) 1480 nm GeO_2–SiO_2 FRL pumped at 1117 nm [60] and (b) 1480 nm P_2O_5–SiO_2 FRL pumped at 1060 nm [63].

recently to fully understand the lasing dynamics among the intermediate Stokes shifts, threshold power, optimal fiber length, and output coupling ratio [66–69].

Raman scattering in silica optical has shown high potential to turn passive optical fiber into highly active gain medium providing efficient gain at arbitrary wavelength within optical communication bands. The main material for Raman gain has been limited to germanosilicate and phosphorosilicate glasses. New material investigation for high Raman gain coefficient, wider Raman shift, and low excess could further expand Raman scattering applications. Modification of waveguide structures such as photonic crystal fibers also demonstrated strong potential to enhance Raman efficiency [70,71] and includes gaseous material [72]. Combination of these two research directions would give a new avenue of Raman scattering devices.

6.7 CONTROL OF BRILLOUIN SCATTERING IN SILICA OPTICAL FIBERS

Among various nonlinearity effects in optical fibers, Brillouin scattering has the lowest power threshold and it is dominantly in the backward direction to the propagating signal. These extraordinary characteristics of Brillouin scattering could cause serious detrimental effects in most of optical signal transmission and fiber lasers, which gave persistent motivations in various fiber designs to suppress the effects. On the other hand, Brillouin scattering shows the most sensitiveness against exterior physical perturbations such as temperature and strain among fiber optic nonlinearities. This sensitiveness has made fiber optic Brillouin sensors very successful in distributed optical sensing industries and markets.

In this section, the general characteristics of Brillouin scattering in optical fiber will be reviewed and then detailed discussions will be followed to address parametric investigations in silica optical fibers to control the main elements in Brillouin scattering process such as Brillouin gain, Brillouin frequency shift, and bandwidth.

Stimulated Brillouin scattering involves three distinctive radiations as schematically shown in Figure 6.24.

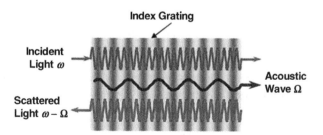

FIGURE 6.24 Schematic diagram for interaction between the acoustic wave and incident light, Brillouin scattered light [73].

The incident pump light at the angular frequency of ω generates an acoustic wave at frequency Ω by the electrostriction process in silica optical fiber [8], which in turn causes a periodic modulation of refractive index in the core, a traveling index grating with the acoustic velocity. This index grating scatters the pump light through Bragg diffraction and the scattered light frequency is downshifted in frequency because of the Doppler effects associated with the moving grating at the acoustic velocity.

SBS is characterized by its spectral distribution in the optical frequency, which is related to decay of acoustic waves in silica optical fiber in the cylindrical shape. The exponential decay of the acoustic waves results in a Lorentzian distribution [74] for the Brillouin gain, $g_B(\nu)$, which is given as follows:

$$g_B(\nu) = g_{B0} \frac{(\Delta\nu_B)^2}{(\nu - \nu_B)^2 + (\Delta\nu_B)^2} \tag{6.36}$$

The peak gain is given by [8]

$$g_{B0} = \frac{2\pi n^7 p_{12}^2}{c\lambda_P^2 \rho_0 V_a \Delta\nu_B} \tag{6.37}$$

where p_{12} is the longitudinal elasto-optic coefficient, ρ_0 is the density, λ_P is the pump wavelength, c is the speed of light in vacuum, V_a is the acoustic velocity, and $\Delta\nu_B$ is the width of Brillouin gain spectrum.

Since the scattered light experiences a Doppler frequency shift, the Brillouin shift ν_B depends on the acoustic velocity V_a and is given by

$$\nu_B = \frac{2n_{\text{eff}} V_a}{\lambda} \tag{6.38}$$

Note that the Brillouin frequency ν_B scales with $1/\lambda$ such that ν_B does change with the pump light wavelength λ_P. Most of the contribution to the acoustic velocity in optical fiber geometry comes from the longitudinal acoustic velocity, V_L, which depends on the density and Young's modulus, E, of silica optical fiber [75]:

$$V_a \approx V_L = \sqrt{\frac{E}{\rho_0}} \tag{6.39}$$

The Brillouin gain spectra width is dependent on the density, viscosity, and the wavelength [76] as follows:

$$\Delta\nu_B = \frac{8\pi\eta}{\rho_0\lambda^2} \tag{6.40}$$

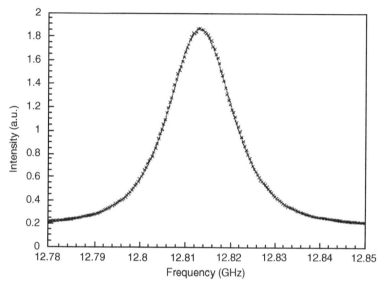

FIGURE 6.25 Typical Brillouin gain spectrum of a single-mode fiber measured at $\lambda_P = 1320\,\text{nm}$ [75].

Room temperature, strain-free Brillouin scattering parameters in conventional single-mode fibers at the pump wavelength of 1550 nm are listed as follows [77]:

$$g_{B0} \simeq 5 \times 10^{-11}\,\text{m/W}$$

$$\nu_B \simeq 11\,\text{GHz}$$

$$\Delta\nu_B \simeq 35\,\text{MHz}$$

Niklès et al. [75] reported SBS gain spectra in conventional single-mode fibers and the typical $g_B(\nu)$ for GeO_2–SiO_2 glass core single-mode fiber is shown in Figure 6.25.

The experimental data shown in crosses in Figure 6.25 perfectly fit the hyperbolic cosine of a Lorentzian function in equation 6.36. Note that the fitted parameters are $\Delta\nu_B = 35.8\,\text{MHz}$ and $\nu_B = 12.8\,\text{GHz}$ for pump light at 1320 nm, which are consistent with values shown above for $\lambda_P = 1550\,\text{nm}$.

Similar to the observation that Raman scattering spectra depend on the glass composition, as shown in Figure 6.8, Brillouin scattering spectral distribution does vary with dopant types and their concentration in silica glass. Tkach et al. [77] first reported that the Brillouin frequency shift ν_B decreases linearly with GeO_2 concentration in the silica glass core. The linear slope in the ν_B shift was estimated as 89 MHz/wt% (=154 MHz/mol%) measured at 1525 nm for GeO_2-doped silica core fibers.

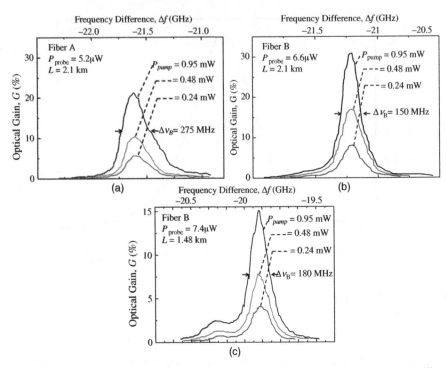

FIGURE 6.26 Brillouin-gain spectra for (a) pure silica fiber, (b) 3.2 mol% GeO_2-doped silica fiber, and (c) 9.9 mol% P_2O_5-doped silica fiber, measured at $\lambda_P = 828$ nm [78].

Shibata et al. [78] followed reporting Brillouin gain spectra for the following three types of optical fibers:

type A: SiO_2 core/F–SiO_2 cladding
type B: GeO_2–SiO_2 core/SiO_2 cladding
type C: P_2O_5–SiO_2 core/P_2O_5–F–SiO_2 cladding

The Brillouin gain spectra measured at $\lambda_P = 828$ nm are shown in Figure 6.26. The Brillouin shifts, ν_B, for the peak gain were found to be 21.6, 21.2, and 20.0 GHz for fibers A, B, and C, respectively. It is noted that the ν_B significantly shifts toward a lower frequency by adding GeO_2 and P_2O_5 in the silica glass core, which is consistent with report of Tkach et al. [77].

Assuming a linear relation between the dopant concentration in the core and the ν_B shift, Shibata et al. [78] reported the ν_B shift coefficients of 125 and 162 MHz/mol% for GeO_2- and P_2O_5-doped silica fibers, respectively.

Brillouin gain measurements for GeO_2- and F-doped fibers were carried out at 1550 nm by Shibata et al. [79]. The linear ν_B shift coefficient was evaluated to be

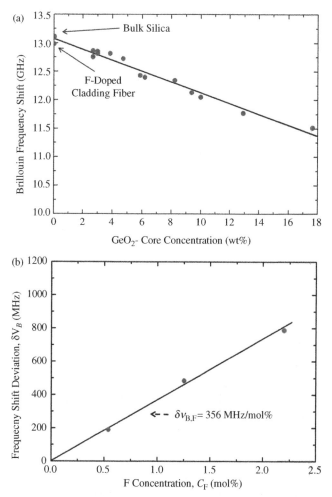

FIGURE 6.27 The Brillouin frequency shift as a function of (a) GeO_2 concentration (wt%) ($\lambda_P = 1320$ nm) [75] and (b) F concentration (mol%) ($\lambda_P = 1550$ nm) [79].

107 MHz/mol% for GeO_2 doping and 356 MHz/mol% for F doping. Note that F doping in the silica core can result in ν_B shift about three times larger than GeO_2 doping. For GeO_2-doped silica core fibers, Niklès et al. [75] reported that ν_B shifts range from 11.5 to 13 GHz at $\lambda_P = 1.32$ μm with a linear slope of 94 MHz/wt%, or equivalent to 162 MHz/mol% for GeO_2 concentration in the core. The results are shown in Figure 6.27.

The location of Brillouin shift, ν_B, is that where the Brillouin gain has its peak, and control of ν_B is a key parameter for Brillouin amplifier, laser, and sensor applications. The dopant concentration in the silica core was found to show a very linear relationship with downward shift of ν_B, which has been consistently confirmed by independent experimental reports [75–79]. Within the conventional range of doping

TABLE 6.7   Brillouin Shift Frequency Coefficient, $\delta\nu_B$, for Various Dopant Types in Silica Optical Fibers

	Dopant						
	$\delta\nu_B$ (MHz/mol%)			$\delta\nu_B$ (MHz/wt%)			
λ_p (nm)	GeO$_2$	P$_2$O$_5$	F	GeO$_2$	P$_2$O$_5$	F	Reference
1525	153			89.0			[77]
828	125	162		72.3	69.5		[78]
1550	107		356	61.8		1113	[79]
1320	162			94			[75]
1550	78		277	45		866	[80]

concentration, we can assume a linear variation of ν_B with the dopant concentration as follows:

$$\nu_B = \nu_B^{SiO_2} - \delta\nu_B^{GeO_2} C_{GeO_2}$$
$$\nu_B = \nu_B^{SiO_2} - \delta\nu_B^{P_2O_5} C_{P_2O_5} \qquad (6.41)$$
$$\nu_B = \nu_B^{SiO_2} - \delta\nu_B^{F} C_F$$

The Brillouin shift frequency coefficient, $\delta\nu_B$, defined in equation 6.41 for various dopant types is summarized in both weight percent (MHz/wt%) and mole percent (MHz/mol%) in Table 6.7. Except the recent report by Shiraki et al. [80], the coefficients $\delta\nu_B^{GeO_2}$ for GeO$_2$ doping are found to be in a good agreement with one another. The discrepancy could be attributed to other parameters such as detailed refractive index profile, its corresponding acoustic index profile, and drawing tension, which will be discussed in the latter part of this section.

Another notable feature observed in the experimental reports is that Brillouin gain bandwidth, $\Delta\nu_B$, changes as a function of GeO$_2$ concentration [75]. It would be highly desirable to increase the gain bandwidth in tunable fiber laser applications, wider amplification band, and multiple sensing capability.

A change of $\Delta\nu_B$ with the linear slope of 1.4 MHz/wt%, equivalent to 2.4 MHz/mol %, was found for GeO$_2$ core concentration (Figure 6.28) [75]. These measurements provide with a high confidence level of 35 MHz value for the Brillouin FWHM linewidth of standard telecommunication fibers (3 wt% GeO$_2$ core doped) at 1320 nm. However, the bandwidth can be also highly affected by nonuniformity in the core refractive index profile, core–cladding interface imperfections, and core diameter. These factors tend to be affected by high GeO$_2$ concentration. It is not fully investigated whether the Brillouin gain bandwidth can be expanded by addition of dopant in a systematic manner.

In contrast to Raman scattering gain increase with GeO$_2$ concentration as shown in Figure 6.9, Brillouin gain in conventional fibers did not show any conspicuous and

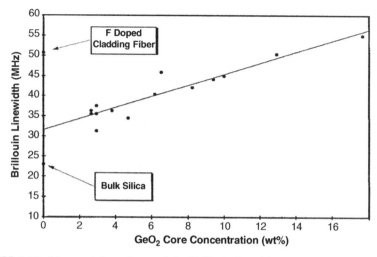

FIGURE 6.28 Measured dependence of the Brillouin linewidth $\Delta\nu_B$ on the GeO_2 concentration in wt% [75].

direct relationship with GeO_2 concentration in previous experimental reports, despite slight increase of g_B for higher GeO_2 concentration [75].

Spontaneous and stimulated Brillouin scattering spectra were measured for modern telecom optical fibers by Yeniay et al. [81]. They reported Brillouin spectra for OFS AllWave (AW) fiber, OFS TrueWave (TW) fiber, DCF, and high Ge-doped (HG) fibers. AW is a conventional SMF with a reduced OH peak attenuation near 1380 nm. TW is a class of nonzero dispersion-shifted fiber. The spontaneous Brillouin spectra measured at $\lambda_P = 1552$ nm are shown in Figure 6.29 and their characteristics are summarized in Table 6.6.

The Brillouin frequency, ν_B, shifted toward a lower frequency as the GeO_2 concentration increases from AW to TW, DCF, and HG. This is consistent with previous reports summarized in Table 6.8.

In Figure 6.29, spontaneous Brillouin scattering spectra show multiple peaks that are related to higher order acoustic modes guided along the optical fibers. As the pump power increases, the Brillouin scattering spectra change as in Figure 6.30. For input powers higher than the SBS threshold, the main resonance peak keeps increasing with the input power, while the other small resonance peaks start to decrease in intensity and disappear eventually. In spontaneous Brillouin scattering, several phonon frequencies with finite lifetimes are coupled to pump photon in the core. However, as the process evolves to a stimulated scattering above the SBS threshold, only one phonon frequency survives with a smaller damping and higher intensity to show unique Lorentzian spectral distribution as in equation 6.36.

As shown in Table 6.3, the threshold for the onset of stimulated Brillouin scattering is estimated by

$$P_B^{th} \simeq 21 \frac{A_{eff}}{g_B L_{eff}} \tag{6.42}$$

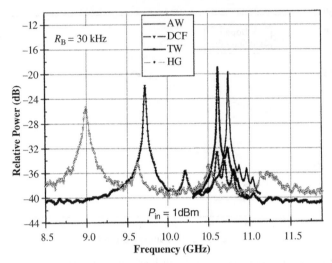

FIGURE 6.29 Spontaneous Brillouin gain spectra of TW, AW, DCF, and HG fibers with a DFB laser at 1552 nm. AW, AllWave fiber; TW, TrueWave fiber; DCF, dispersion-compensating fiber; HG, high Ge-doped fiber [81].

Despite similar peak gain g_B, DCF and HG fibers showed higher threshold than AW and TW by several factors. This is attributed to nonuniformity in the core diameter and the refractive index profile along the fiber in the case of high GeO_2 contents with complex waveguide structures.

Koyamada et al. [82] reported theoretical simulation to estimate Brillouin scattering gain spectra by directly calculating gain coefficients for the interaction between the guided LP_{01} optical mode and the ith longitudinal acoustic mode:

$$g_i = p_{12}\omega_P\beta_i \int_0^\infty X^2(r)Y_i(r)r\,dr \qquad (6.43)$$

Here p_{12} is the longitudinal elasto-optic coefficient, ω_P is the angular frequency of the LP_{01} mode of the pump light, and β_i is the propagation constant for the ith acoustic mode. $X(r)$ and $Y_i(r)$ are the amplitude function over the optical fiber cross-section for the LP_{01} mode optical electric field and ith acoustic field, respectively, and they can be

TABLE 6.8   Brillouin Characteristics of Conventional Telecom Fibers

Fiber Type	GeO_2 (mol%)	L_{eff} (km)	Attenuation (dB/km)	P_{th} (dBm)	ν_B (GHz)
AW	3.6	15.3	0.186	5.2	10.81
TW	5.7	16.1	0.180	4.8	10.85
DCF	18	4.4	0.64	8.2	9.75
HG	20	0.3	4.5	19	8.9

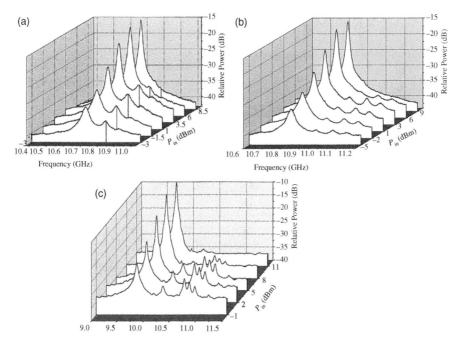

FIGURE 6.30 Brillouin gain spectra evolution from spontaneous to stimulated Brillouin scattering for (a) TW, (b) AW, and (c) DCF fibers with a DFB laser for various input power levels at 1552 nm. AW, AllWave fiber; TW, TrueWave fiber; DCF, dispersion-compensating fiber; HG, high Ge-doped fiber [81].

found by solving corresponding wave equations with the core/cladding boundary conditions.

The gain spectrum $g_B(\nu)$ can be estimated by summing all the Lorentzian contributions for the ith acoustic mode as follows:

$$g_B(\nu) = \sum_i g_i^2 \frac{\Gamma_i}{2\pi^2\{(\nu - \nu_p + f_i)^2 + (\Gamma_i/2\pi)^2\}} \tag{6.44}$$

Here Γ_i and f_i are the lifetime and acoustic frequency of the ith acoustic mode, respectively.

In previous experimental measurements as in Refs [75,83,84], linear relationships between the dopant concentration (C_{GeO_2}, C_F) in the silica core and the following parameters can be assumed at the wavelength of 1.55 μm:

$$n = 1.458(1 + 1.0 \times 10^{-3}C_{GeO_2} - 3.3 \times 10^{-3}C_F) \tag{6.45}$$

$$V_L = 5944(1 - 7.2 \times 10^{-3}C_{GeO_2} - 2.7 \times 10^{-2}C_F) \quad [\text{m/s}] \tag{6.46}$$

$$V_S = 3749(1 - 6.4 \times 10^{-3}C_{GeO_2} - 2.7 \times 10^{-2}C_F) \quad [\text{m/s}] \tag{6.47}$$

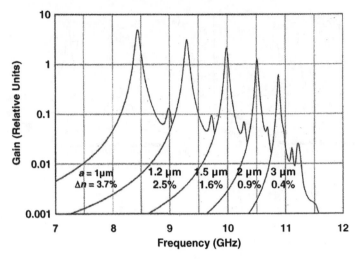

FIGURE 6.31 Simulated Brillouin gain spectra for five types of GeO$_2$-doped core and pure silica cladding fibers [82].

$$\rho_0 = 2202(1 + 6.4 \times 10^{-3} C_{GeO_2} - 3.4 \times 10^{-3} C_F) \quad [kg/m^3] \qquad (6.48)$$

$$\frac{1}{\Gamma} = \frac{10^{-6}}{2\pi(17.5 + 0.71 \times C_{GeO_2})} \quad [s] \qquad (6.49)$$

Here n is the refractive index at 1.55 μm, and V_L and V_S are the longitudinal velocity and shear velocity, respectively. ρ_0 is the mass density and $1/\Gamma$ is the acoustic mode lifetime. Figure 6.31 shows the simulated Brillouin scattering gain spectra of five different fibers, each of which has a common normalized frequency $V = 1.6$ but a different core radius a, and a different refractive index difference between the core and cladding, Δ. Here step-index profiles with GeO$_2$-doped silica core and silica cladding were assumed. The normalized frequency is defined as

$$V = \frac{2\pi a}{\lambda} n\sqrt{2\Delta} \qquad (6.50)$$

where a is the core radius, λ is the wavelength of light in vacuum, n is the refractive index of the core, and Δ is the relative refractive index difference between the core and cladding.

In Figure 6.31, the peak position of the Brillouin gain spectrum, ν_B, moves toward a lower frequency when the index difference increases and the core radius decreases. This is because the longitudinal velocity, V_L, becomes lower in the core when the GeO$_2$ concentration increases. This is consistent with equation 6.38 and experimental observation summarized in Table 6.7 and Figure 6.27a.

Assuming a perfect uniformity in the waveguide structure along the fiber length, it is predicted that the peak Brillouin gain coefficient, g_B, increases as a decreases and

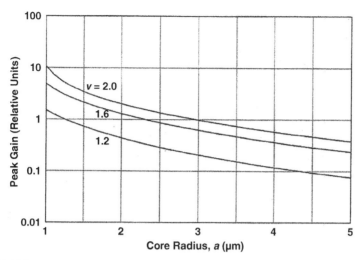

FIGURE 6.32   Peak Brillouin gain coefficients as functions of the core radius under constant normalized frequencies of $v = 1.2$, 1.6, and 2.0 for fibers with GeO_2-doped core and pure silica cladding [82].

Δ increases as in Figure 6.32. The relationship between the peak Brillouin gain coefficients and the core radius under constant normalized frequencies of 1.2, 1.6, and 2.0 is plotted in Figure 6.32. In the simulation, the peak Brillouin gain coefficient increases as the core radius decreases and the normalized frequency increases or equivalently GeO_2 concentration increases. The g_B scales approximately with a^{-2}.

Even though this correlation between g_B and a, Δ, has not been experimentally confirmed due the nonuniformity in high GeO_2 concentration fibers, this information is important because the SBS threshold is approximately in inverse proportion to the peak Brillouin gain coefficient. Control of g_B in optical fiber is still an open area for further research.

There exists another important parameter that determines the Brillouin frequency shift in optical fibers, the fiber drawing tension. Zou et al. [85] reported dependence of Brillouin frequency shift, v_B, in optical fibers on the residual elastic and inelastic strains induced by different fiber draw tensions. In experiments, v_B was found to be linearly proportional to the draw tension as shown in Figure 6.33a such that a higher tension resulted in a lower v_B in optical fibers drawn from the same preform with a step-index profile as in Figure 6.33b. The linear coefficient was measured to be $-42.0\,MHz/100\,g$, which agrees with the theoretical value of $-41.96\,MHz/100\,g$. Theoretical analysis further showed that the elastic strain in fiber core influences predominantly the v_B due to the second-order nonlinearity of Young's modulus while the effect of the inelastic strain in fiber cladding is less than 1.0%.

In dense WDM systems and high power fiber lasers, SBS threshold is a fundamental limiting factor to increase the optical power level carried over the fiber. Intensive research efforts have been focused on the SBS threshold in high capacity transmission fibers and high power laser cavity fibers to alleviate the detrimental SBS

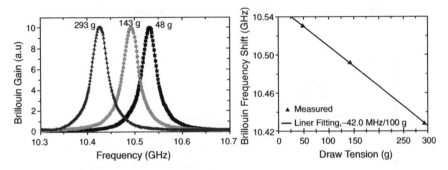

FIGURE 6.33 Brillouin gain spectra for fiber draw at different tensions: (a) measured (symbolic points) and Lorentzian fitted BGS (solid curves) of the fiber samples; (b) triangle symbols are measured data and solid line is the least squares linear fitting [85].

effect that transfers the power of forward incident light to unwanted backward-propagating light at a downshifted frequency by ν_B.

Shiraki et al. [80] reported a novel technique to increase the SBS threshold by introducing an axially nonuniform refractive index profile or equivalently introducing wide variation of ν_B. The fiber had a gradual increase of F concentration across the core and inner cladding from one end to the other, which required the following fabrication steps. First, a soot preform with a GeO_2-doped silica core and pure silica cladding was fabricated by the vapor axial deposition (VAD) method. The soot was then consolidated into a transparent glass by moving the soot body axially upward in a high temperature furnace. The consolidation atmosphere contained F and its concentration was varied in a manner synchronized with the axial preform movement, which varies F dopant concentration along the longitudinal direction of the consolidated preform. The refractive index profiles along the fabricated preform are shown in Figure 6.34a.

F dopant concentration in the fiber preform varies along its length. The variation in concentration was about 0.2%, while the relative index difference Δ between the core

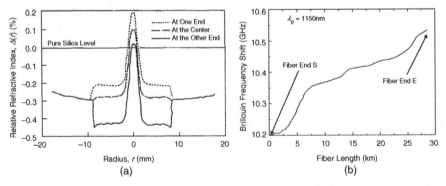

FIGURE 6.34 Suppression of stimulated Brillouin scattering by axially nonuniform refractive index profile distribution: (a) refractive index profiles of the preform at both ends and near the center; (b) distribution of Brillouin frequency shift along the fiber length [80].

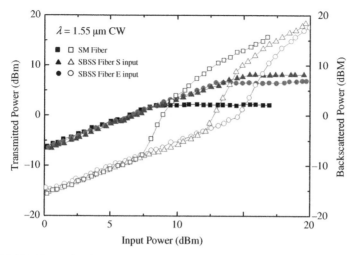

FIGURE 6.35 Transmitted and backscattered powers as a function of input power for SBS suppression fiber and conventional 32 km long single-mode fiber [80]. Here the solid and open symbols represent the transmitted and back-scattered power, respectively.

and cladding was kept at about 0.4%. A 28.5 km long fiber was drawn from this preform and the distribution of ν_B along the fiber is shown in Figure 6.34b. Due to this nonuniform distribution of ν_B, the SBS threshold increased by 7 dB in this fiber and the experimental results are compared with conventional SMF in Figure 6.35.

An alternative method to suppress SBS has been reported by Li et al. [86] by utilizing unique antiguiding property for the acoustic waves in Al_2O_3-doped silica along with precise control of the overlap between optical field and acoustic field around the core. Instead of rough estimation of the SBS threshold as in equation 6.34, new parameters have been introduced to understand the role of waveguide structure and its material composition [87–89] and the expression for the threshold power is given as follows:

$$P_B^{th} \propto \frac{KA_{eff}\alpha_u}{G(\nu_B, L)\overline{I}_u^{ao}} \qquad (6.51)$$

where α_u is the acoustic attenuation coefficient for the acoustic mode of order u, A_{eff} is optical effective mode area, $G(\nu_B, L)$ is the effective gain coefficient at the peak frequency, and K is the polarization factor. $\overline{I}_u^{ao}$ is the normalized overlap integral between the electric and acoustic fields given by:

$$\overline{I}_u^{ao} = \frac{\left(\int E_oE_o^*\rho_u^* r\, dr\, d\theta\right)^2}{\left(\int E_oE_o^*\right)^2 r\, dr\, d\theta\right)\left(\int (\rho_u\rho_u^*)r\, dr\, d\theta\right)} \qquad (6.52)$$

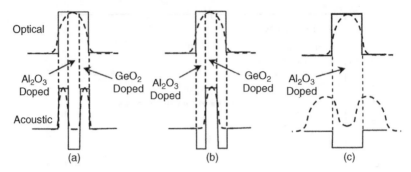

FIGURE 6.36 (a–c) Dopant designs that can reduce the overlap between the optical and acoustic fields [86]. Top row is the optical field and bottom row is acoustic field.

where E_0 is the optical field associated with the fundamental mode and ρ_u is the field of a longitudinal acoustic eigenmode of order u. The above equation indicates that, in addition to the mode area, A_{eff}, the SBS threshold can be further controlled by increasing the acoustic loss (α_u), or by decreasing the overlap integral ($\overline{I}_u^{ao}$) and the maximum gain coefficient ($G(\nu_{max}, L)$). The overlap integral can be controlled by fiber refractive index profile design and acoustic velocity profile design, which can be effectively modified by glass composition near the core region.

It is observed that Al_2O_3 in silica glass has a unique antiguiding property for acoustic wave [86] and it serves as an acoustic index-decreasing element, which is very similar to fluorine's role in optical refractive index. Three different cases for acoustic index profiles are shown in Figure 6.36. Note that the optical index profiles are identical such that the optical field distribution follows the LP_{01} mode in conventional SMFs. These profiles can be achieved by adding GeO_2 and Al_2O_3 in silica glass. Figure 6.36a is the case where Al_2O_3 is confined to the center of the core and it is surrounded by a GeO_2-doped ring. This profile guides the acoustic field only along the GeO_2-doped ring. Figure 6.36b is the opposite case of (a); the center of the core is GeO_2 doped and its surrounding ring is doped with Al_2O_3, which guides the acoustic wave only over the central core region. Figure 6.36c describes more radical case where the core is doped by only Al_2O_3 to push the acoustic field outward from the center.

These profiles schematically illustrate that the overlap integral ($\overline{I}_u^{ao}$) between the optical and acoustic fields can be efficiently controlled by selective doping of Al_2O_3 and GeO_2 over the core region. This concept has been successfully implemented in a high power Yb-doped double-clad fiber laser [86], whose core composition and its SBS threshold are shown in Figure 6.37. The schematic index profiles for both optical and acoustic waves are shown in the inset. The concentration of Al_2O_3 is linearly ramped down from the center of the core to the edge of the core to provide an effective acoustic index shown as dashed line in the inset. SBS threshold increase by 5 dB was reported in this configuration.

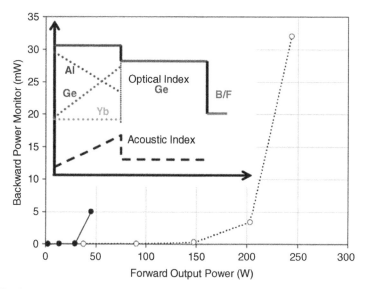

FIGURE 6.37 Experimental observation of SBS threshold increase in a double-clad fiber laser with Yb, Ge, and Al codoped core [86].

In Brillouin scattering, there exist consistent methods to control the Brillouin frequency shift

(1) doping GeO_2, P_2O_5, and F in the core as summarized in Table 6.7,
(2) controlling the drawing tension as in Figure 6.35.

However, there is only a theoretical prediction to increase the peak Brillouin gain by GeO_2 doping and proper design of the core diameter as in Figure 6.34. In terms of the magnitude enhancement of the SBS gain, there would be ample research opportunities. Al_2O_3 doping in the core has allowed a very intelligent waveguide design capability for both optical and acoustic waves to provide a wide range of overlap control between them. This new technique could be applied in both directions—suppressing SBS or increasing SBS gain—which also provides an open area for fiber optic research.

6.8 REVIEW ON FIBER BRILLOUIN SENSORS AND RECENT NOVEL APPLICATIONS

As schematically shown in Figure 6.5, the position of Brillouin scattering peak (ν_B) in both Stokes and anti-Stokes components shifts with the changes in temperature (T) and strain (ε). The Brillouin frequency shift (ν_B) is determined by the Bragg condition and expressed as

$$\nu_B = \frac{2 n_{eff} V_a}{\lambda} \tag{6.38}$$

where n_{eff} is the effective refractive index of the guided optical mode and λ is the optical wavelength in vacuum. V_a is the acoustic velocity, which is approximated as the longitudinal acoustic velocity, V_L, in the cylindrical optical fiber geometry. V_L is determined by the Young's modulus (E) and the density (ρ) as follows [90]:

$$V_a \approx V_L = \sqrt{\frac{E}{\rho_0}} \tag{6.39}$$

All of the parameters except λ in equation 6.38 are affected by both temperature and strain, to provide a linear shift in ν_B. The Brillouin frequency ν_B in optical fibers changes under the influence of strain (ε) and temperature (T) in a highly linear manner as follows:

$$\nu_B(\varepsilon) = \nu_B(\varepsilon_r)[1 + C_\varepsilon(\varepsilon - \varepsilon_r)] \tag{6.53}$$

$$\nu_B(T) = \nu_B(T_r)[1 + C_T(T - T_r)] \tag{6.54}$$

where ε_r and T_r are the reference strain and temperature. In conventional single-mode fibers, the linearity coefficients for the strain and temperature are $C_\varepsilon = 4.6$ and $C_T = 9.4 \times 10^{-5}$ (for the temperature in Kelvin), respectively [91,92].

The linear dependence of ν_B on the applied strain and the temperature change can be also expressed in a relative scale as follows:

$$\nu_B - \nu_B^r = A \cdot \delta\varepsilon + B \cdot \delta T$$
$$\frac{\nu_B - \nu_B^r}{\nu_B^r} = A' \cdot \delta\varepsilon + B' \cdot \delta T \tag{6.55}$$

where ν_B^r is the Brillouin frequency shift measured at room temperature (25 °C) and in the strain-free state. $A' (10^{-6}/\mu\varepsilon)$ is the normalized strain coefficient and $B' (10^{-6}/°C)$ is the normalized temperature coefficient.

Zou et al. [93] analyzed these normalized coefficients for both pure SiO_2 glass and GeO_2-doped silica glass to deduce the effects of GeO_2 in the Brillouin sensing capability. According to equations and 6.55, the strain and temperature normalized coefficients can be dissolved into three parts, respectively:

$$A' = \left(\frac{1}{n_{\text{eff}}}\frac{\delta n_{\text{eff}}}{\delta\varepsilon}\right) + \left(\frac{-1}{2\rho_0}\frac{\delta\rho_0}{\delta\varepsilon}\right) + \left(\frac{-1}{2E}\frac{\delta E}{\delta\varepsilon}\right) = A'_n + A'_\rho + A'_E \tag{6.56}$$

$$B' = \left(\frac{1}{n_{\text{eff}}}\frac{\delta n_{\text{eff}}}{\delta T}\right) + \left(\frac{-1}{2\rho_0}\frac{\delta\rho_0}{\delta T}\right) + \left(\frac{-1}{2E}\frac{\delta E}{\delta T}\right) = B'_n + B'_\rho + B'_E \tag{6.57}$$

Each of the three parts in the above equations is determined by relative change rates in the effective index (n_{eff}), density (ρ), and Young's modulus (E), under the applied

TABLE 6.9 Strain and Temperature Normalized Coefficients for Brillouin Frequency Shift in Pure Silica [93]

$A'_{SiO_2} = A'_n + A'_p + A'_E = 3.785(10^{-6}/\mu\varepsilon)$ [93]		
$A'_n \simeq -\frac{pn^2}{2} = -0.230$	$n_{eff} \simeq n(1.55\,\mu m) = 1.444$	Refractive index
	$p = 0.22$	Elasto-optic coefficient [85]
$A'_p = \frac{1-2\gamma}{2} = 0.315$	$\gamma = 0.186$	Poisson ratio [90]
$A'_E = \frac{s}{2} = 3.70$	$s = 7.4$	Strain-induced nonlinear coefficient of E [94]
$B'_{SiO_2} = B'_n + B'_p + B'_E \simeq 111.575(10^{-6}/^\circ C)$ [93]		
$B'_n = 8.310$	–	[95,96]
$B'_p = \frac{3\alpha}{2} = 0.765$	$\alpha = 0.51 \times 10^{-6}\,^\circ C^{-1}$	Thermal expansion coefficient [97]
$B'_E = \frac{\xi}{2} = 102.5$	$\xi = 2.05 \times 10^{-6}\,^\circ C^{-1}$	Thermally induced nonlinear coefficient of E [98,99]

strain or the temperature change. A'_n and B'_n are determined by the elasto-optic and thermo-optic effects. A'_p and B'_p are subject to the strain-induced distortion and the thermal expansion, respectively. A'_E and B'_E are decided by the strain-induced second-order nonlinearity of Young's modulus and the thermal-induced second-order nonlinearity of Young's modulus, respectively.

The parameters in equations 6.56 and 6.57 are summarized in Table 6.9 for pure SiO_2 glass. Theoretical values for strain coefficients and temperature coefficients are obtained as follows:

$$A'_{SiO_2} = 3.785(10^{-6}/\mu\varepsilon)$$

$$B'_{SiO_2} = 111.575(10^{-6}/^\circ C)$$

The A' and B' were estimated for optical fibers with various GeO_2 concentrations by measuring the shifts in ν_B and the experimental results are plotted as a function of GeO_2 concentration in Figure 6.38a and b. The least square linear fittings to the measured data gave the linear dependence slopes of 1.48%/mol% and 1.61%/mol% for A' and B', respectively. The normalized coefficients A' and B' relatively decreased by 1.48 and 1.61% for 1 mol% increase of GeO_2 concentration in the core, that is, an incremental relative index Δ of 0.1%:

$$A' = A'_{SiO_2}(1 - 0.0148 \times C_{GeO_2}) \qquad (6.58)$$

$$B' = B'_{SiO_2}(1 - 0.0161 \times C_{GeO_2})$$

The normalized coefficients of 0 mol% GeO_2-doped silica, pure silica, are extracted from the linearly fitted curves in Figure 6.38, $A' = 3.906$ and

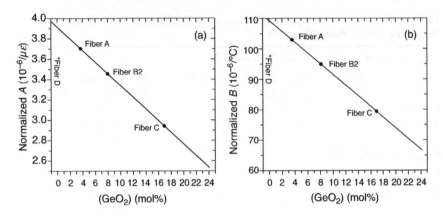

FIGURE 6.38 Normalized (a) strain coefficient A' and (b) temperature coefficient B' as a function of GeO_2 concentration. The solid lines are the least squares linear fittings to the experimental results (circles) [93].

$B' = 108.961$. They are in a good agreement with the values obtained for bulk SiO_2 as in Table 6.9.

It is noted that addition of GeO_2 actually decreases the normalized coefficients A' and B' as in Figure 6.38, which in turn reduces the sensitivity of ν_B shift over temperature and strain. However, high GeO_2 concentration can locate ν_B away from the Rayleigh scattering as in Figure 6.27, so that the background noise can be reduced for sensing applications. Therefore, the Stokes sensing frequency location ($\nu_S = \nu_P - \nu_B$) and its sensitivity A' and B' can be precisely controlled by appropriate fiber design.

Brillouin bandwidth ($\Delta\nu_B$) and electric power signal-to-noise ratio (SNR) considerations determine the minimum detectable resolution in ν_B [100]:

$$\delta\nu_B \simeq \frac{\Delta\nu_B}{\sqrt{2}(\text{SNR})^{1/4}} \tag{6.59}$$

By the relations in equations 6.53 and 6.54, the minimum detectable changes in strain and temperature are then given as follows:

$$\delta\varepsilon = \frac{\delta\nu_B}{C_e\nu_B(\varepsilon_r)}$$

$$\delta T = \frac{\delta\nu_B}{C_e\nu_B(T_r)} \tag{6.60}$$

Using a typical value of 35 MHz for $\Delta\nu_B$ in conventional SMFs, a strain change of 10^{-4} and a temperature change of 1 K can be measured for SNR of 25 and 52 dB, respectively.

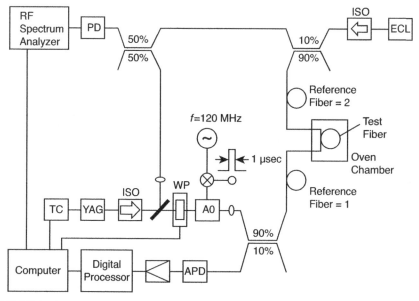

FIGURE 6.39 Experimental arrangement for measurement of temperature distribution. WP, half-wave plate; PD, photodiode; TC, temperature controller [101].

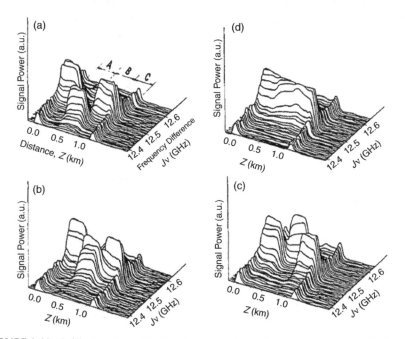

FIGURE 6.40 Brillouin gain distribution for different ν_B values in the test fiber and reference fibers #1 and #2. A, B, and C denote reference fiber #1, the test fiber, and reference fiber #2, respectively. The temperature of the reference fibers, T_r, is fixed at 28.00 °C. The temperature of the test fiber, T, is (a) − 28.50 °C, (b) 1.40 °C, (c) 61.40 °C, and (d) 28.00 °C [100].

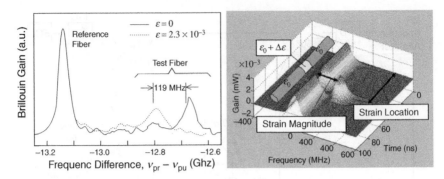

FIGURE 6.41 Distributed strain sensing using Brillouin scattering. (a) Strain dependence of the Brillouin frequency shift [91]; (b) strain point location using a BOTDA system [103].

Based on optical time domain reflectometry technique, the spatial resolution in Brillouin scattering can be also estimated as follows:

$$\delta z = \frac{c' W}{2} \tag{6.61}$$

where c' and W are the light speed in the optical fiber and the pulse duration of light, respectively. According to equation 6.58, a spatial resolution of 10 cm and 100 m is estimated for the light pulse width of 1 ns and 1 μs, respectively.

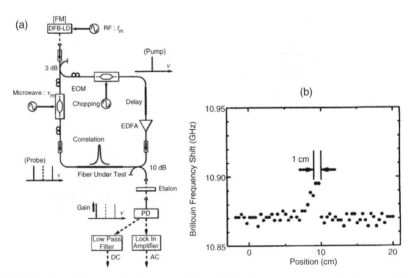

FIGURE 6.42 (a) System for distributed fiber Brillouin strain sensing with the correlation-based CW technique; (b) measurement result of Brillouin frequency shift distribution along the fiber with the strain-loaded section of 1 cm.

Based on linear shifts in ν_B for the strain and temperature changes, various distributed fiber optic sensors have been demonstrated, which enables strain and temperature reading at an arbitrary location along the optical fiber.

One of the most frequently referred experiments on distributed temperature sensing was demonstrated by Kurashima et al. [101] and their experimental setup is shown in Figure 6.39. As the temperature of the test fiber located in the middle section is changed from -28.5 to $61.4\,°C$, shifts in ν_B were measured to be $1.25\,MHz/°C$ as in Figure 6.40, which is consistent with equation 6.54.

In the case of distributive sensing of strain along the fiber, Horiguchi and Tateda [102] reported ν_B shift of $119\,MHz$ for 2.3×10^{-3} strain, as shown in Figure 6.41a. This value is consistent with equation 6.53. Based on these results, Brillouin optical time domain analysis (BOTDA) technique has been developed and its schematic output is shown in Figure 6.41b, where the location and the amount of strain are simultaneously measured along the fiber with a high spatial resolution [103].

The spatial resolution of Brillouin sensing system is furthermore developed and $\sim 1\,cm$ resolution has been reported for strain measurement by Hotate and Tanaka [104] as shown in Figure 6.42.

Distributed sensing techniques using nonlinear scattering such as Brillouin and Raman process are being intensively developed for structural health monitoring in civil structures and transportation vehicles and monitoring fiber optic communication network breaks. Detailed description of fiber optic sensors is beyond the scope of this book and the readers are referred to excellent books [105–108] and review papers [28–30].

REFERENCES

[1] A. R. Chraplyvy, "Limitations on lightwave communications imposed by optical fiber nonlinearities," Journal of Lightwave Technology, vol. 8, no. 10, pp. 1548–1557, 1990.

[2] B. B. Tiwari, V. Prakash, V. Tripathi, and N. Malaviya, "Nonlinear effects in optical fiber transmission system," IETE Technical Review, vol. 16, no. 5–6, pp. 461–479, 1999.

[3] E.-H. Lee, K. H. Kim, and H. K. Lee, "Nonlinear effects in optical fiber: advantages and disadvantages for high capacity all-optical communication application," Optical and Quantum Electronics, vol. 34, no. 12, pp. 1167–1174, 2002.

[4] M. Wu and W. I. Way, "Fiber nonlinearity limitations in ultra-dense WDM systems," Journal of Lightwave Technology, vol. 22, no. 6, pp. 1483–1498, 2004.

[5] J. Toulouse, "Optical nonlinearities in fibers: review, recent examples, and systems applications," Journal of Lightwave Technology, vol. 23, no. 11, pp. 3625–3641, 2005.

[6] ITU-T G650. 2 "Definitions and test methods for statistical and non-linear related attributes of single-mode fibre and cable," 2007.

[7] ITU-T G. 665 "Generic characteristics of Raman amplifiers and Raman amplified subsystems," 2005.

[8] G. P. Agrawal, "Nonlinear fiber optics,", 4th edition,Academic Press 2006.

[9] G. P. Agrawal, "Applications of nonlinear fiber optics,", 2nd edition,Academic Press 2008.

[10] A. Hasegawa, "Optical soliton in fibers," Springer Tracts in Modern Physics, vol. 116, pp. 1–74, 1989.

[11] M. E. Marhic, "Fiber optical parametric amplifiers, oscillators, and related device," Cambridge University Press 2007.

[12] D. Cotter, R. J. Manning, K. J. Blow, A. D. Ellis, A. E. Kelly, D. Nesset, I. D. Phillips, A. J. Poustie, and D. C. Rogers, "Nonlinear optics for high-speed digital information processing," Science, vol. 286, no. 5444, pp. 1523–1528, 1999.

[13] S. Radic and C. J. McKinstrie, "Optical amplification and signal processing in highly nonlinear optical fiber," IEICE Transactions on Electronics, vol. 88, no. 5, pp. 859–869, 2005.

[14] J. M. Dudley and J. R. Taylor, "Supercontinuum generation in optical fibers," Cambridge University Press 2010.

[15] L. Thévenaz, "Slow and fast light in optical fibres," Nature Photonics, vol. 2, pp. 474–481, 2008.

[16] E. Bhahaa, E. A. Saleh, and M. C. Teich, "Fundamentals of photonics,", Chapter 19, John Wiley & Sons, Inc., 1991.

[17] C. C. Wang, "Empirical relation between the linear and the third order nonlinear optical susceptibilities," Physical Review B, vol. 2, no. 6, pp. 2045–2048, 1970.

[18] T. M. Monro and H. Ebendorff-Heidepriem, "Progress in microstructured optical fibers," Annual Review of Material Research, vol. 36, pp. 467–495, 2006.

[19] N. L. Boling, A. J. Glass, and A. Owyoung, "Empirical relationships for predicting nonlinear refractive index changes in optical solids," IEEE Journal of Quantum Electronics, vol. QE-14, no. 8, pp. 601–608, 1978.

[20] K. Nakajima and M. Ohashi, "Dopant dependence of effective nonlinear refractive index in GeO_2- and F-doped core single mode fibers," IEEE Photonics Technology Letters, vol. 14, no. 4, pp. 492–494, 2002.

[21] T. Kato, Y. Suetsugu, and M. Nishimura, "Estimation of nonlinear refractive index in various silica-based glasses for optical fibers," Optics Letters, vol. 20, no. 22, pp. 2279–2281, 1995.

[22] J. Hansryd, P. A. Andrekson, M. Westlund, J. Li, and P.-O. Hedekvist, "Fiber-based optical parametric amplifiers and their applications," IEEE Journal of Selected Topics in Quantum Electronics, vol. 8, no. 3, pp. 506–520, 2002.

[23] Y. Namihira, "Relationship between nonlinear effective area and mode field diameter for dispersion shifted fibres," Electronics Letters, vol. 30, no. 3, pp. 262–264, 1994.

[24] R. H. Stolen and E. P. Ippen, "Raman gain in glass optical waveguides," Applied Physics Letters, vol. 22, pp. 276–278, 1973.

[25] E. P. Ippen and R. H. Stolen, "Stimulated Brillouin scattering in optical fibers," Applied Physics Letters, vol. 21, pp. 539–541, 1972.

[26] R. G. Smith, "Optical power handling capacity of low loss optical fibers as determined by stimulated Raman and Brillouin scattering," Applied Optics, vol. 11, pp. 2489–2494, 1972.

[27] B. D. Gupta, B. Das, "Fiber Optic Sensors: Principles and Applications" New India Publishing, 2006.

[28] A. D. Kersey, "A review of recent developments in fiber optic sensor technology," Optical Fiber Technology, vol. 2, pp. 291–317, 1996.

[29] B. Lee, "Review of the present status of optical fiber sensors," Optical Fiber Technology, vol. 9, pp. 57–79, 2003.

[30] O. S. Wolfbeis, "Fiber-optic chemical sensors and biosensors," Analytical Chemistry, vol. 76, no. 12, pp. 3269–3284, 2004.

[31] C. V. Raman and K. S. Krishnan, "A new type of secondary radiation," Nature, vol. 121, p. 501, 1928.

[32] J. Bromage, "Raman amplification for fiber communications systems," Journal of Lightwave Technology, vol. 22, no. 1, pp. 79–93, 2004.

[33] R. H. Stolen, C. Lee, and R. K. Jain, "Development of the stimulated Raman spectrum in single-mode fibers," Journal of Optical Society of America B, vol. 1, no. 4, pp. 652–657, 1984.

[34] M. E. Lines, "Raman-gain estimates for high-gain optical fibres," Journal of Applied Physics, vol. 62, pp. 4363–4370, 1987.

[35] N. Shibata, M. Horigudhi, and T. Edahiro, "Raman spectra of binary high-silica glasses and fibers containing GeO_2, P_2O_5 and B_2O_3," Journal of Non-Crystalline Solids, vol. 45, pp. 115–126, 1981.

[36] F. L. Galeener, J. C. Mikkelsen, R. H. Geils, and W. J. Mosby, "The relative Raman cross sections of vitreous SiO_2, GeO_2, B_2O_3 and P_2O_5," Applied Physics Letters, vol. 32, no. 1, pp. 34–36, 1978.

[37] S. T. Davey, D. L. Williams, B. J. Ainslie, W. J. M. Rothwell, and B. Wakefield, "Optical gain spectrum of GeO_2–SiO_2 Raman fibre amplifiers," IEE Proceedings, vol. 136, no. 6, pp. 301–305, 1989.

[38] J. Bromage, K. Rottwitt, and M. E. Lines, "A method to predict the Raman gain spectra of germanosilicate fibers with arbitrary index profiles," IEEE Photonics Technology Letters, vol. 14, no. 1, pp. 24–26, 2002.

[39] S. K. Sharma, D. W. Matson, J. A. Philpotts, and T. L. Roush, "Raman study of the structure of glasses along the join SiO_2–GeO_2," Journal of Non-Crystalline Solids, vol. 68, pp. 99–114, 1984.

[40] E. M. Dianov, "Advances in Raman fibers," Journal of Lightwave Technology, vol. 20, no. 8, pp. 1457–1462, 2002.

[41] E. M. Dianov, V. M. Mashinsky, V. B. Neustruev, O. D. Sazhin, A. N. Guryanov, V. F. Khopin, and N. N. Vechkanov, "Origin of excess loss in single-mode optical fibers with high GeO-doped silica core," Optical Fiber Technology, vol. 3, pp. 77–86, 1977.

[42] E. M. Dianov, M. V. Grekov, I. A. Bufetov, V. M. Mashinksy, O. D. Suzhin, A. M. Prokhorov, G. G. Devyatykh, A. N. Guryanov, and V. F. Khopin, "Highly efficient 1.3 m Raman fiber amplifier," Electronics Letters, vol. 34, pp. 669–670, 1998.

[43] K. Suzuki, K. Noguchi, and N. Uesugi, "Selective stimulated Raman scattering in highly P_2O_5-doped silica single-mode fibers," Optics Letters, vol. 11, pp. 656–658, 1986.

[44] K. Suzuki and M. Nakazawa, "Raman amplification in P_2O_5-doped optical fiber," Optics Letters, vol. 13, pp. 666–668, 1988.

[45] E. M. Dianov, M. V. Grekov, I. A. Bufetov, S. A. Vasiliev, O. I. Medvedkov, V. G. Plotnichenko, V. V. Koltashev, A. V. Belov, M. M. Bubnov, S. L. Semjonov, and A. M. Prokhorov, "CW high power 1.24 μm and 1.48 μm Raman lasers based on low loss phosphosilicate fiber," Electronics Letters, vol. 33, pp. 1542–1544, 1997.

[46] J.-C. Bouteiller, "Raman fiber lasers for optical communication application," Annals of Telecommunications, vol. 58, no. 9–10 pp. 1342–1363, 2003.

[47] V. I. Karpov, E. M. Dianov, A. S. Kurokov, V. M. Paramonov, V. N. Protopopov, M. E. Bachynski, and W. R. L. Clements, "LD-pumped 1.48 μm laser based on Yb-doped

double-clad fiber and phosphorosilicate-fiber Raman converter," Optical Fiber Communication Conference, Technical Digest, pp. 202–204, 1999, paper WM3.

[48] M. Prabhu, N. S. Kim, L. Jianren, and K. Ueda, "Output characteristics of high-power continuous wave Raman fiber laser at 1484 nm using phosphosilicate fiber," Optical Review, vol. 7, no. 5, pp. 455–461, 2000.

[49] T. Schneider, "Nonlinear optics in telecommunications,", Springer 2004.

[50] M. N. Islam (ed), "Raman amplifiers for telecommunications 2: sub-systems and systems,", Springer 2004.

[51] C. Headley and G. P. Agrawal, "Raman amplification in fiber optical communication systems,", Elsevier Academic Press 2005.

[52] M. Ikeda, "Stimulated Raman amplification characteristics in long span single mode silica fibers," Optics Communications, vol. 39, no. 3, pp. 148–152, 1981.

[53] T. Naito, T. Tanaka, K. Torii, N. Shimojoh, H. Nakamoto, and M. Suyama, "A broadband distributed Raman amplifier for bandwidth beyond 100 nm," Optical Fiber Communications Conference, A. Sawchuk (ed), OSA Trends in Optics and Photonics, Vol. 70, Optical Society of America, 2002, paper TuR1.

[54] K. Fujimura, A. Oguri, T. Nakajima, Y. Emori, S. Namiki, and M. Sakano, "Applying a numerical simulation technique to the design of WDM pumped Raman amplifiers, and methods for the automatic determination of pump powers," Furukawa Review, no. 25, pp. 1–8, 2004.

[55] V. E. Perlin and H. G. Winful, "On distributed Raman amplification for ultrabroad-band long-haul WDM systems," Journal of Lightwave Technology, vol. 20, pp. 409–416, 2002.

[56] Y. Emori, K. Tanaka, and S. Namiki, "100 nm bandwidth flat-gain Raman amplifiers pumped and gain-equalised by 12-wavelength-channel WDM laser diode unit," Electronics Letters, vol. 35, pp. 1355–1356, 1999.

[57] H. Kidorf, K. Rottwitt, M. Nissov, M. Ma, and E. Rabarijaona, "Pump interactions in a 100-nm bandwidth Raman amplifier," IEEE Photonics Technology Letters, vol. 11, pp. 530–532, 1999.

[58] S. Namiki and Y. Emori, "Ultrabroad-band Raman amplifiers pumped and gain-equalized by wavelength division multiplexed high power laser diodes," IEEE Journal on Selected Topics in Quantum Electronics, vol. 7, no. 1, pp. 3–16, 2001.

[59] S. G. Grubb, T. Erdogan, V. Mizrahi, T. Strasser, W. Y. Cheung, W. A. Reed, P. J. Lemaire, A. E. Miller, S. G. Kosinski, G. Nykolak, P. C. Beccer, and D. W. Peckham, "1.3 μm cascaded Raman amplifier in germanosilicate fibers," Optical Amplifiers and Their Applications (OAA), 1994, paper PD-3.

[60] S. G. Grubb, T. Strasser, W. Y. Cheung, W. A. Reed, V. Mizrahi, T. Erdogan, P. J. Lemaire, A. M. Vendarkar, and D. J. DiGiovanni, "High power 1.48 μm cascaded Raman laser in germanosilicate fibers," OSA Topical Meeting: Optical Amplifiers and Their Applications (OAA), 1995, paper SaA4-1.

[61] S. G. Grubb and A. J. Stentz, "Fiber Raman lasers emit at many wavelengths," Laser Focus World, 1996.

[62] V. I. Karpov, E. M. Dianov, V. M. Paramonov, O. I. Medvedkov, M. M. Bubnov, S. L. Semyonov, S. A. Vasiliev, V. N. Protopopov, O. N. Egorova, V. F. Hopin, A. N. Guryanov, M. P. Bachynski, and W. R. L. Clements, "Laser-diode-pumped phosphosilicate-fiber Raman laser with an output power of 1 W at 1.48 μm," Optics Letters, vol. 24, no. 13, pp. 887–889, 1999.

[63] N. S. Kim, M. Prabhu, C. Li, J. Song, and K. Ueda, "1239/1484 nm cascaded phosphosilicate Raman fiber laser with CW output power of 1.36 W at 1484 nm pumped by CW Yb-doped double-clad fiber laser at 1064 nm and spectral continuum generation," Optics Communications, vol. 176, pp. 219–222, 2000.

[64] P. B. Hansen, A. J. Stentz, L. Eskilden, S. G. Grubb, T. A. Strasser, and J. R. Pedrazzani, "High sensitivity 1.3 μm optically pre-amplified receiver using Raman amplification," Electronics Letters, vol. 32, no. 23, pp. 2164–2165, 1996.

[65] M. N. Islam, "Raman amplifiers for telecommunications," IEEE Journal of Selected Topics in Quantum Electronics, vol. 8, no. 3, pp. 548–559, 2002.

[66] M. Rini, I. Cristiani, and V. Degiorgio, "Numerical modeling and optimization of cascaded CW Raman fiber lasers," IEEE Journal of Quantum Electronics, vol. 36, no. 10, pp. 1117–1122, 2000.

[67] S. D. Jackson and P. H. Muir, "Theory and numerical simulation of nth-order cascaded Raman fiber lasers," Journal of Optical Society of America B, vol. 18, no. 9, pp. 1297–1305, 2001.

[68] B. Burgoyne, N. Godbout, and S. Lacroix, "Theoretical analysis of nth-order cascaded continuous-wave Raman fiber lasers. I. Model and resolution," Journal of Optical Society of America B, vol. 22, no. 4, pp. 764–771, 2005.

[69] B. Burgoyne, N. Godbout, and S. Lacroix, "Theoretical analysis of nth-order cascaded continuous-wave Raman fiber lasers. I. Optimization and design rules," Journal of Optical Society of America B, vol. 22, no. 4, pp. 772–776, 2005.

[70] J. C. Travers, S. V. Popov, and J. R. Taylor, "Efficient continuous-wave holey fiber Raman laser," Applied Physics Letters, vol. 87, no. 3, p. 031106, 2005.

[71] M. Fuochi, F. Poli, A. Cucinotta, and L. Vincetti, "Study of Raman amplification properties in triangular photonic crystal fibers," Journal of Lightwave Technology, vol. 21, pp. 2247–2254, 2003.

[72] F. Benabid, J. C. Knight, G. Antonopoulos, and P. St. J. Russell, "Stimulated Raman scattering in hydrogen-filled hollow-core photonic crystal fiber," Science, vol. 298, no. 5592, pp. 399–402, 2002.

[73] M. J. Li, "Managing nonlinearity in optical fiber for high power lasers," SPIE News Room, July 28 2006, DOI: 10.1117/2.1200607. 0317.

[74] D. Heiman, D. S. Hamilton, and R. W. Hellwarth, "Brillouin scattering measurements on optical glasses," Physical Review B, vol. 19, p. 6583, 1979.

[75] M. Niklès, L. Thévenaz, and P. A. Robert, "Brillouin gain spectrum characterization in single-mode optical fibers," Journal of Lightwave Technology, vol. 15, no. 10, pp. 1842–1851, 1997.

[76] K. Hotate and T. Hasegawa, "Measurement of Brillouin gain spectrum distribution along an optical fiber using a correlation based technique-proposal, experiments, and simulation," IEICE Transactions in Electronics, vol. E38-C, no. 3, pp. 405–412, 2000.

[77] R. W. Tkach, A. R. Charaplyvy, and R. M. Derosier, "Spontaneous Brillouin scattering for single-mode optical-fibre characterization," Electronics Letters, vol. 22, no. 19, pp. 1011–1013, 1986.

[78] N. Shibata, R. G. Waarts, and R. P. Braun, "Brillouin-gain spectra for single-mode fibers having pure-silica, GeO_2-doped, and P_2O_5-doped cores," Optics Letters, vol. 12, no. 4, pp. 269–271, 1987.

[79] N. Shibata, K. Okamoto, and Y. Azuma, "Longitudinal acoustic modes and Brillouin gain spectra for GeO$_2$ doped core single mode fibers," Journal of Optical Society of America B, vol. 6, no. 6, pp. 1167–1174, 1989.

[80] K. Shiraki, M. Ohashi, and M. Tateda, "SBS threshold of a fiber with a Brillouin frequency shift distribution," Journal of Lightwave Technology, vol. 14, no. 1, pp. 50–57, 1996.

[81] A. Yeniay, J.-M. Delavaux, and J. Toulouse, "Spontaneous and stimulated Brillouin scattering gain spectra in optical fibers," Journal of Lightwave Technology, vol. 20, no. 8, pp. 1425–1432, 2002.

[82] Y. Koyamada, S. Sato, S. Nakamura, H. Sotobayashi, and W. Chujo, "Simulating and designing Brillouin gain spectrum in single-mode fibers," Journal of Lightwave Technology, vol. 22, no. 2, pp. 631–639, 2004.

[83] N. Lagakos, J. A. Bucaro, and R. Hughes, "Acoustic sensitivity predictions of single-mode optical fibers using Brillouin scattering," Applied Optics, vol. 19, pp. 3668–3670, 1980.

[84] K. Shiraki and M. Ohashi, "Sound velocity measurement based on guided acoustic-wave Brillouin scattering," IEEE Photonics Technology Letters, vol. 4, pp. 1177–1180, 1992.

[85] W. Zou, Z. He, A. D. Yablon, and K. Hotate, "Dependence of Brillouin frequency shift in optical fibers on draw-induced residual elastic and inelastic strains," IEEE Photonics Technology Letters, vol. 19, no. 18, pp. 1389–1391, 2007.

[86] M.-J. Li, X. Chen, J. Wang, S. Gray, A. Liu, J. A. Demeritt, A. B. Ruffin, A. M. Crowley, D. T. Walton, and L. A. Zenteno, "Al/Ge co-doped large mode area fiber with high SBS threshold," Optics Express, vol. 15, no. 13, pp. 8290–8299, 2007.

[87] M.-J. Li, X. Chen, J. Wang, A. Liu, S. Gray, D. T. Walton, A. B. Ruffin, J. Demeritt, and L. Zenteno, "Fiber designs for higher power lasers," Proceedings of SPIE, vol. 6469, p. 64690H, 2007.

[88] A. Boh Ruffin, M.-J. Li, X. Chen, A. Kobyakov, and F. Annunziata, "Brillouin gain analysis for fibers with different refractive indices," Optics Letters, vol. 30, pp. 3123–3125, 2005.

[89] A. Kobyakov, S. Kumar, D. Q. Chowdhury, A. B. Ruffin, M. Sauer, S. R. Bickham, and R. Mishra, "Design concept for optical fibers with enhanced SBS threshold," Optics Express, vol. 13, pp. 5338–5346, 2005.

[90] S. P. Timoshenko and J. N. Goodier, "Theory of elasticity," 3rd edition, New York: McGraw-Hill, 1970.

[91] T. Horiguchi, T. Kurashima, and M. Tateda, "Tensile strain dependence of Brillouin frequency shift in silica optical fibers," IEEE Photonics Technology Letters, vol. 1, no. 5, pp. 107–108, 1989.

[92] T. Kurashima, T. Horiguchi, and M. Tateda, "Thermal effects of Brillouin gain spectra in silica optical fibers," IEEE Photonics Technology Letters, vol. 2, no. 10, pp. 718–720, 1990.

[93] W. Zou, Z. He, and K. Hotate, "Investigation of strain- and temperature-dependences of Brillouin frequency shifts in GeO$_2$-doped optical fibers," Journal of Lightwave Technology, vol. 26, no. 13, pp. 1854–1861, 2008.

[94] A. Bertholds and R. Dandliker, "Deformation of single-mode optical fibers under static longitudinal stress," Journal of Lightwave Technology, vol. LT-5, no. 7, pp. 895–900, 1987.

[95] I. H. Malitson, "Inter-specimen comparison of the refractive index of fused silica," Journal of Optical Society of America, vol. 55, no. 10, pp. 1205–1209, 1965.

[96] T. Toyoda and M. Yabe, "The temperature dependence of the refractive indices of fused silica and crystal quartz," Journal of Physics D, vol. 16, no. 5, pp. L97–L100, 1983.

[97] P. K. Bachmann, D. U. Wiechert, and T. P. M. Meeuwsen, "Thermal expansion coefficients of doped and undoped silica prepared by means of PCVD," Journal of Materials Science, vol. 23, no. 7, pp. 2584–2588, 1988.

[98] A. S. Pine, "Brillouin scattering study of acoustic attenuation in fused quartz," Physical Review, vol. 185, no. 3, pp. 1187–1193, 1969.

[99] J. A. Bucaro and H. D. Dardy, "High-temperature Brillouin scattering in fused quartz," Journal of Applied Physics, vol. 45, no. 12, pp. 5324–5329, 1974.

[100] T. Horiguchi, K. Shimizu, T. Kurashima, M. Tateda, and Y. Koyamada, "Development of a distributed sensing technique using Brillouin scattering," Journal of Lightwave Technology, vol. 13, no. 7, pp. 1296–1302, 1995.

[101] T. Kurashima, T. Horiguchi, and M. Tateda, "Distributed-temperature sensing using stimulated Brillouin scattering in optical silica fibers," Optics Letters, vol. 15, no. 18, pp. 1038–1040, 1990.

[102] T. Horiguchi and M. Tateda, "BOTDA-nondestructive measurement of single mode optical fiber attenuation characteristics using Brillouin interaction: theory," Journal of Lightwave Technology, vol. 7, no. 8, pp. 1170–1176, 1989.

[103] Neubrex Co. Ltd., http://www.neubrex.com/htm/technology/kouseido_1.htm.

[104] K. Hotate and M. Tanaka, "Distributed fiber Brillouin strain sensing with 1-cm spatial resolution by correlation-based continuous-wave technique," IEEE Photonics Technology Letters, vol. 14, no. 2, pp. 179–181, 2002.

[105] E. Udd (ed), "Fiber optic sensors: an introduction for engineers and scientists,", Wiley Sciences in Pure and Applied Optics, 1991.

[106] K. T. V. Grattan and B. T. Meggitt (eds), "Optical fiber sensor technology: fundamentals," Kluwer Academic Publisher, 2000.

[106] J. M. Lopez-Higuera (ed), "Handbook of optical fibre sensing technology," Wiley, 2002.

[108] S. Yin, P. B. Ruffin, and F. T. S. Yu (eds), "Fiber optic sensors," 2nd edition, CRC Press, 2008.

CHAPTER SEVEN

Birefringence Control in Optical Fibers

7.1 PHYSICAL PARAMETERS FOR THE POLARIZATION CHARACTERIZATION IN OPTICAL FIBERS

In optics the direction of the electric field has a specific name, polarization. This might be confusing because the net electric dipole moment per unit volume, $\vec{P}$, bears the same name, polarization. In most cases, these two different physical quantities can be distinguished as "light polarization or equivalently electromagnetic wave polarization" and "electric polarization, $\vec{P}$." In this chapter we will focus on the direction of electric field, the light polarization in optical fibers and the way to characterize its states, and finally methods to control its properties in optical fibers.

Let us begin with basic nomenclature of polarization optics, assuming a monochromatic transverse electromagnetic wave propagating along z direction in vacuum. The electric field is confined in the x–y plane and its components can be described as

$$\vec{E} = E_x \hat{x} + E_y \hat{y}$$
$$E_x = A_x \exp[j(\beta z - \omega t)] \tag{7.1}$$
$$E_y = A_y \exp[j(\beta z - \omega t + \Delta)]$$

The phase retardance between E_x and E_y is denoted as Δ. And β is the propagation constant at the frequency of ω. The locus of the electric field direction on the x–y plane, or equivalently polarization state, will be determined by the phase retardance, Δ, and the amplitude ratio A_x/A_y. When E_x and E_y have the same amplitude $A_x = A_y$, the

Silica Optical Fiber Technology for Devices and Components: Design, Fabrication, and International Standards, First Edition. By Kyunghwan Oh and Un-Chul Paek.

TABLE 7.1   Summary of States of Polarization for Various Retardance Values

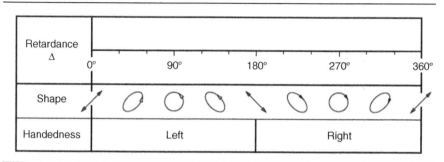

states of the polarization states are summarized in Table 7.1. As the retardance increases from 0 to 180°, the polarization state evolves from a linear polarization at 45° angle to a left elliptic, to left circular, and then to a linear polarization at 135° angle. For the retardance larger than 180°, the polarization changes from left to right handedness.

The same terminologies for the polarization states are being used in optical fibers. In a perfectly circular optical fiber waveguide, the fundamental mode L_{P01} mode is in a linear polarization state and in fact it is doubly degenerate modes, HE_{11x} and HE_{11y}, whose directions of the electric field are orthogonal to each other. In a perfectly cylindrical geometry, the polarization state of the guided mode does not change as it propagates when there is no external perturbation as schematically shown in Figure 7.1a. The incident vertical linear polarization would retain its direction along the fiber. But when there exist external perturbations along the fiber length, the retardance between electric fields in x and y directions will build up according to equation 7.1, to change the polarization states as schematically shown in Figure 7.1b. External perturbations include bending, pressure, tension, and twist, which naturally occur in fiber cabling and deployment. In some special cases external electric or magnetic fields can be applied to optical fiber as well. Deviations from the perfect circular symmetry such as the ellipticity in the core and/or the cladding also generate the retardance internally, which is not avoidable to a certain degree even in the present state-of-the-art technology.

Due to random nature of these internal and external perturbations, the output polarization states in conventional single-mode fibers cannot be predicted in an analytic manner and they are only described statistically. However, the demand to control the polarization states of the guided modes along the optical fiber is ever increasing in various optical science and engineering fields, such as recently revived coherent optical communication for ultralarge transmission capacity [2,3], and various high-precision optical sensing [4,5], biomedical imaging [6,7], and fiber coupling with polarization-sensitive optical devices [8,9].

In order to cope with these demands, polarization maintaining fibers (PMFs) have been developed since late 1970s. PMF research and development have explored high potential in the cladding engineering by imbedding various structures near the core to

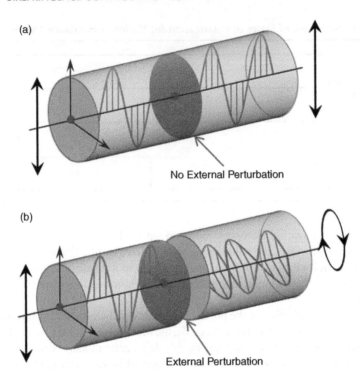

FIGURE 7.1 Schematic diagrams for evolution of the state of polarization in a perfect cylindrically symmetric optical fiber: (a) the vertical polarization is preserved along the propagation direction when there is no external perturbation; (b) the polarization state changes depending on the amount of retardance induced by the applied external perturbation [1].

induce a high level of internal anisotropy within the optical fiber. This high anisotropy in optical fibers will then break the degeneracy of the L_{P01} mode to provide different propagation constants for x and y polarizations, β_x and β_y, respectively. In other words, we make two different modes propagate instead of degenerate modes. Therefore, in PMFs the two orthogonal electric components are described by

$$
\begin{aligned}
\vec{E} &= E_x\hat{x} + E_y\hat{y} \\
E_x &= A_x \exp[j(\beta_x z - \omega t)] \\
E_y &= A_y \exp[j(\beta_y z - \omega t)]\exp(j\Delta_{ext}) \\
\delta\beta &= |\beta_x - \beta_y| \neq 0
\end{aligned}
\tag{7.2}
$$

Polarization maintaining property can be expected if the retardance due to the difference in the propagation constant over distance L, $\delta\beta L$, is significantly larger than the external perturbation contribution, Δ_{ext}. This case is schematically illustrated in Figure 7.2.

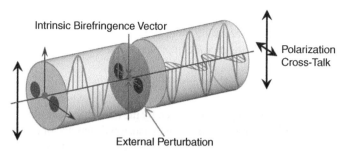

FIGURE 7.2 Schematic diagram to describe polarization preservation in polarization maintaining fiber. By the external perturbation some portion of light in the initial polarization state, the vertical polarization in this case, is coupled to the other orthogonal polarization state [1].

In Figure 7.2 the incident light polarization is aligned to the vertical direction of the fiber. External perturbation along the fiber will induce the mode coupling from the initial vertical polarization mode to the horizontal polarization mode, to result in polarization cross-talk. Once two orthogonal polarization modes exist together, there will be beating between them. However, the amount of deviation from the initial vertical polarization can be suppressed by reducing the coupling between the orthogonal polarization modes. Therefore, in the PMF design, the main issue is to increase $\delta\beta$ and minimize the coupling between the orthogonal polarization modes.

As shown in Table 7.1 there are two pairs of orthogonal polarization modes. In the linear polarization two orthogonal polarization states are represented as linear arrows, the retardance angle $0°$ and $180°$, which can be also named as horizontal and vertical polarization states. In contrast to these linear polarization states, we do have right- and left-handed circular polarizations that are represented as circles with opposite rotation directions. In most of PMFs we are dealing with linear polarizations and the difference in their propagation constant, $\delta\beta = |\beta_x - \beta_y|$, is referred as linear birefringence. Circular birefringence will be discussed in the latter part of this chapter.

In this section, basic parameters used in PMF characterization will be reviewed along with discussion on their physical meanings.

The polarization properties in a PMF can be modeled as a discrete linearly birefringent element whose retardance $R(z)$ is proportional to fiber length z:

$$R(z) = \delta\beta z \tag{7.3}$$

Two commonly used parameters that are useful in describing fibers with length-invariant properties are the linear birefringence B and the polarization mode beat length L_p at which the polarization state is periodically repeated:

$$B = \frac{\lambda}{2\pi}\delta\beta = \frac{\lambda}{2\pi}\frac{R(z)}{z} = \delta n_{\text{eff}}(\lambda) \tag{7.4}$$

$$L_p = \frac{2\pi}{\delta\beta} = \frac{\lambda}{B} \tag{7.5}$$

Here λ is the free-space wavelength of light and δn_{eff} is the difference in effective indices of two orthogonal polarization modes.

In general, the modal birefringence B comprises three anisotropy contributions [10]:

$$B = B_{\text{G}} + B_{\text{S}}^{\text{i}} + B_{\text{S}}^{\text{e}} \qquad (7.6)$$

Here B_{G} is the geometrical component induced by the core shape anisotropy. B_{S}^{i} is the self-stress component induced by the thermal expansion difference of the asymmetric core and the outer stress component. B_{S}^{e} is induced by the stress-applying parts imbedded in the cladding.

Birefringence of conventional single mode is in the range of $B = 10^{-6}$ to 10^{-5}. In recent high-birefringence fibers the nominal value of modal birefringence, B, exceeds 10^{-4}. In the case of low-birefringence fibers, B is in the range of 10^{-9} [11].

The physical meaning of the mode beat length is illustrated in Figure 7.3. When a linearly polarized incident light is aligned to 45° with respect to horizontal axes as in Figure 7.3, it excites both vertical and horizontal polarization modes. Due to the difference in the propagation constant, $\delta\beta$, the state of polarization evolves in a very similar manner as in Table 7.1. The propagation length corresponding to the retardance of 2π is the beat length, L_{p}.

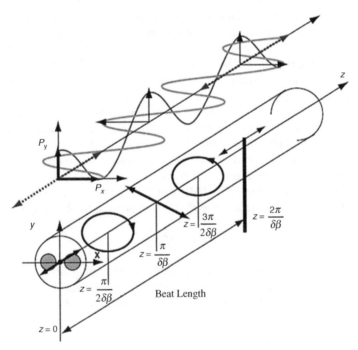

FIGURE 7.3 Evolution of polarization states in a linear birefringent optical fiber. The incident polarization state is aligned at 45° exciting both horizontal and vertical polarization modes. Due to the accumulated retardance, the polarization state evolves in a cyclic manner and its period along the z direction is beat length [12].

Typical value of the beat length L_p for recent high-birefringence fibers is a few millimeters. Conventional single-mode fibers have a typical beat length of a few centimeters and low-birefringence fiber beat length is as long as a few tens of meters.

In linear high-birefringence fibers, the two orthogonal axes with different propagation constants are usually called "fast" and "slow" axes, depending on the magnitude of the phase velocity ($v = c/n_{eff}$), or equivalently the effective indices (n_{eff}), and the propagation constant ($\beta = 2\pi n_{eff}/\lambda$) as follows:

$$v^{slow}\left(= \frac{c}{n_{eff}^{slow}}\right) < v^{fast}\left(= \frac{c}{n_{eff}^{fast}}\right)$$

$$n_{eff}^{slow} > n_{eff}^{fast} \qquad (7.7)$$

$$\beta^{slow}\left(= \frac{2\pi n_{eff}^{slow}}{\lambda}\right) > \beta^{slow}\left(= \frac{2\pi n_{eff}^{fast}}{\lambda}\right)$$

In the slow axis, the phase velocity is slower because the effective index is larger than the fast axis. Usually slow axis is taken as the horizontal axis in most PM fibers by convention.

Another important pair of physical parameters to characterize PMF is the crosstalk (CT) and h parameter, which are used to describe the amount of power coupling between the two orthogonal polarization states as described in Figure 7.4. In the figure a certain amount of power in the horizontal linear polarization is launched to a PMF. After a certain propagation length, the power is coupled to the vertical axis due to internal imperfections within the PMF or external perturbations. The CT and h parameter for a PMF length L are defined as follows:

$$CT = 10 \log\left(\frac{P_y}{P_x}\right)$$

$$h = \frac{\tan^{-1}(P_y/P_x)}{L} \approx \frac{1}{L}\left(\frac{P_y}{P_x}\right) \qquad (7.8)$$

Typical values of CT in present PMFs are in the range of -30 to $-35\,dB$ for $100\,m$, which corresponds to h of 10^{-5} to $10^{-6}\,m^{-1}$.

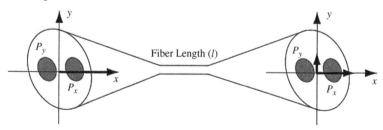

FIGURE 7.4 Schematic diagram for polarization cross-talk in a PMF. After propagation of fiber length l, the initial power $P_x(z=0)$ along the x axis is partly coupled to y axis to result in $P_x(z=l)$ and $P_y(z=l)$.

Thus far, we only considered linear birefringence, where the propagation constant of the light linearly polarized in the slow axis differs from that in the fast axis. In circular birefringence fiber, we can similarly assume two different propagation constants for the lights whose states of polarization are either the right-handed or the left-handed circular polarization. In the case of low linear birefringence fiber, we can induce the circular birefringence by simply twisting the fiber [13,14]. Twist introduces a torsional stress around the core and this in turn leads to an optical activity in proportion to the twist by the photoelastic effect. Assuming the absence of linear birefringence, a fiber under a uniform twist rate ξ can be modeled as a discrete polarization rotation element with rotation angle $\Omega(z)$ that increases linearly with fiber length:

$$\Omega(z) = g'\xi z = \alpha z \tag{7.9}$$

where $g' = 0.073$ for silica [14,15] and α is the optical rotation per unit length. The two orthogonal polarization modes are now left and right circularly polarized lights and they experience the difference in the propagation constants:

$$\delta\beta_{\text{circ}} = 2g'\xi = 2\alpha \tag{7.10}$$

In a low linear birefringence fiber in the absence of twist, the magnitude of α will determine its circular birefringence.

7.2 REPRESENTATION OF THE STATE OF POLARIZATION IN OPTICAL FIBER USING POINCARÉ SPHERE

The state of polarization and its evolution over the propagation distance in an optical fiber can be very efficiently represented in a visual manner using a Poincaré sphere [16]. The Poincaré sphere is a graphical tool in real, three-dimensional space that allows convenient description of polarized signals and of polarization transformations caused by propagation through devices. Any state of polarization can be uniquely represented by a point on or within a unit sphere centered on a rectangular (x, y, z) coordinate system. The coordinates of the point are the three normalized Stokes parameters (S_1, S_2, S_3) describing the state of polarization.

Let us assume a plane wave propagating in z direction as in equation 7.1, whose electric field components are represented in real number as

$$\vec{E}(z, t) = E_x\hat{x} + E_y\hat{y}$$
$$E_x(z, t) = A_x \cos(\beta z - \omega t) \tag{7.11}$$
$$E_y(z, t) = A_y \cos(\beta z - \omega t + \Delta)$$

We have already confirmed that the electric field locus can be described by an ellipse whose equation is given by

$$\frac{E_x^2(z,t)}{A_x^2} + \frac{E_y^2(z,t)}{A_y^2} - 2\frac{E_x(z,t)E_y(z,t)}{A_xA_y}\cos\Delta = \sin^2(\Delta) \qquad (7.12)$$

If we take the time average of equation 7.12, we will get

$$(A_x^2 + A_y^2)^2 - (A_x^2 - A_y^2)^2 - (2A_xA_y \cos\Delta)^2 = (2A_xA_y \sin\Delta)^2 \qquad (7.13)$$

Stokes parameters for a plane wave are defined as follows:

$$\begin{aligned}
S_0 &= A_x^2 + A_y^2 \\
S_1 &= A_x^2 - A_y^2 \\
S_2 &= 2A_xA_y \cos\Delta \\
S_3 &= 2A_xA_y \sin\Delta
\end{aligned} \qquad (7.14)$$

The Stokes vector whose components are S_i as defined above can be represented by two angle parameters, χ and ψ, on a unit circle:

$$\vec{S} = \begin{bmatrix} S_0 \\ S_1 \\ S_2 \\ S_3 \end{bmatrix} = S_0 \begin{bmatrix} 1 \\ \cos(2\chi)\cos(2\psi) \\ \cos(2\chi)\sin(2\psi) \\ \sin(2\chi) \end{bmatrix} \qquad (7.15)$$

$$\begin{aligned}
\tan(2\psi) &= \frac{S_2}{S_1}, \quad 0 \le \psi \le \pi \\
\sin(2\chi) &= \frac{S_3}{S_0}, \quad -\frac{\pi}{4} \le \chi \le \frac{\pi}{4}
\end{aligned} \qquad (7.16)$$

By taking S_1, S_2, and S_3 as the rectangular coordinates in three-dimensional space, the state of polarization is projected as a point on a sphere of unit radius, which is the Poincaré sphere as shown in Figure 7.5.

Now let us consider some of states of polarization using the Stokes vector and corresponding points in the Poincaré sphere.

Linearly horizontally polarized light holds $A_y = 0$ and it is represented as

$$S = I_0 \begin{bmatrix} 1 \\ 1 \\ 0 \\ 0 \end{bmatrix}, \quad I_0 = A_x^2 \qquad (7.17)$$

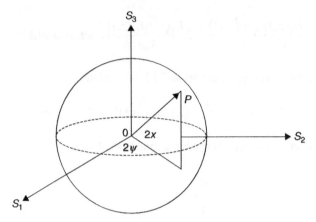

FIGURE 7.5 Poincaré sphere to represent the state of polarization P. Note that the sphere is drawn on the (S_1, S_2, S_3) rectangular coordinate system.

Linearly vertically polarized light holds $A_x = 0$ and it is represented as

$$S = I_0 \begin{bmatrix} 1 \\ -1 \\ 0 \\ 0 \end{bmatrix}, \quad I_0 = A_y^2 \tag{7.18}$$

Similarly $+45°$ polarized light has $A_x = A_y$ and $\Delta = 0$, and they are represented as

$$S = I_0 \begin{bmatrix} 1 \\ 0 \\ 1 \\ 0 \end{bmatrix}, \quad I_0 = A_x^2 + A_y^2 \tag{7.19}$$

In the case of $-45°$ polarized light, we have $A_x = A_y$ and $\Delta = \pi$ and its Stokes vector is

$$S = I_0 \begin{bmatrix} 1 \\ 0 \\ -1 \\ 0 \end{bmatrix}, \quad I_0 = A_x^2 + A_y^2 \tag{7.20}$$

For right-handed circular polarized light, we have $A_x = A_y$ and $\Delta = \pi/2$ and its Stokes vector is

$$S = I_0 \begin{bmatrix} 1 \\ 0 \\ 0 \\ 1 \end{bmatrix}, \quad I_0 = A_x^2 + A_y^2 \tag{7.21}$$

For left-handed circular polarized light, we have $A_x = A_y$ and $\Delta = 3\pi/2$ and its Stokes vector is

$$S = I_0 \begin{bmatrix} 1 \\ 0 \\ 0 \\ -1 \end{bmatrix}, \quad I_0 = A_x^2 + A_y^2 \tag{7.22}$$

These polarization states are marked on the Poincaré sphere as shown in Figure 7.6.

Because a state of polarization is represented by a point, a continuous evolution of polarization can be represented as a continuous path on the Poincaré sphere. A path can also record the polarization history of a signal, for example, in response to changing strain applied to a birefringent fiber. Ulrich and Simon [14] reported the Poincaré presentation of various birefringent fibers and impact of twist. In Figure 7.7, we have three types of typical birefringence.

In the case of linear birefringence as shown in Figure 7.7a, the polarization state makes a circular trajectory about a point on the equator, which represents the linear polarization state. This is the same case we have discussed in Figure 7.2, where the external perturbation induced the cross-talk between two orthogonal linear polarization states to result in a cyclic evolution around the initial linear polarization state.

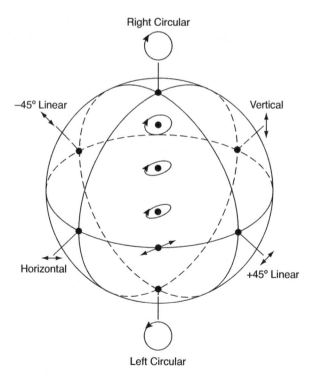

FIGURE 7.6 Representation of various states of polarization on the Poincaré sphere.

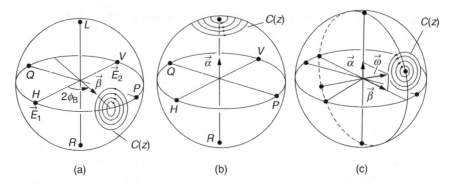

FIGURE 7.7 Evolution of polarization, $C(z)$, for various kinds of birefringence: (a) linear, (b) circular, and (c) elliptic.

More detailed discussion on the other polarization state evolution is referred to report of Ulrich and Simon [14].

7.3 CLASSIFICATIONS OF LINEAR POLARIZATION MAINTAINING FIBERS

In terms of linear polarization, polarization maintaining fibers can be first categorized into high-birefringence fiber (HBF) and single-polarization fiber (SPF). HBF supports two orthogonal linear polarization modes over a wide wavelength range and in the spectral domain it is not a single-mode but a dual-mode fiber.

HBF can be further categorized by the origin of its birefringence

$$B = B_G + B_S^i + B_S^e \tag{7.6}$$

Here B_G is the geometrical component induced by the core shape anisotropy. B_S^i is the self-stress component induced by the thermal expansion difference of the asymmetric core and the outer stress component, which is related to interior of the core. B_S^e is induced by the stress-applying parts imbedded in the cladding, which is exterior to the core.

B_G is directly related to the anisotropy in the core shape and elliptic core fiber, side-pit fiber, and side tunnel fiber belong to this category. The schematic cross-sections of these B_G-HBFs, geometrically induced high-birefringence fibers, are shown in Figure 7.8.

It is noted that these B_G-HBFs have contribution from B_S^i as well. In elliptic core fibers [17–20], the anisotropy in the core is obviously induced by the length difference in the major axis, a, and the minor axis, b, which is parameterized by the core eccentricity $\varepsilon = (a/b) - 1$. In the side-pit fibers [21] the core composition is no longer uniform with GeO_2 doping, but low refractive index "pits" that are usually composed of B_2O_3-doped silica are added on both sides. The core shape is subsequently deformed. Similar concept was applied to side tunnel fibers [22], where the side pits were replaced by side air tunnels.

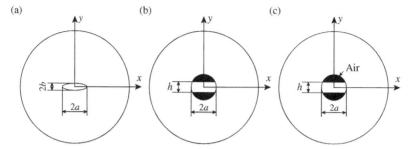

FIGURE 7.8 Schematic cross-section of B_G-HBFs, geometrically induced high-birefringence fibers: (a) elliptic core fiber, (b) side-pit fiber, and (c) side tunnel fiber. The slow and fast axes are along x and y directions, respectively.

It is noteworthy that side-pit fibers further evolve to PANDA fibers, and side tunnel fibers are directly related to recent hole-assisted single-polarization fibers, which will be discussed in the following sections. Despite relatively simple waveguide structures and easy fabrication, all of these B_G-HBFs suffered from high optical loss and furthermore their noncircular core has caused significant amount of splice loss with conventional single-mode fibers. Due to these shortcomings, B_G-HBFs are not being preferred in long-distance applications:

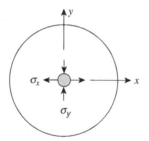

$$B_S = C(\sigma_x - \sigma_y)$$
$$C = 3.36 \times 10^{-5} \text{ nm}^2/\text{kg} \tag{7.23}$$

There is another avenue of generating birefringence in optical fiber maintaining the circular shape of the core, B_S-stress-induced birefringence, which is in fact the mainstream in PMFs. In the B_S-HBFs stress-applying parts (SAPs) are introduced near the circular core to imbed anisotropy in mechanical stress based on the thermal expansion difference between the core glass and SAPs. These anisotropic stress distributions subsequently induce the anisotropy in effective indices of guided modes in two orthogonal directions by photoelastic effect, which is described by equation 7.23.

Here C is the photoelastic coefficient of silica optical fiber and σ_x and σ_y are the stress components along x and y axes.

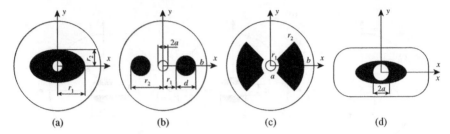

| (a) | (b) | (c) | (d) |

FIGURE 7.9   Schematic cross-section of B_S-HBFs, stress-induced high-birefringence fibers: (a) elliptic clad fiber, (b) PANDA fiber, (c) bow-tie fiber, and (d) flat clad fiber. The slow and fast axes are along x and y directions, respectively. Note that the horizontal axes are the slow axes and the vertical axes are the fast ones.

B_S-HBFs include elliptic clad fiber, PANDA fiber, bow-tie fiber, and flat clad fiber, whose schematic cross-sections are shown in Figure 7.9. Here the SAPs are illustrated in the dark areas and they are located near the circular core. In most cases SAP is made of B_2O_3–SiO_2 glass due to its high thermal expansion coefficient and low refractive index.

In elliptic clad fibers [23–25], the B_2O_3–SiO_2 glass inner cladding is deformed to an ellipse with an ellipticity of $\varepsilon = (r_1/r_2) - 1$ as in Figure 7.9a, to provide an anisotropic mechanical stress over the circular GeO_2–SiO_2 glass core. PANDA fiber [1,12,26–29] is the most favored commercial PMF and its mechanical anisotropy is provided by two circular B_2O_3–SiO_2 glass SAPs on both sides of the core. It has flexible waveguide parameters such as r_1, r_2, and doping concentration of SAP as shown in Figure 7.9b to control the optimal birefringence with a low optical loss. Bow-tie fibers [30–33] have an advantage to finish all the processes in MCVD and the bow-tie-shaped SAPs provide anisotropic stress. The angle of SAP is in most cases $90°$ and its dimensions are controlled by r_1 and r_2 as in Figure 7.9c. Flat clad fibers [34–37] are very close to the elliptic clad fiber for the inner cladding SAP structure but they add more anisotropy by noncircular cladding. Mechanical pressure is applied to the circular preform to deform both the outer cladding and low viscosity inner cladding, maintaining the circular core as in Figure 7.9d. Flat clad fibers can have practical advantages in orienting the fiber axes. In addition to these fiber structures, recently various HBFs are being reported based on microstructured air–silica fibers and we will discuss them separately in this chapter.

The HBFs discussed thus far carry dual modes in the spectral domain. Single-polarization fibers (SPFs), which transmit only one of two orthogonally polarized HE_{11} modes, would be strongly preferred for long-distance coherent communications applications, due to absence of polarization cross-talk, polarization-dependent loss, and polarization dispersion [38–46]. The most common SPF types are listed in Figure 7.10, along with their schematic structures as reported by Li et al. [38].

Making SPF requires more complicated fiber designs than HBFs, with high birefringence and accurate manufacturing technology to control the fiber fundamental mode cutoff wavelengths. The design goal of SPF is to guide only one polarization

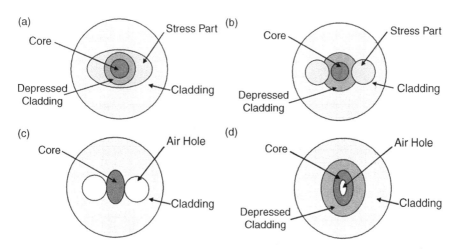

FIGURE 7.10   Single-polarization fiber (SPF) designs using stress birefringence. (a) B_2O_3-doped silica elliptic stress-applying cladding; (b) two circular boron-doped stress-applying parts, SPF designs using hole-assisted structures; (c) dual air hole; (d) elliptic central air hole [38]. Note that horizontal axis in "a" and "b" and the vertical axis in "c" and "d" are the slow axes.

mode in the slow axis by providing an efficient loss mechanism to the other polarization mode in a highly selective manner. This purpose can be accomplished by designing a fiber so that the effective index of the unwanted polarization mode in the fast axis is below the index of the cladding, while the effective refractive index of the preferred polarization in the slow axis is above that of the cladding in a certain spectral range of interests.

Selecting cutoff wavelength can be achieved by introducing a low refractive index inner cladding, which is usually called W-type fiber [47–49]. The depressed inner cladding selectively provides a cutoff for the HE_{11} mode in the fast axis, whose effective index is lower than the mode in the slow axis, leaving behind only the HE_{11} mode in the slow axis guided. Therefore, the SPFs shown in Figure 7.10a, b, and d have depressed inner cladding and that in Figure 7.10c has equivalent air holes near the core that serves the same purpose.

Two different approaches have been adopted in SPF fabrications. One approach relying on the stress-induced birefringence from B_2O_3–SiO_2 glass stress-applying part combined with depressed inner cladding is shown in Figure 7.10a and b, which was experimentally demonstrated in Refs [42,44,45]. Figure 7.10a shows one design with B_2O_3–SiO_2 glass as a stress-applying part, which is very similar to Figure 7.9a except the depressed inner cladding. Figure 7.10b is another design with PANDA structure. In both designs, the circular core is surrounded by a lower index inner depressed cladding doped with fluorine, where differential cutoff wavelengths are provided for the two polarization modes.

Noda et al. [50] have reported a very comprehensive review on PMFs and proposed a systematic classification of PMFs similar to the chart shown in Figure 7.11.

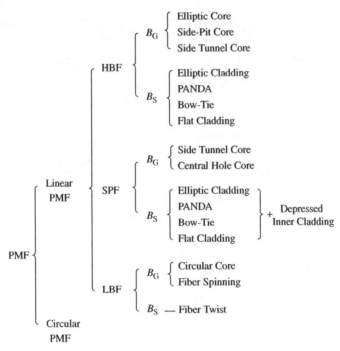

FIGURE 7.11 Classification of polarization maintaining fibers (PMF), high-birefringence fiber (HBF), single-polarization fiber (SPF), low-birefringence fiber (LBF), geometrical Birefringence (B_G), and stress-induced birefringence (B_S).

7.4 FABRICATION METHODS FOR HIGH BIREFRINGENCE FIBERS

Fabrication processes of some of typical HBFs are schematically shown in Figure 7.12, which are variations of modified chemical vapor deposition (MCVD), outside vapor deposition (OVD), and vapor axial deposition (VAD) processes described in Chapter 2.

The easiest way to fabricate the elliptic core fibers or elliptic cladding fibers is by drawing the flat cladding preform as shown in Figure 7.12a [17]. The thickness of removed preform cladding will determine the ellipticity of the core in the fiber. A higher temperature for fiber drawing is necessary in order to make the fiber cladding round by surface tension, which will induce the desired anisotropy in the core.

The reduced pressure collapsing method is shown in Figure 7.12b and this has been widely used for producing elliptic cladding, elliptic jacket [23–25], and elliptic core preforms with a large ellipticity. In MCVD process for conventional single-mode fibers, a positive pressure is applied during the substrate tube collapse to minimize ellipticity in the core and inner cladding, yet in this technique a reduced pressure is maintained to intentionally introduce anisotropy.

Bow-tie fibers are also based on modification of MCVD process, where the B_2O_3-doped silica layers are etched away at two substrate tube positions with $180°$ angular

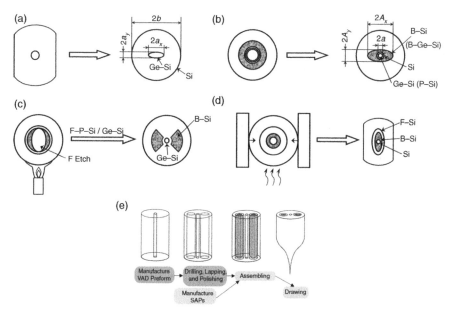

FIGURE 7.12 Schematic diagram of fabrication process for some of typical high-birefringence fibers (HBFs): (a) elliptic core fiber, (b) elliptic cladding fiber, (c) bow-tie fiber, (d) flat cladding fiber, and (e) PANDA fiber [50].

separation by keeping the substrate tube stationary (see Figure 7.12c). Usually high temperature fluorine gas-phase etching process is used to provide high axial uniformity and a large cross-sectional area for the stress-applying parts [32]. The first layer of $F-P_2O_5-SiO_2$ glass serves as a reliable stopper for fluorine-liberated gas etching. Over the etched B_2O_3-doped silica layers higher viscosity core layers of GeO_2-doped silica glass are deposited and then collapsed in conventional MCVD process.

Flat clad fibers are fabricated by a pressing method as seen in Figure 7.12d, which is based on the same mechanism of the easier deformation of the boron-doped cladding than of the core [35]. In contrast to the elliptic core case in Figure 7.12a, the drawing temperature of this fiber should be kept low enough to maintain the flat cladding. In most cases a high tension is accompanied during the drawing process. This structure has an advantage over the others of the self-alignment of the shape to the birefringence principal axis. Alignment procedures are easier when making fiber couplers or coiling fibers.

PANDA fiber fabrication process is illustrated in Figure 7.12e. Single-mode fiber preform is first fabricated by conventional VAD technique. The positions where the SAPs are to be located are drilled in circles and their inner surfaces are polished. The B_2O_3-doped silica glass SAPs are prepared separately in rods using either MCVD or VAD. SAPs and drilled preform are assembled and then drawn to fiber with an appropriate pressure control [29].

Except the elliptic core fiber fabrication process, all the others are using B_2O_3-doped silica glass as the stress-applying parts. By using heavily doped $B_2O_3-SiO_2$

glass, we can flexibly control the viscosity and thermal expansion coefficient of the SAP layers. An important factor determining the ellipticity is the difference of the viscosity, that is, softening temperature difference between the support tube and the doped layer. The temperature difference ΔT between the room temperature and softening temperature of the doped silica glass at a constant viscosity is given [51] as the first linear approximation by

$$\Delta T = \Delta T_0 + m(Q_m/Q_0 - 1)(T_r + \Delta T_0) \tag{7.24}$$

Here ΔT_0 is the temperature difference between the room temperature T_r and softening temperature of the pure silica glass. Q_0 and Q_m are the activation energies of the pure silica glass and the doped silica glass, respectively. m is the dopant concentration in mole percent. Here $Q_m/Q_0 = 0.984$ and 0.955 for B_2O_3–SiO_2 glass SAP and GeO_2–SiO_2 glass core, respectively, at a viscosity of $10^{14.5}$ P. The pure SiO_2 glass cladding and the GeO_2–SiO_2 glass core float in the B_2O_3–SiO_2 glass SAPs collapsing. The thermal expansion coefficient for the B_2O_3–SiO_2 glass and GeO_2–SiO_2 glass is given [51]:

$$\alpha = (1 - m)\alpha_{SiO_2} + m\alpha_d \tag{7.25}$$

where m is mole percent of doping concentration and α_{SiO_2} and α_d are thermal expansion coefficients of SiO_2 glass and dopant oxide, respectively. The thermal expansion coefficients of important oxide glasses are listed in Table 7.2.

Usually the doping concentration in the GeO_2–SiO_2 glass core is 4–5 mol% and in the B_2O_3–SiO_2 glass SAP the concentration could reach $\sim$10 mol%, which generates the anisotropic mechanical stress around the core.

7.5 CONTROL OF BIREFRINGENCE BY WAVEGUIDE DESIGN IN BIREFRINGENT FIBERS

In this section, we will review the waveguide parameters for some of typical PMFs and their impacts over birefringence control. We are going to discuss four types of HBFs whose birefringence can be analytically calculated by formulae: (1) elliptic core fiber, (2) elliptic cladding fiber, (3) bow-tie fiber, and (4) PANDA fiber.

Following the initial stress analysis made by Kaminow [52,53] using the slab waveguide model, various types of numerical analyses for the stress components have been presented for HBFs. Stress components in elliptic core and elliptic cladding fibers are analyzed in Refs [54–58] and the stress components induced by the SAPs located at both sides of the core such as in PANDA and bow-tie fibers have been analyzed in Refs [59–61].

TABLE 7.2 Thermal Expansion Coefficients of SiO_2 and Dopants used in PMFs

	SiO_2	GeO_2	B_2O_3	P_2O_5
α (10^{-7} °C^{-1})	5.4	70	100	140

7.5.1 Elliptic Core Fiber

In an elliptic core fiber shown in Figure 7.8a, the modal birefringence B at the center of the core, $(0, 0)$, is given in an analytic form [59,61] as follows:

$$B_0 = \frac{\Delta\alpha^{core}\,\Delta T^{core}\,CE}{2(1-\nu)}\frac{a-b}{a+b}\left[2 - \frac{3ab(a+b)^2}{r_{clad}^4}\right] \tag{7.26}$$

$$B_0 = B_m B_0* \tag{7.27}$$

$$B_m = \frac{\Delta\alpha^{core}\,\Delta T^{core}\,EC}{2(1-\nu)}, \qquad B_0* = \frac{a-b}{a+b}\left[2 - \frac{3ab(a+b)^2}{r_{clad}^4}\right] \tag{7.28}$$

Here ΔT^{core} is the temperature difference between the room temperature and softening temperature of the core glass, and $\Delta\alpha^{core}$ is the difference of thermal coefficient between the core and the cladding. E and C are the Young's modulus and the photoelastic constant of silica glass, respectively. ν is Poisson's ratio of silica glass. Therefore, B_m is dependent only on the material properties. The physical constants for B_m are summarized in Table 7.3.

In contrast, B_0* represents the effects of elliptic core geometry at the center, where a and b are the major and minor axes of the elliptic core and r_{clad} is the fiber cladding radius.

Tsai et al. [61] also calculated the average core birefringence, $\bar{B}$, which is given as

$$\bar{B} = \frac{C\iint_{core}(\sigma_x - \sigma_y)dA}{\iint_{core}dA} = B_m\bar{B}* \tag{7.29}$$

It is expected that $B_0 \approx \bar{B}$ and $B_0* \approx \bar{B}*$ when the elliptic core is small in comparison to the cladding. The two values calculated in Ref. [61] are plotted as a function of $(a-b)/(a+b)$ in Figure 7.13.

In the elliptic core fiber, the birefringence related to the geometrical effects, $\bar{B}*$ and B_0*, has its maximum value at a certain optimal choice of the ellipticity for a given core dimension. It is found that birefringence of the elliptic core fiber linearly increases with its core ellipticity up to a certain optimal value and then decreases. It is also noted that a smaller core provides a higher birefringence value.

For the material-dependent birefringence, B_m, it is also noteworthy that $\Delta\alpha^{core}$ and ΔT^{core} terms are proportional to GeO_2 dopant concentration in the elliptic core glass, which is in turn proportional to relative refractive index difference, Δ, as

TABLE 7.3 Physical Constants for the Material Contribution in Birefringence [62]

ΔT (°C)	1000–1500
E (kg/mm^2)	7830
C (10^{-6} mm^2/N)	3.57
ν	0.186

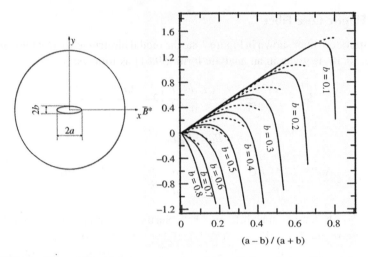

FIGURE 7.13   Plots of $\bar{B}^*$ and B_0^* as a function of $(a-b)/(a+b)$ in solid and dotted lines, respectively, for various elliptic core sizes, b. Here the major and minor axes of the elliptic core, a, and b are normalized to the cladding radius r_{clad} [61].

shown in Chapter 2. Therefore, the material-dependent birefringence B_m is proportional to Δ^2 as Payne et al. reported [11].

Therefore, in elliptic core fiber, birefringence increases with (1) a larger refractive index between the core and cladding ($\propto \Delta^2$), (2) a larger ellipticity, and (3) a smaller core diameter. Elliptic fibers show a typical birefringence B of $4-5 \times 10^{-4}$ at the wavelength of 850 nm and the relative refractive index Δ was ~4% as reported in Refs [17–20]. Elliptic core fiber is very easy to fabricate but the requirements of high Δ, high ε, and small core usually result in very high optical loss over 80 dB/km, which limits the elliptic core fibers within short span applications.

Total birefringence is composed of three contributions as described in Section 7.1:

$$B = B_G + B_S^i + B_S^e \qquad (7.6)$$

Sakai and Kimura [58] calculated the dispersion characteristics in the elliptic core fiber birefringence. In the elliptic core fiber $B = B_G + B_S^i$, with $B_S^e = 0$, and it is expressed as [58]

$$B_G = n\varepsilon\Delta^2 G(v) \qquad (7.30)$$

$$G(v) = \frac{w^4}{v^4}\left[u^2 + (u^2 - v^2)\left\{\frac{J_0(u)}{J_1(u)}\right\}^2 + uw^2\left\{\frac{J_0(u)}{J_1(u)}\right\}^3\right] \qquad (7.31)$$

Here the waveguide parameters are defined as $u = k(n_1^2 - n_{eff}^2)^{1/2}\bar{a}$, $w = k(n_{eff}^2 - n_2^2)^{1/2}\bar{a}$, $v = (u^2 + w^2)^{1/2} = k\bar{a}n_1\sqrt{2\Delta}$, $\Delta = (n_1 - n_2)/n_1$, $\bar{a} = (a+b)/2$, and $\varepsilon = 1 - (b/a)$. n_1 and n_2 are the refractive indices of the core and cladding, respectively. n_{eff} is the effective index of the guided mode, and u and w are the transverse propagation parameters as defined in Chapter 2. v is the normalized frequency

$$B_S^i = \frac{w^2}{v^2} \varepsilon B_m \qquad (7.32)$$

where B_m is as defined in equation 7.28.

Because the elliptic core itself is where most light is guided through, the birefringence is inevitably highly dispersive, or, in other words, highly dependent on wavelength in elliptic core fibers. The dispersive characteristics are shown in Figure 7.14 [50] for the elliptic core fiber in comparison to PANDA fiber. It is noted that the elliptic core fiber's birefringence, $B = B_G + B_S^i$, originated from the anisotropy in the core results in highly dispersive characteristics. But in the case of PANDA fiber the core and SAP are separate and unless the optical field is very weakly guiding or equivalently it has a small v value, its birefringence, $B = B_S^e$, is almost independent of wavelength.

7.5.2   Elliptic Cladding Fiber

In an elliptic cladding fiber as shown in Figure 7.9a, we assume the core is a perfect circle with a uniform doping concentration. The modal birefringence B at the center of the core, $(0, 0)$, is given in an analytic form [59,61] as follows:

$$B_0 = \frac{\Delta\alpha^{SAP}\,\Delta T^{SAP}CE}{2(1-\nu)}\frac{r_1-r_2}{r_1+r_2}\left[2 - \frac{3r_1r_2(r_1+r_2)^2}{r_{clad}^4}\right] \qquad (7.33)$$

$$B_0 = B_m B_0{}^* \qquad (7.34)$$

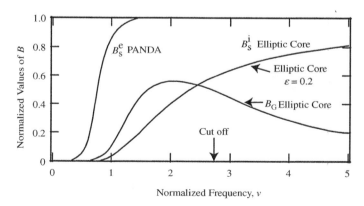

FIGURE 7.14 Dispersive characteristics of three birefringence terms, B_G, B_S^i, and B_S^e, in PANDA fiber and elliptic core fiber, as a function of normalized frequency [50].

$$B_m = \frac{\Delta\alpha^{SAP} \, \Delta T^{SAP} EC}{2(1-\nu)}, \qquad B_0^* = \frac{r_1 - r_2}{r_1 + r_2}\left[2 - \frac{3r_1 r_2 (r_1 + r_2)^2}{r_{clad}^4}\right] \qquad (7.35)$$

Here ΔT^{SAP} is the temperature difference between the room temperature and softening temperature of the SAP glass, and $\Delta\alpha^{SAP}$ is the difference of thermal coefficient between the SAP and the cladding. Note that r_1 and r_2 are the major and minor axes of the inner cladding (see Figure 7.9a). Equation 7.33 for the elliptic cladding fiber is in fact in the identical form as the elliptic core fiber case (equation 7.26) because the perfectly circular and uniform core does not contribute any birefringence.

In the elliptic cladding fiber the stress distribution around the core is almost uniform and we have $B_G = B_S^i = 0$ and $B = B_S^e$. Therefore, the elliptic cladding fiber would show a nondispersive behavior as PANDA fiber in Figure 7.14.

The average birefringence at the core of the elliptic cladding fiber defined in equation 7.29 is plotted in Figure 7.15.

In comparison to elliptic core fiber in Figure 7.13, the birefringence in the elliptic cladding is in general larger and the $\bar{B}^*$ and B_0^* are almost identical. It is noteworthy in equation 7.33 that $\Delta\alpha^{SAP}$ and ΔT^{SAP} terms are proportional to B_2O_3 dopant concentration in the elliptic cladding glass, which is now independent of the relative refractive index difference, Δ, of the core. Therefore, high dopant concentration in the elliptic cladding can be attempted without sacrificing the optical loss as in the case of elliptic core fibers.

Elliptic cladding could find advantages, therefore, over elliptic core fiber in the following aspects: (1) birefringence is only due to stress components, $B = B_S^e$, which is nondispersive near the cutoff conditions; (2) birefringence can be increased with

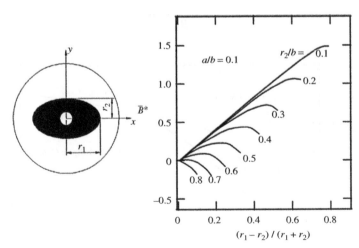

FIGURE 7.15 Plots of $\bar{B}^*$ as a function of $(r_1 - r_2)/(r_1 + r_2)$ for various elliptic cladding sizes, r_2. Here the core radius is 0.1 times the cladding radius b [61].

further doping of elliptic cladding independently keeping the core in the optimal loss condition. However, the fabrication process is rather complicated and the yield of qualified fiber is significantly lower than other types of PMFs. Heavy doping of B_2O_3 near the core significantly increases optical loss [29] and therefore there is in fact limitation of doping level in the elliptic cladding.

7.5.3 Bow-Tie Fibers

In a bow-tie fiber shown in Figure 7.9c, the modal birefringence B at the center of the core, (0, 0), is given in an analytic form [60,61] as follows:

$$B = \frac{2}{\pi} B_m \sin(\phi) \left\{ 2 \ln \frac{r_2}{r_1} - \frac{3}{2b^4} (r_2^4 - r_1^4) \right\} \qquad (7.36)$$

$$B_0^* = \frac{2 \sin(\phi)}{\pi} \left\{ 2 \ln \frac{r_2}{r_1} - \frac{3}{2b^4} (r_2^4 - r_1^4) \right\} \qquad (7.37)$$

where r_1 and r_2 are the inner and outer radii of the SAP, respectively. ϕ is the angle of the SAP spread out and usually it is $\pi/2$. Note that b is the cladding radius as in Figure 7.9c.

In bow-tie fibers, we have $B_G = B_S^i = 0$ and $B = B_S^e$. Therefore, the bow-tie fiber shows a nondispersive behavior as the PANDA fiber in Figure 7.14. The average birefringence at the core of the bow-tie fiber defined in equation 7.29 is plotted in Figure 7.16.

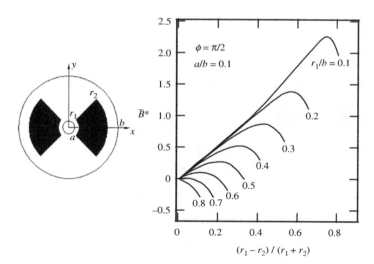

FIGURE 7.16 Plots of $\bar{B}^*$ as a function of $(r_1 - r_2)/(r_1 + r_2)$ for various inner radius of SAP, r_1. Here the core radius is 0.1 times the cladding radius b [61].

Assuming the small core relative to the cladding, the values in $\bar{B}^*$ and B_0^* were almost identical as in the case of elliptic cladding. It is also clear that the birefringence will have maximum when $\phi = \pi/2$.

7.5.4   PANDA Fibers

In a PANDA fiber shown in Figure 7.9b, the modal birefringence B at the center of the core, $(0, 0)$, is given in an analytic form [60,61] as follows:

$$B = 4B_m \left(\frac{r_2 - r_1}{r_2 + r_1} \right) \left\{ 1 - \frac{3}{b^4} (r_2 - r_1)^4 \right\} \tag{7.38}$$

$$B_0^* = 4 \left(\frac{r_2 - r_1}{r_2 + r_1} \right) \left\{ 1 - \frac{3}{b^4} (r_2 - r_1)^4 \right\} \tag{7.39}$$

where r_1 and r_2 are the inner and outer radii of the SAP, respectively. Note that b is the cladding radius as in Figure 7.9b.

In PANDA fibers, we have $B_G = B_S^i = 0$ and $B = B_S^e$. Therefore, the PANDA fiber shows a nondispersive behavior as in Figure 7.14. The average birefringence at the core of the PANDA fiber defined in equation 7.29 is plotted in Figure 7.17.

One of notable features of PANDA fiber is that the birefringence increases relatively slowly as a function of the inner radius of SAP, r_2/b, in comparison to other types of PMF. In other words, the SAP should be located in the optimal proximity to the core to have the net effect in the birefringence.

Sasaki [29] reported detailed PANDA fiber structure design considerations to obtain high birefringence, low loss, and low cross-talk over a long length. As we have

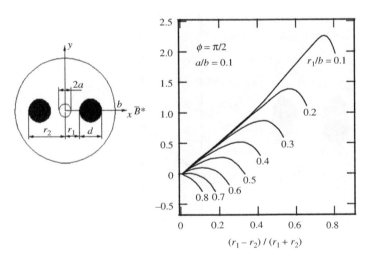

FIGURE 7.17   Plots of $\bar{B}^*$ as a function of $(r_2 - r_1)/(r_1 + r_2)$ for various inner radius of SAP, r_1. Here the core radius is 0.1 times the cladding radius b [61].

considered in this section and previous ones, there is conflicting trade-off between high birefringence and low loss as long as we use B_2O_3-doped SAPs. Furthermore, the maintenance of high birefringence over a long length is another critical fabrication process issue because the SAPs are very vulnerable to external perturbation even in the current fiber drawing process.

Long-distance PMF is of high interest in current coherent communications and the performance requirements in PMFs are as follows:

(1) CT must be less than $-10\,\text{dB}$ over a length of 200 km.
(2) Transmission loss must be less than 0.20 dB/km in the 1.5 μm wavelength region.

These system performance requirements can be interpreted in fiber manufacturing process requirements as follows [29]:

(1) Modal birefringence B must be more than 3×10^{-4}.
(2) Additional loss due to B_2O_3-doped SAPs must be less than 0.05 dB/km at 1.56 μm.
(3) Δ of GeO_2-doped core is kept at 0.3% and concentration of B_2O_3 is kept at 15 mol%.

The SAP distance r_1 and its diameter d in Figure 7.9b were optimized to satisfy the above conditions.

In Figure 7.17, the range of SAP distance in terms of (r_1/a) is described for the first two requirements: (a) birefringence B over 3×10^{-4} and (b) additional loss less than 0.05 dB/km at 1.56 μm. The optimal SAP distance in PANDA fiber was found to be in the following range:

$$3.4 < r_1/a < 4.5 \tag{7.40}$$

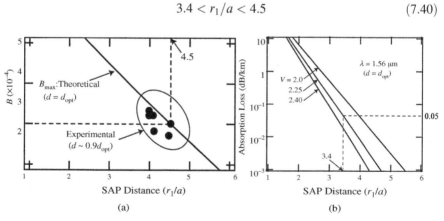

FIGURE 7.18 Finding range of SAP distance (r_1/a) by considering (a) birefringence B over 3×10^{-4} requirement and (b) additional loss less than 0.05 dB/km at 1.56 μm. Note that r_1 is the inner SAP distance and a is the core radius of PANDA fiber shown in Figure 7.9b [29].

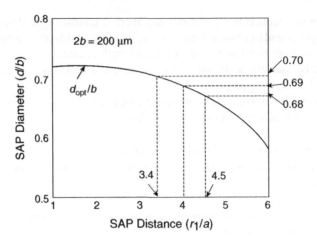

FIGURE 7.19 Finding range of SAP thickness d/b from the optimal d/b versus r_1/a curve. Here d_{opt} is designed for maximum B [29].

For this range of SAP distance the optimal range of SAP thickness in reference to the cladding diameter b, (d/b), was found from the trace of (d_{opt}/b) versus (r_1/a), as shown in Figure 7.18. Here d_{opt} is the design parameter for maximum birefringence B. For the range of r_1/a given in equation 7.40, the SAP thickness in PANDA fiber was found to be in the following range (Figure 7.19):

$$0.68 < d/b < 0.70 \qquad (7.41)$$

The present commercial PANDA fibers are designed with SAP parameters in these ranges.

The waveguide parameters of optimized PANDA fiber and its birefringence characteristics are summarized in Table 7.4. These values are being widely accepted by commercial PANDA fiber manufacturers.

After this experimental report by Sasaki and following development in the mass production process, PANDA fiber became the standard HBF and being used in various long-distance applications. In Table 7.5, commercially available HBFs and their optical characteristics are summarized, based on data sheets.

7.6 SINGLE-POLARIZATION FIBERS

In Section 7.3, we have briefly introduced four typical types of SPFs as shown in Figure 7.10. SPFs that transmit only one of two orthogonally polarized HE_{11} modes could find clear advantages over HBF for long-distance coherent communications and highly sensitive sensing applications. SPF can, in principle, get rid of polarization cross-talk, polarization-dependent loss, and polarization dispersion [38–46]. Recent SPFs have been reviewed by Li et al. [38] and according to the classification

TABLE 7.4 Waveguide Parameters and Characteristics of a Low-Loss PANDA Fiber Measured at the Wavelength of $\lambda = 1.56\,\mu m$ [29]

Cladding diameter ($2b$, μm)	200
Normalized frequency (v)	2.25
Relative core refractive index (Δ, %)	0.3
Core radius (a, μm)	10
SAP distance (r_1/a)	4.1
SAP thickness (d/b)	0.61
SAP B_2O_3 concentration (mol%)	15
Modal birefringence (B)	3.2×10^{-4}
Cross-talk (CT, dB) at 5 km length	-27
h parameter (m^{-1})	4.0×10^{-7}
Minimum loss (dB/km)	0.22

chart in Figure 7.11, SPFs can be divided into two groups depending on the origin of birefringence: (1) geometrical birefringence B_G and (2) stress-induced birefringence B_S.

It is noteworthy that SPFs are in fact birefringent and guiding two orthogonal polarization modes in the wavelength region out of the single-polarization bandwidth (SP-BW), where only the HE$_{11}$ mode along the slow axis is guided. The SP-BW is usually defined to be the wavelength range over which the fundamental mode of one polarization state, usually along the fast axis, is attenuated by at least 25 dB while the other orthogonal state suffers less than 1 dB/km increase in its base attenuation [45].

TABLE 7.5 Optical Characteristics of High-Birefringent Optical Fiber Products

Product	λ (nm)	MFD (μm)	L_P (mm)	B ($\times 10^{-4}$)	CT (dB)	α (dB/km)
Corning	PM 1550	1550	10.5	3.0–5.0	3.1–5.2	-30	0.5
	PM 850	850	5.5	1.0–2.0	4.3–8.5	-30	3.0
Sumitomo	PM-155	1550	10.5	≤5.2	≥3.0	-25	0.5
						-35	1.0
	PM-085	850	5.1	≤2.8	≥3.0	-30	3.0
Fibercore	HB1500	1550	7.9	≤2.0	≥7.8	N/A	2.0
	HB800	830	4.2	≤2.0	≥4.3	N/A	5.0
Fujikura	SM15-PS-U40A	1550	10.5	3.0–5.0	3.1–5.2	-30	0.5
	SM85-PS-U40A	850	5.5	1.0–2.0	4.3–8.5	-30	3.0
Furukawa-OFS	CL Truephase 1550 400	1550	10.5	≤4.5	≥3.4	-30	1.0
	CL Truephase 980 400	980	6.6	≤2.8	≥3.5	-30	2.5

Here λ is the operating wavelength, MFD is mode field diameter, L_P is beat length, B is modal birefringence, CT is cross-talk measured at 100 m, and α is optical attenuation at the operating wavelength λ. Fibercore fibers are bow-tie types and other fibers are PANDAs.

Therefore, the key design issues in SPFs are (1) locating the SP-BW at the desired spectral location with a desired bandwidth, (2) inducing the highly selective attenuation to the polarization mode along the fast axis, and (3) maintaining the minimal added loss for the guided polarization mode along the slow axis within the SP-BW.

These design issues have been successfully solved by introducing the mode cutoff in the fundamental HE_{11} mode. In conventional SMFs, the HE_{11} mode does not have cutoff and only the higher order modes have their cutoffs at unique normalized frequencies as described in Chapter 2. This guidance of HE_{11} can be significantly modified in a special waveguide structure named as W-type fiber [47–49], where the HE_{11} cutoff can be flexibly adjusted. Initially W-type fibers have been intensively investigated to control the chromatic dispersion of the fundamental mode well apart from the HE_{11} cutoff wavelength, as discussed in Chapters 2 and 3. Most of current dispersion-compensating fibers and some of dispersion-shifted fibers are based on the W-type waveguide design.

Typical refractive index profile of a W-type fiber is shown in Figure 7.20a. It is a step-index structure with the core–inner cladding–outer cladding layers. In a shorter wavelength, as indicated as λ_1 in the figure, the fundamental HE_{11} mode is tightly confined in the core and its effective index $n_{eff}(\lambda_1)$ is well above the outer cladding index, n_3. In this case the effect of the outer cladding is minimal. However, as the wavelength increases, the effective index gets close to the outer cladding index and the guided light rapidly expands out of the core, as in the case of $\lambda = \lambda_2$ in the figure. When the wavelength further increase to λ_c, the mode satisfies the cutoff condition as follows and the HE_{11} mode is no longer guided along the core:

$$n_{eff}(\lambda_c) = n_3 \tag{7.42}$$

A schematic transmission spectrum of a W-type fiber is shown in Figure 7.20b, where the long wavelength cutoff, λ_c, is marked beyond which transmission rapidly

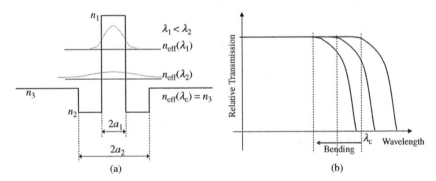

(a)   (b)

FIGURE 7.20 Schematic diagram of a W-type fiber waveguide and the HE_{11} mode cutoff therein. (a) Refractive index profile of a W-type fiber with the core–inner cladding–outer cladding structure. The refractive indices of the core, inner cladding, and outer cladding are n_1, n_2, and n_3, respectively. (b) Schematic transmission spectra of W-type fiber with the HE_{11} mode cutoff at the longer wavelength, λ_c, which can be shifted toward a shorter wavelength by bending.

decreases. It is noted that the HE_{11} mode cutoff length position can be flexibly controlled by applying bending and waveguide engineering using the parameters in Figure 7.20a.

This concept of W-type fiber is combined with HBF with the stress-induced birefringence B_S to introduce separate HE_{11} mode cutoffs in the fast and slow axes at different spectral positions, λ_c^F and λ_c^S ($\lambda_c^F < \lambda_c^S$). The schematic diagram for this B_S type of SPF is illustrated in Figure 7.21.

The birefringence of this type of fibers is induced by the SAP with the refractive index of n_{SAP} near the core. The refractive index of the SAP is adjusted to be in between the inner cladding index and outer cladding index. The HE_{11} modes in the fast and slow axes have effective indices $n_{eff}^F(\lambda) < n_{eff}^S(\lambda)$, respectively, at a given wavelength. The birefringence is given as $B_S(\lambda) = n_{eff}^S(\lambda) - n_{eff}^F(\lambda)$. In this case the mode cutoff condition is imposed by the SAP index as described in Figure 7.21a. Note that the light in the HE_{11} mode in the core is leaked to SAP at the cutoff condition, where the following conditions are satisfied:

$$n_{eff}^F(\lambda_c^F) = n_{SAP}$$
$$n_{eff}^S(\lambda_c^S) = n_{SAP} \qquad (7.43)$$
$$\lambda_c^F < \lambda_c^S$$

W-type waveguide structure, therefore, provides separate HE_{11} cutoffs for the polarization modes along the slow and fast axes. Schematic transmission spectra are shown in Figure 7.21b. The polarization mode at the fast axis is first cut off with rapidly increasing attenuation beyond λ_c^F, leaving behind only one polarization mode along the slow axis up to λ_c^S. Therefore, we could achieve single-polarization single-mode guidance near the wavelength range of $\lambda_c^F < \lambda < \lambda_c^S$. Considering the conventional 25 dB extinction ratio requirement, actual single-polarization bandwidth is narrower, as illustrated in Figure 7.21b.

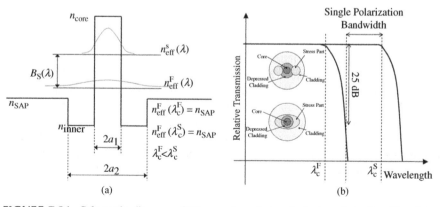

FIGURE 7.21 Schematic diagram of W-type stress-induced single-polarization fiber (B_S-SPF): (a) refractive index profile along with the polarization mode cutoff conditions and (b) transmission spectra showing two separate cutoffs for the polarization modes in the fast and slow axes. Two types of this B_S-SPF are shown in the inset, which are Figure 7.10a and b.

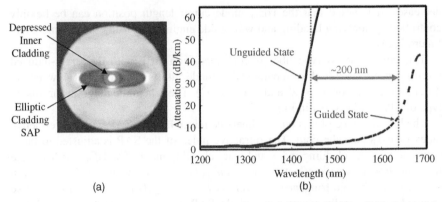

FIGURE 7.22 A commercially available B_S-SPF based on elliptic cladding SAP along with the depressed inner cladding: (a) cross-section of the fiber and (b) transmission spectra defining the single-polarization bandwidth of 200 nm centered near 1550 nm [63].

Waveguide structures of this B_S-SPF are shown in the insets of Figure 7.21b, which correspond to Figure 7.10a and b. They are in fact elliptic cladding fiber and PANDA fiber with depressed inner cladding, which allows the cutoffs for the polarization modes. Single-polarization bandwidth of ~45 nm was demonstrated in an SPF design similar to Figure 7.10a [42]. Further expansion of the bandwidth of 100 nm was reported in Ref. [44]. SPF based on PANDA structure as in Figure 7.10b showed a single-polarization window of 80 nm [45].

Recently commercial SPF, Verrillon PZF [63], was developed based on elliptic cladding SAP with the depressed inner cladding structure similar to Figure 7.10a, which showed a wide SP-BW over 200 nm centered near 1550 nm. The cross-section of the fiber and the transmission spectra are shown in Figure 7.22. The fiber also showed a high extinction ratio over 40 dB within the SP-BW to ensure the single-polarization mode guidance.

Another class of SPF is based on geometrical birefringence B_G combined with differential mode cutoff for the fundamental polarization modes. The principles of this SPF are illustrated in Figure 7.23.

In this B_G-SPF, two different W-type waveguide structures are provided in a single fiber for the two orthogonal polarization modes, which are shown as the solid and dotted lines in Figure 7.23a for the slow and fast axes, respectively. Note that the core diameter $2a_1^S$ of the slow axis profile is larger than $2a_1^F$ of the fast axis. The depressed inner cladding layer along the slow axis is narrower ($a_2^S < a_2^F$) and shallower ($n_2^S < n_2^F$) than that along the fast axis. These geometrical differences result in a large birefringence, B, and two separate HE_{11} mode cutoffs for the slow and fast axes as follows:

$$n_{eff}^F(\lambda_c^F) = n_3^F$$
$$n_{eff}^S(\lambda_c^S) = n_3^S \qquad (7.44)$$
$$\lambda_c^F < \lambda_c^S$$

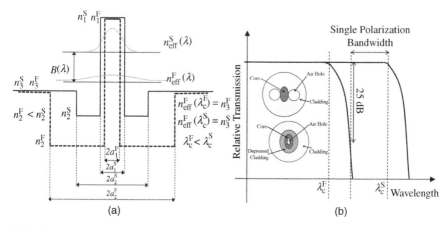

FIGURE 7.23 Schematic diagram of W-type geometrical birefringence single-polarization fiber (B_G-SPF): (a) refractive index profile along with the polarization mode cutoff conditions and (b) transmission spectra showing two separate cutoffs for the polarization modes in the fast and slow axes. Two types of this B_G-SPF are shown in the inset, which are Figure 7.10c and d. In the diagram the vertical axis is the slow axis.

This type of B_G-SPFs was reported by Hinata et al. [46] in the name of "hollow pit" and more recently by Nolan et al. [64,65] from Corning. In these fibers the anisotropy was achieved by the elliptic core and W-type depressed inner cladding was provided by two air holes on both sides of the core. The schematic waveguide structures are shown in Figure 7.24.

Corning recently added SPFs in its product lineups, SP1060, SP1310, and SP1550 that are single-polarization single-mode at the wavelength near 1060, 1310, and 1550 nm, respectively [66]. Their waveguide structures are similar to Figure 7.24b consisted of the elliptic core along with adjacent side air holes. These air holes provide an effective W-type depressed cladding. The cross-section of the fiber and the transmission spectra are shown in Figure 7.25.

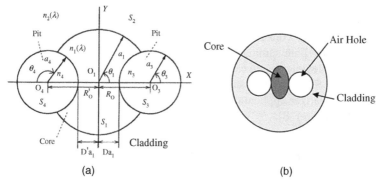

FIGURE 7.24 Waveguide structure of B_G-SPFs with side air holes: (a) hollow pit SPF, where side holes are drilled at both sides of the circular core [46], and (b) hole-assisted SPF, where side holes are drilled just outside of the elliptic core [64,65].

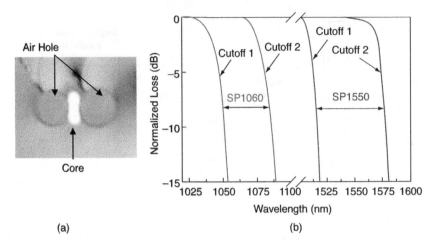

(a)   (b)

FIGURE 7.25 A commercially available B_G-SPF based on the elliptic core and adjacent air holes that serve as an effective depressed inner cladding: (a) cross-section of the fiber [38] and (b) transmission spectra defining the single-polarization bandwidth near 1060 and 1550 nm [66].

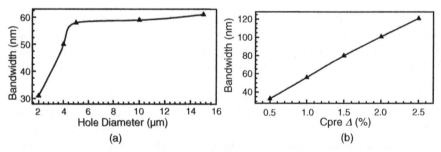

(a)   (b)

FIGURE 7.26 (a) Single-polarization bandwidth as a function of air hole size. The core Δ is 1%, and the aspect ratio of the major and minor axes of the elliptic core is 1.5. (b) SP-BW as a function of core Δ. The air hole diameter is 5 μm; the aspect ratio of the core is 3.0 [64].

This particular fiber showed very flexible control of SP-BW from 30 to 120 nm by varying air hole size, core Δ, and core ellipticity. And the experimental results reported in Ref. [64] are summarized in Figure 7.26.

Despite these flexible design capability and excellent polarization characteristics, the hole-assisted SPs, however, suffer from high optical loss and they would serve better in fiber polarizer applications. General characteristics of this fiber are summarized in Table 7.6 [38].

7.7 LOW LINEAR BIREFRINGENCE FIBERS

A high birefringence and subsequently a large group delay between two orthogonal polarization modes in an optical fiber would not be always welcomed in optical applications, especially in high–data rate communication system over 40 Gbps and

TABLE 7.6   Experimental Characteristics of Hole-Assisted Single-Polarization Fibers [38]

Fiber	Δ (%)	λ_c^F (nm)	λ_c^S (nm)	SP-BW (nm)	L_P (mm)	Loss (dB/m)
1	1.1	1148	1177	30	3.0	0.03
2	1.1	1067	1098	31	2.8	0.03
3	1.1	911	935	24	–	0.08
4	2	972	1014	42	1.1	1.76
5	2	1478	1533	55	–	0.10

polarimetric sensors. In high–data rate optical communication system, polarization mode dispersion (PMD), which is a statistical average of differential group delay (DGD) between two polarization modes as described in Chapter 2, should be minimized to resolve the digital signal in the bit period of ~20 ps [67]. In polarimetric optical fiber sensors based on Faraday effects, variations of output polarization states under external perturbation such as magnetic field, electric field, and acoustic wave should be maximized rather than minimized as in HBFs [11,15].

Therefore, there is also increasing technical demand to control the birefringence in optical fiber as low as possible, opposite to HBF and SPF discussed in previous sections. In this section we will describe three types of techniques for fabricating low-birefringence fibers (LBFs): (1) fiber twisting, (2) preform spinning, and (3) fiber spinning methods, which are schematically illustrated in Figure 7.27.

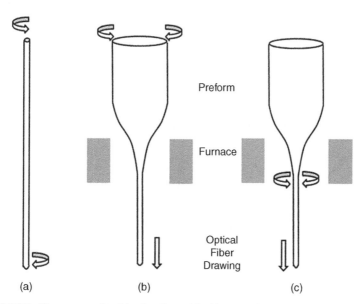

FIGURE 7.27   Three types of making low linear birefringence fibers (LBFs): (a) fiber twisting, (b) preform spinning, and (c) fiber spinning.

In order to achieve a low birefringence, main efforts were initially focused on optical fiber preform manufacturing processes, which were aimed at reducing (1) the geometrical birefringence B_G by maintaining a highly circular core and (2) the stress-induced anisotropy B_S by decreasing the residual core stress. Despite success in reducing B_G in the preform fabrication processes, it was soon found that the internal stress anisotropy B_S is not controllable beyond a certain level. The residual stress results from differential thermal contraction between core, cladding, and substrate materials on cooling to room temperature, and is frozen into fiber along with the tension-induced mechanical stress during the optical fiber drawing process. Therefore, preform fabrication process alone cannot reduce the birefringence to a desired level and eventually the development efforts have shifted toward fiber drawing process.

Three principal techniques in LBFs are illustrated in Figure 7.27 and we will discuss fiber twist, preform spinning, and fiber spinning techniques in this section.

When a linearly birefringent fiber is twisted [14], at a rate of ξ rad/m, the azimuth of the fiber elliptic cross-section undergoes precession. Then the twisted fiber can be considered as composed of individual local sections having a length of a quarter twist-period and a linear-birefringence that alternates in sign. Thus, although each local section may have a relatively high linear birefringence, net effect along the twisted fiber is compensated by the next rotated section. By twisting fiber, a low retardance $R(z)$ as defined in equation 7.3 could be achieved, which oscillates with a small amplitude along the fiber length. Fiber twisting can be easily used to effectively reduce the fiber linear birefringence, but it simultaneously introduces an elasto-optic rotation α as defined in equation 7.9. The interaction between the local linear and circular birefringence in a twisted optical fiber lead results in a net retardance $R(z)$ and a net rotation $\Omega(z)$ given as follows [14,15]:

$$R(z) = 2\sin^{-1}\left[\frac{1}{(1+q^2)^{1/2}}\sin\gamma z\right] \tag{7.45}$$

$$\Omega(z) = \xi z + \tan^{-1}\left[-\frac{q}{(1+q^2)^{1/2}}\tan\gamma z\right] \tag{7.46}$$

$$q = \frac{2(\xi - \alpha)}{\delta\beta} \tag{7.47}$$

$$\gamma = \frac{1}{2}(\delta\beta^2 + 4(\xi - \alpha)^2)^{1/2} \tag{7.48}$$

$$\alpha = g'\xi \tag{7.49}$$

In the twisted fiber, the main parameter to control the net linear birefringence, or equivalently $R(z)$, is $\delta\beta/\xi$. Depending on the magnitude of this value, both $R(z)$ and $\Omega(z)$ show very different behavior.

In the case of "small twist," we have $\xi \ll \delta\beta$ and net retardance and rotation grow linearly with fiber length z as follows:

$$R(z) \approx \delta\beta z$$
$$\Omega(z) \approx \xi z \tag{7.50}$$

Note that twist effect in this case is negligible and the fiber maintains its intrinsic linear birefringence of $\delta\beta$. This situation is illustrated in Figure 7.28 for $\delta\beta/\xi = 4000$ and we can confirm a linear growth of the retardance with the fiber length.

In the case of "large twist," $\xi \gg \delta\beta$, we have completely different behavior shown as follows:

$$R(z) \approx \frac{\delta\beta}{\xi}\sin\xi z$$
$$\Omega(z) \approx \alpha z \tag{7.51}$$

In this case the fiber retardance is considerably reduced by the twist and becomes oscillatory about zero with the amplitude of $\delta\beta/\xi$ as illustrated in Figure 7.28 for $\delta\beta/\xi = 0.04$. Launching linearly polarized light into this twisted fiber at any azimuth results in near-linear output, rotated with respect to the input by an angle of αz. In fact, the state of polarization along the fiber consists of an oscillation between very slightly left and right elliptically polarized light that rotates in orientation.

Despite its effectiveness, fiber twisting cannot be applied to long-length fiber and fiber cable. Individual tracking and control of twist for each fiber strand in recent high fiber count cables are impractical. Fiber twisting technique is not used in practical applications.

Barlow et al. [15] reported a preform spinning method as illustrated in Figure 7.27b. In this technique, optical fiber is drawn from a preform that is

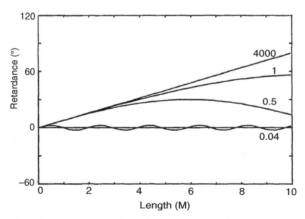

FIGURE 7.28 Calculated retardance of a twisted optical fiber as a function of length for various values of $\delta\beta/\xi$ (0.04, 0.5, 1, and 4000). Here an intrinsic linear birefringence $\delta\beta = 0.044\pi\,\text{rad/m}$ is assumed [15].

simultaneously spun, in order to freeze an effective fiber twist rate of ξ rad/m into the pulled optical fiber. No photoelastic effect will be present as the twisting occurs in the furnace hot zone, where the glass viscosity is sufficiently low to prevent the support of shear stress. Thus, $\alpha = 0$ in this case. Assuming a large twist rate $\xi \gg \delta\beta$, the preform spun fiber will show the following characteristics:

$$R(z) \approx \frac{\delta\beta}{\xi}\sin\xi z$$

$$\Omega(z) \approx 0$$
(7.52)

The fiber now has very small values of both retardation and rotation and, in the limit of large spin, behaves as a perfectly isotropic waveguide capable of transmitting any polarization state unchanged.

Typical fabrication conditions for preform spun fiber are: preform spinning rate of $\sim$600 rotation/min at fiber drawing speed of $\sim$1 m/s. Most of preforms had the diameter of 10–20 mm. At these conditions, the net retardance $R(z)$ less than 1 °/m and optical rotation less than 0.5 ° are usually provided, which are near the measurement and detection limits.

Preform spinning technique indeed has provided the first continuous way of fiber fabrication to freeze the twist effect into optical fiber during the drawing process. However, in recent manufacturing process, the preform diameter is over 80 mm and the fiber drawing speed exceeds 25 m/s, which requires more than 15,000 rpm spinning rate. It is very difficult to rotate the large preform at very high speeds. A more practical technique, described, for example, by Hart et al. [68], was to spin the fiber directly instead of the preform as illustrated in Figure 7.27c. Note that the preform is linearly fed toward the furnace just as in conventional drawing process but in the neck-down region fiber is spun. The torque for the fiber spinning is provided in the far end side of fiber spooling, where the fiber is rapidly oscillated in a certain plane. This oscillation is transferred to rotation of optical fiber in the neck-down region. Fiber spinning method can be readily applied to modern high speed drawing technique and is being widely used to manufacture low polarization fiber and more importantly low PMD fibers.

Another key advantage of fiber spinning technique is the fact that various spinning profiles can be easily adopted, which was not possible in preform spinning techniques. In recent fiber fabrication process, it is found that uniform spin, keeping the same twist rate of ξ rad/m over the entire fiber length, cannot reduce PMD effectively and spin rate should be modulated along the fiber length in terms of spin amplitude and spin frequency [67]. In Figure 7.29 various spin profiles are shown.

In mathematical form, these spin profiles can be expressed as follows:

$$\text{Constant spinning}: \quad \xi(z) = \xi_0$$
(7.53)

$$\text{Sinusoidal spinning}: \quad \xi = \xi_0 \sin(2\pi f_0 z)$$
(7.54)

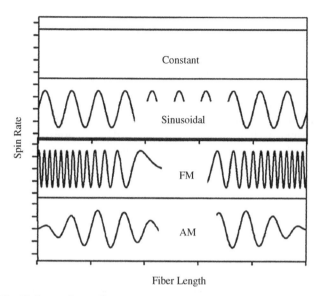

FIGURE 7.29 Various spin profiles used in fiber spinning technique. FM, frequency modulation; AM, amplitude modulation [67].

$$\text{FM spinning}: \quad \xi(z) = \xi_0 \sin\left\{ 2\pi z \left[f_0 + f_m \sin\left(\frac{2\pi z}{\Lambda} \right) \right] \right\} \tag{7.55}$$

$$\text{AM spinning}: \quad \xi = \xi_0 \sin(2\pi f z) \sin\left(\frac{2\pi z}{\Lambda} \right) \tag{7.56}$$

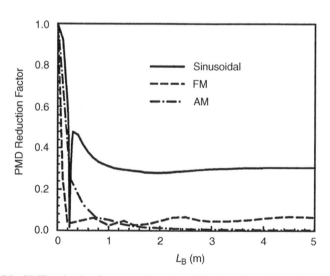

FIGURE 7.30 PMD reduction factor as a function of fiber beat length for sinusoidal, FM, and AM spin profiles. The spin magnitude for all cases has been chosen to be 3.0 turns/m [67].

There have been extensive research reports on the effects of spin profiles over birefringence and PMD [69–71]. Detailed description of those analyses and ongoing investigations are beyond the scope of this book and one comparative result shown in Figure 7.30 will be discussed in order to emphasize the importance of spin profile in the fiber spinning technique.

It can be found that PMD reduction factor (PMDRF) remains low for a wide range of beat length. For a sinusoidal spin profile that has only one spatial frequency, even though it is possible to adjust the spin magnitude so that the corresponding PMDRF at a given beat length is low, such low PMDRF cannot sustain over a wide beat length range, for example, from 0.5 m to a few meters. On the other hand, FM and AM spin profiles have multiple Fourier frequency components. As a result, good PMD reduction can be achieved over a wide range of fiber beat lengths. They can also reduce the variation in fiber PMD distribution.

REFERENCES

[1] Fujikura PANDA fiber products and basics of PM fibers, http://www.fujikura.co.uk/pdf/fibre_optics_panda_fibre_presentation.pdf

[2] T. Okoshi and K. Kikuchi, "Coherent optical fiber communications," KTK Scientific Publishers, 1988.

[3] H. Sun, K. T. Wu, and K. Roberts, "Real-time measurement of a 40 Gb/s coherent system," Optic Express, vol. 16, no. 2, pp. 873–879, 2008.

[4] A. Vengsarkar, W. C. Michie, L. Jankovic, B. Culshaw, and R. O. Claus, "Fiber-optic dual-technique sensor for simultaneous measurement of strain and temperature," Journal of Lightwave Technology, vol. 12, no. 1, pp. 170–177, 1994.

[5] W. J. Bock, T. R. Wolinski, and A. Barwicz, "Development of a polarimetric optical fiber sensor for electronic measurement of high pressure," IEEE Transactions on Instrumentation and Measurement, vol. 39, no. 5, pp. 715–721, 1990.

[6] S. G. Demos and R. R. Alfano, "Optical polarization imaging," Applied Optics, vol. 36, no. 1, pp. 150–155, 1997.

[7] B. H. Park, C. Saxer, S. M. Srinivas, J. S. Nelson, and J. F. de Boer, "*In vivo* burn depth determination by high-speed fiber-based polarization sensitive optical coherence tomography," Journal of Biomedical Optics, vol. 6, no. 4, pp. 474–479, 2001.

[8] E. L. Wooten, K. M. Kissa, A. Yi-Yan, E. J. Murphy, D. A. Lafaw, P. F. Hallemeier, D. Maack, D. V. Attanasio, D. J. Fritz, G. J. McBrien, and D. E. Bossi, "A review of lithium niobate modulators for fiber-optic communications systems," IEEE Journal of Selected Topics in Quantum Electronics, vol. 6, no. 1, pp. 69–82, 2000.

[9] F. Di Pasquale and F. Meli, "New Raman pump module for reducing pump–signal four-wave-mixing interaction in co-pumped distributed Raman amplifiers," Journal of Lightwave Technology, vol. 21, no. 8, pp. 1742–1748, 2003.

[10] S. C. Rashleigh, "Origins and control of polarization effects in single-mode fibers," Journal of Lightwave Technology, vol. LT-1, no. 2, pp. 312–331, 1983.

[11] D. N. Payne, A. J. Barlow, and J. J. R. Hansen, "Development of low- and high-birefringence optical fibers," IEEE Journal of Quantum Electronics, vol. QE-18, no. 4, pp. 477–488, 1982.

[12] S. Arai, H. Saitou, N. Oyama, T. Nakamura, K. Yokomizo, and K. Aiso, "Polarization maintaining fiber," Furukawa Electric Review, no. 109, pp. 5–10, 2002.

[13] A. Papp and H. Harms, "Polarisation optics of index-gradient optical waveguide fibers," Applied Optics, vol. 14, pp. 2406–2411, 1975.

[14] R. Ulrich and A. Simon, "Polarisation optics of twisted single mode fibers," Applied Optics, vol. 18, pp. 2241–2251, 1979.

[15] A. J. Barlow, J. J. Ramskov Hansen, and D. N. Payne, "Birefringence and polarisation mode-dispersion in spun single-mode fibers," Applied Optics, vol. 20, pp. 2962–2968, 1981.

[16] E. Collett, "Polarized light in fiber optics," SPIE Press, 2003.

[17] R. B. Dyott, "Elliptical fiber waveguides," Artech House, London, UK, 1995.

[18] R. B. Dyott, J. R. Cozens, and D. G. Morris, "Preservation of polarization in optical-fiber waveguides with elliptical cores," Electronics Letters, vol. 15, no. 13, pp. 380–382, 1979.

[19] S. C. Rashleigh and M. J. Marrone, "Polarization holding in elliptical-core birefringent fibers," IEEE Journal of Quantum Electronics, vol. QE-18, no. 10, pp. 1515–1523, 1982.

[20] N. Shibata, M. Tateda, S. Seikai, and N. Uchida, "Birefringence and polarization-mode dispersion caused by thermal stress in single mode fibers with various core ellipticities," IEEE Journal of Quantum Electronics, vol. QE-19, no. 8, pp. 1223–1227, 1983.

[21] T. Hosaka, K. Okamoto, Y. Sasaki, and T. Edahiro, "Single-mode fibers with asymmetrical refractive-index pits on both sides of the core," Electronics Letters, vol. 17, no. 5, pp. 191–193, 1981.

[22] T. Okoshi, K. Oyamada, M. Nishimura, and H. Yokota, "Side tunnel fiber: an approach to polarization-maintaining optical waveguiding scheme," Electronics Letters, vol. 18, no. 19, pp. 824–826, 1982.

[23] R. H. Stolen, V. Ramaswamy, P. Kaiser, and W. Pliebel, "Linear polarization in birefringent single-mode fibers," Applied Physics Letters, vol. 33, no. 8, pp. 699–701, 1978.

[24] V. Ramaswamy, R. H. Stolen, M. D. Divine, and W. Pliebel, "Birefringence in elliptically clad borosilicate single-mode fibers," Applied Optics, vol. 18, no. 24, pp. 4080–4084, 1979.

[25] E. M. Dianov, A. B. Grudinin, A. N. Gurjanov, D. D. Gusovsky, Z. E. Harutjunjian, S. V. Ignatjev, and O. B. Smirnov, "Circular core polarization-maintaining optical fibers with elliptical stress-induced cladding," Journal of Lightwave Technology, vol. 10, pp. 118–124, 1992.

[26] T. Hosaka, K. Okamoto, T. Miya, Y. Sasaki, and T. Edahiro, "Low loss single polarization fibers with asymmetrical strain birefringence," Electronics Letters, vol. 17, pp. 530–531, 1981.

[27] N. Shibata, Y. Sasaki, K. Okamoto, and T. Hosaka, "Fabrication of polarization-maintaining and absorption-reducing fibers," Journal of Lightwave Technology, vol. LT-1, pp. 38–43, 1983.

[28] Y. Sasaki, T. Hosaka, K. Takada, and J. Noda, "8 km-long polarization-maintaining fiber with highly stable polarization state," Electronics Letters, vol. 19, pp. 792–794, 1983.

[29] Y. Sasaki, "Long-length low-loss polarization maintaining fibers," Journal of Lightwave Technology, vol. JT-5, no. 9, pp. 1139–1146, 1987.

[30] M. P. Varnham, D. N. Payne, R. D. Birch, and E. J. Tarbox, "Single-polarization operation of highly birefringent bow-tie optical fibers," Electronics Letters, vol. 19, no. 7, pp. 246–247, 1983.

[31] R. H. Stolen, R. E. Howard, and W. Pleibel, "Substrate tube lithography for optical fibers," Electronics Letters, vol. 18, no. 18, pp. 764–765, 1982.

[32] R. D. Birch, M. P. Varnham, D. N. Payne, and E. J. Tarbox, "Fabrication of polarization-maintaining fibers using gas-phase etching," Electronics Letters, vol. 18, no. 24, pp. 1036–1038, 1982.

[33] Hi-bi:bow-tie fiber, Fibercore Inc., http://www.fibercore.com.

[34] J. R. Simpson, R. H. Stolen, F. M. Sears, W. Pleibel, J. B. MacChesney, and R. E. Howard, "A single-polarization fiber," Journal of Lightwave Technology, vol. LT-1, no. 2, pp. 370–373, 1983.

[35] R. H. Stolen, W. Pleibel, and J. R. Simpson, "High-birefringence optical fibers by preform deformation," Journal of Lightwave Technology, vol. LT-2, no. 5, pp. 639–641, 1984.

[36] M. W. Shute, Sr. and C. S. Brown, "A study of the polarization properties of a rectangular polarization-maintaining fiber," Journal of Lightwave Technology, vol. 7, pp. 2013–2017, 1989.

[37] Y. Liu, B. M. A. Rahman, and K. T. V. Grattan, "Analysis of the birefringence properties of optical fibers made by a preform deformation technology," Journal of Lightwave Technology, vol. 13, no. 2, pp. 142–147, 1995.

[38] M.-J. Li, D. A. Nolan, G. E. Berkey, X. Chen, J. Koh, D. T. Walton, J. Wang, W. A. Wood, and L. A. Zenteno, "High performance single polarization optical fibers," Passive Components and Fiber-Based Devices, Proceedings of SPIE, vol. 5623, pp. 612–621, 2005, doi: 10.1117/12.580336.

[39] T. Okoshi and K. Oyamada, "Single polarization single mode optical fiber with refractive index pits on both sides of core," Electronics Letters, vol. 16, no. 18, pp. 712–713, 1980.

[40] T. Hosaka, K. Okamoto, T. Miya, Y. Sasaki, and T. Edahiro, "Low-loss single polarisation fibres with asymmetrical strain birefringence," Electronics Letters, vol. 17, no. 15, pp. 530–531, 1981.

[41] M. P. Varnham, D. N. Payne, R. D. Birch, and E. J. Tarbox, "Single-polarisation operation of highly birefringent bow-tie optical fibres," Electronics Letters, vol. 19, no. 7, pp. 246–247, 1983.

[42] K. Okamoto, "Single-polarization operation in highly birefringent optical fibers," Applied Optics, vol. 23, no. 15, pp. 2638–2642, 1984.

[43] K. S. Chiang, "Stress-induced birefringence fibers designed for single polarization single-mode operation," Journal of Lightwave Technology, vol. 7, pp. 436–441, 1989.

[44] K. Tajima, M. Ohasi, and Y. Sasaki, " A new single polarization optical fiber," Journal of Lightwave Technology, vol. 7, no. 10, pp. 1499–1503, 1989.

[45] M. J. Messerly, J. R. Onstott, and R. C. Mikkelson, "A broad-band single polarization optical fiber," Journal of Lightwave Technology, vol. 9, pp. 817–820, 1991.

[46] T. Hinata, S. Furukawa, N. Namatame, and S. Nakajima, "A single-polarization optical fiber of hollow pit type with zero total dispersion at wavelength of 1.55 m," Journal of Lightwave Technology, vol. 12, pp. 1921–1925, 1994.

[47] L. G. Cohen, D. Marcuse, and W. L. Mammel, "Radiating leaky-mode losses in single-mode lightguides with depressed-index claddings," Transactions on Microwave Theory and Techniques, vol. MTT-30, no. 10, pp. 1455–1460, 1982.

[48] H.-J. Hagemann, H. Lade, J. Wamier, and D. Wiechert, "The performance of depressed-cladding single-mode fibers with different b/a ratio," Journal of Lightwave Technology, vol. 9, no. 6, pp. 689–694, 1991.

[49] J. Auge, C. Brehm, L. Jeunhomme, and C. L. Sergent, "Parametric study of depressed inner cladding single-mode fibers," Journal of Lightwave Technology, vol. LT-3, no. 4, pp. 767–772, 1985.

[50] J. Noda, K. Okamoto, and Y. Sasaki, "Polarization maintaining fibers and their applications," Journal of Lightwave Technology, vol. LT-4, no. 8, pp. 1071–1089, 1986.

[51] N. Shibata, Y. Sasaki, and T. Hosaka, "Structure design for polarization-maintaining and absorption-reducing optical fibers," Review of the Electrical Communication Laboratory, vol. 31, no. 3, pp. 393–399, 1983.

[52] I. P. Kaminow and V. Ramaswamy, "Single-polarization optical fibers: slab model," Applied Physics Letters, vol. 34, no. 4, pp. 268–270, 1979.

[53] I. P. Kaminow, "Polarization in optical fibers," IEEE Journal of Quantum Electronics, vol. QE-17, no. 1, pp. 15–22, 1981.

[54] T. Katsuyama, H. Matsumura, and T. Suganuma, "Low-loss single-polarization fibers," Applied Optics, vol. 22, no. 11, pp. 1741–1747, 1983.

[55] P. L. Chu, "Thermal stress-induced birefringence in single-mode elliptical optical fiber," Electronics Letters, vol. 18, no. 1, pp. 45–47, 1982.

[56] J. Sakai and T. Kimura, "Birefringence caused by thermal stress in elliptically deformed core optical fibers," IEEE Journal of Quantum Electronics, vol. QE-18, no. 11, pp. 1899–1909, 1982.

[57] W. Eickhoff, "Stress-induced single-polarization fiber," Optics Letters, vol. 7, no. 12, pp. 629–631, 1982.

[58] J. Sakai and T. Kimura, "Birefringence and polarization characteristics of single mode optical fibers under elastics deformation," Journal of Quantum Electronics, vol. QE-17, no. 6, pp. 1041–1051, 1981.

[59] M. P. Varnham, D. N. Payne, A. J. Barlow, and R. D. Birch, "Analytical solution for the birefringence produced by thermal stress in polarization-maintaining optical fibers," Journal of Lightwave Technology, vol. LT-1, no. 2, pp. 332–339, 1983.

[60] P. L. Chu and R. A. Sammut, "Analytical method for calculation of stresses and material birefringence in polarization-maintaining optical fiber," Journal of Lightwave Technology, vol. LT-2, no. 5, pp. 650–662, 1984.

[61] K. H. Tsai, K.-S. Kim, and T. F. Morse, "General solutions for stress-induced polarization in optical fibers," Journal of Lightwave Technology, vol. 9, no. 1, pp. 7–17, 1991.

[62] H. Aben and C. Guillemet, "Photoelasticity of glass," Springer-Verlag, Berlin, Germany, 1993.

[62] Single polarization fiber (PZF) data sheet, Verrillon, http://www.verrillon.com.

[64] D. A. Nolan, G. E. Berkey, M.-J. Li, X. Chen, W. A. Wood, and L. A. Zenteno, "Single polarization fiber with high extinction ratio," Optics Letters, vol. 29, no. 16, pp. 1855–1857, 2004.

[65] D. A. Nolan, M.-J. Li, X. Chen, and J. Koh, "Single polarization fibers and applications," *Optical Fiber Communication Conference Technical Digest*, 2006, paper OWA1.

[65] Single polarization specialty fibers, Corning, http://www.corning.com.

[67] D. A. Nolan, M.-J. Li, and X. Chen, "Fibers with low polarization-mode dispersion," Journal of Lightwave Technology, vol. 22, no. 4, pp. 1066–1077, 2004.

[68] A. C. Hart, Jr., R. G. Huff, and K. L. Walker, "Method of making a fiber having low polarization mode dispersion due to a permanent spin," U.S. Patent 5,298,047 March 29, 1994.

[69] M. J. Li and D. A. Nolan, "Fiber spin-profile designs for producing fibers with low polarization mode dispersion," Optics Letters, vol. 23, pp. 1659–1661, 1998.

[70] R. E. Schuh, X. Shan, and A. S. Siddiqui, "Polarization mode dispersion in spun fibers with different linear birefringence and spinning parameters," Journal of Lightwave Technology, vol. 16, p. 1583, 1998.

[71] A. Galtarossa, L. Palmieri, and A. Pizzinat, "Optimized spinning design for low PMD fibers: an analytical approach," Journal of Lightwave Technology, vol. 19, p. 1502, 2001.

CHAPTER EIGHT

Optical Fibers Based on Air–Silica Guiding Structure

8.1 REVIEW OF AIR–SILICA GUIDANCE IN OPTICAL FIBERS

In silica optical fibers, the refractive index control is one of key technologies in mode confinements, chromatic dispersion controls, and birefringence manipulation. The control should be in both ways—increasing and decreasing the refractive index. In comparison to relatively sufficient and efficient doping of the index-raising GeO_2 into silica, the index-lowering agents in silica glass such as B_2O_3 and F have shown severe limitation of doping concentration due to the huge excess loss for high B_2O_3 and chemically unstable glass forming for high F. The lowest possible refractive index is that of air and holey regions have been incorporated in optical fiber to maximize the refractive index contrast, and various air–silica guiding structures have been proposed in various forms [1–3].

However, it is since 1996 when Knight et al. [4] first reported a transverse periodic air hole structure in silica cladding that intensive and explosive research efforts have begun for innovative nano- and micro-scale air–silica structured optical fibers and their applications. There are various names for these air–silica guiding optical fibers such as photonic crystal fibers (ASHF), microstructured optical fiber (MOF), and holey fiber (HF). In this chapter, we will use the term air-silica holey fiber (ASHF) to be more inclusive in classification of similar fibers.

These ASHFs have opened a new avenue of guided wave optics and fiber devices demonstrating elegant and innovative optical properties that have not been attainable in prior all-silica fibers. According to Russell [5] ASHFs have provided unprecedented and unique optical properties such as (1) a high refractive index difference, Δ, that enabled flexible control in the group velocity dispersion, birefringence,

Silica Optical Fiber Technology for Devices and Components: Design, Fabrication, and International Standards, First Edition. By Kyunghwan Oh and Un-Chul Paek.
© 2012 John Wiley & Sons, Inc. Published 2012 by John Wiley & Sons, Inc.

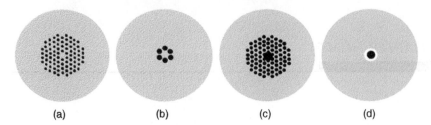

(a)   (b)   (c)   (d)

FIGURE 8.1 Four types of ASHF to be discussed in this chapter: (a) effective index guiding (EIG), (b) air hole cladding (AHC), (c) photonic band gap (PBG), and (d) hollow ring core (HRC). Here black regions indicate air, gray parts are pure silica, and brighter gray in (d) is GeO_2-doped silica.

nonlinearity, and bending performance and (2) hollow core guidance that effectively enhances light–gas and light–liquid interactions with a high overlap.

There have been excellent review papers [5–12] and books [13–15] on ASHFs and their applications in recent years. It would be beyond the scope of this chapter to discuss all of the reported ASHFs and the related applications. In order to make this chapter consistent in terms of its subject, air–silica guiding structure, we confine our discussion to four types of ASHF as shown in Figure 8.1.

These four distinctive classes, although they act as waveguides, do exhibit different optical properties leading to different applications. Especially types (a)–(c) have been extensively reviewed in Refs [5–15], yet type (d) has not been sufficiently represented despite its high potentials in practical applications. We will discuss mainly the effective index guiding fiber before concentrating on the more unique hollow ring core fiber. In terms of applications, we will further confine ourselves to modal guidance, chromatic dispersion, and polarization. Further detailed discussions on other issues are referred to Refs [5–15].

8.2 FABRICATION TECHNIQUE—STACK AND DRAW METHOD

ASHFs have been fabricated in various techniques and they have two distinctive steps: (1) preform fabrication and (2) fiber drawing. This is in fact not much different from conventional single-mode fibers (SMFs) but the ASHF preforms are not made by chemical vapor deposition (CVD) process as in SMFs. ASHF preforms can be prepared in several manners including stacking of glass capillaries and rods [4,6], extrusion [16,17], sol–gel casting [18], and drilling.

In most of silica glass ASHFs, so-called "stack and drawing" techniques are widely used and extrusion has been successfully applied in multicomponent soft glasses. ASHF preforms are most commonly fabricated by hand stacking an array of silica capillary tubes or solid rods into the desired pattern, and then fusing the stack into a solid preform. Recent improvements of this fabrication process have not only served to bring down losses but also greatly increased the diversity of structural designs. Recently, the method has been further stabilized by introducing a cane-pulling process, where the first stack of tubes and rods is drawn to intermediate canes of a few millimeters diameter. These intermediate canes can be further stacked to provide a complicated air–silica cross-section. Consequently, new ASHF designs are continuously appearing, and it will probably be a few years yet before the field can be said to have matured. We will briefly overview the stack and draw technique.

The typical starting point for an effective index guiding ASHF or an air cladding ASHF is organizing an array of hollow capillary silica tubes bundled around a pure silica rod replacing the center capillary. For PBG fibers, one or more capillary tubes may simply be left out in the center of preform in order to create a hollow defect core.

After stacking, the capillaries and rods may be held together by thin wires and fused together during an intermediate drawing process, where the preform is drawn into preform canes. Such an intermediate step may be introduced to provide a large number of preform canes for the development and optimization of the later drawing of the ASHFs to their final dimensions. During the drawing process, the outer lying tubes/rods may experience some distortion, but the core region and its nearest surroundings generally retain to a large degree the desired morphology.

The drawing of ASHF preform is generally performed in a conventional drawing tower operating at a relatively low temperature of around 1990°C. The reason for drawing at this temperature level is that surface tension otherwise may be found to collapse the air holes. The key element in the drawing of ASHF is the ability to maintain the highly regular structure of the preform all the way down to fiber dimensions and a dedicated pressure control is necessary.

Typical outer fiber diameter is 125 μm, but diameters from 80 to around 700 μm are routinely fabricated. This fiber maintains the structure of the preform but now on a microscopic scale. Standard protective UV curable polymer coatings are applied to the bare fibers in order to improve handling characteristics. The several steps of ASHF fabrication are summarized in Figure 8.2.

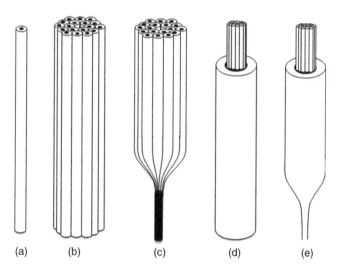

(a)   (b)   (c)   (d)   (e)

FIGURE 8.2 Schematic diagram for "stack and drawing" process for ASHF: (a) preparation of silica tubing whose diameter is in the range of 1–2 mm, (b) stacking capillaries in a hexagonal shape, (c) drawing the stacked capillaries into a capillary cane maintaining the air holes, (d) assembling the capillary cane with a jacketing tube, and (e) drawing the final preform into air–silica optical fiber along with appropriate pressure control.

8.3   EFFECTIVE INDEX GUIDING AIR–SILICA OPTICAL FIBERS

The structure reported by Knight et al. [4] in 1996 is the first of the air–silica optical fibers with periodic hole structures in the transverse cladding plane. The structure is composed of central silica defect imbedded in a triangular air hole lattice as shown in Figure 8.3.

Due to its central silica defect, this fiber is named as solid core photonic crystal fiber (PCF) as well. The term photonic crystal fiber is used to denote a waveguide that is periodically patterned in two dimensions with the characteristic length or period being on the order of the wavelength of light being guided. The initial report on the PCF structure, however, showed guiding properties that are governed by effective step-index optical fiber. The periodic hole in the cladding provides an effective index lower than the central silica core and the light is guided by modified total internal reflection at the high refractive index silica core and the lower effective index holey cladding, very similar to conventional SMFs. Therefore, we will classify these types as effective index guiding air–silica holey fibers (EIG-ASHFs). It is noteworthy though that the dispersion of air–silica cladding is totally different from bulk silica and it can be controlled by the air hole periodicity, air hole size, and the symmetry of arrangement. This cladding dispersion engineering capability in EIG-ASHF has provided a fundamental stepping stone toward innovative novel fibers.

Soon after the initial report of EIG-ASHF, various modifications have been reported in five major areas: (1) so-called "endless single-mode" operation from visible to IR [19,20], (2) single mode with a very large mode area [21–23], (3) flexible

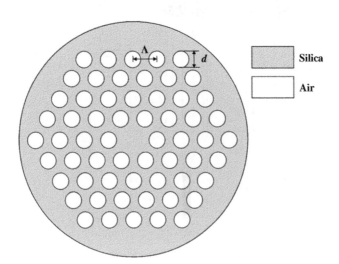

FIGURE 8.3   Schematic diagram of the most common EIG-ASHF design platform in the triangular lattice. Note that at the center, a hole is replaced by the silica defect, which serves as the core.

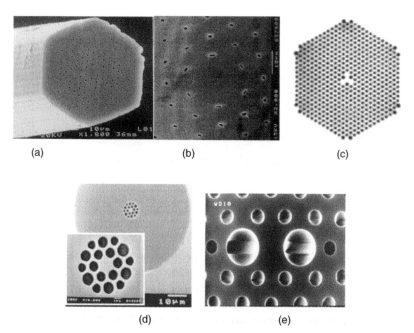

FIGURE 8.4 Cross-section photographs of major effective index guiding air–silica optical fibers (EIG-ASHFs): (a) endless single-mode fiber [19], (b) large mode area fiber [21], (c) dispersion-flattened fiber [29], (d) highly nonlinear fiber [32], and (e) high-birefringence fiber [35].

chromatic dispersion control over wide bands [24–31], (4) high nonlinearity [32–34], and (5) high birefringence [35–39].

Examples of such a structure are given in Figure 8.4. Here the defect manifests itself as a missing cylindrical hole in an array, thus forming a guiding core surrounded by cladding that consists of periodic holes. These holes act to lower the effective refractive index in the cladding region providing efficient trapping of light modes along the central core region. In contrast to conventional SMF where the core engineering has been traditionally emphasized, EIG-ASHFs require sophisticated cladding engineering based on two parameters: the air hole diameter (d) and the air hole pitch (Λ) as indicated in Figure 8.3.

8.3.1 Modal Guidance—Endless Single Mode, Single Mode, Multimode

One of the earliest discoveries in ASHF was the fact that it can maintain the single-mode state in a very wide spectral range from visual to infrared, which brought a new concept, "endless single mode" [19]. This peculiar modal guidance in EIG-ASHFs is attributed to unique dispersion characteristics in the holey cladding, which can be tailored by the air hole diameter (d) and the air hole pitch (Λ).

In conventional step-index single-mode fibers (SI-SMFs), the modal guidance is described by an important parameter, normalized frequency, V, which is defined as

$$V = \frac{2\pi a}{\lambda} \sqrt{n_{co}^2 - n_{cl}^2}$$

$$V_{cutoff} = 2.405$$

(8.1)

where a is core radius and n_{co} and n_{cl} are refractive indices of the core and the cladding, respectively. Note that the core and the cladding indices in SI-SMF are determined by the constituent material and they are very slowly varying over the wavelength region of interests as discussed in Chapter 2. The first excited mode cutoff is defined by $V = 2.405$, which defines the boundary between the single-mode and multimode operation.

In the case of EIG-ASHFs, the cladding is periodic air–silica structure and the core is formed by the silica defect in the center, as shown in Figure 8.5.

One notable feature of EIG-ASHFs is the fact that boundary of the core region could be ambiguous as shown in Figure 8.5, and there have been various definitions in the core diameter in the numerical analysis ranging, from Λ, $\Lambda/\sqrt{3}$, and $\Lambda/2$ to $\Lambda - d/2$ [40–43].

Similar to SI-SMFs, V parameter has been defined for EIG-ASHFs in two different versions [42,43] as follows:

$$\begin{cases} V = \frac{2\pi\Lambda}{\lambda} \sqrt{n_{FM}^2(\lambda) - n_{FSM}^2(\lambda)} \\ V_{cutoff} = \pi \end{cases}$$

(8.2)

$$\begin{cases} V = \frac{2\pi a_{eff}}{\lambda} \sqrt{n_{co}^2 - n_{FSM}^2(\lambda)} \\ V_{cutoff} = 2.405 \end{cases}$$

(8.3)

Mortensen et al. [42] defined V parameter of EIG-AOSF as in equation 8.2, where $n_{FM}(\lambda)$ is the wavelength-dependent effective index of the fundamental mode (FM) and $n_{FSM}(\lambda)$ is the corresponding effective index of the first cladding mode in the

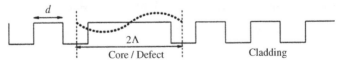

FIGURE 8.5 Refractive index profile of an EIG-ASHF along the symmetry axis. The top and bottom levels correspond to the refractive indices of the silica and the air, respectively [40].

infinite periodic cladding structure often denoted the fundamental space-filling mode (FSM). This definition of V parameter took Λ as the core radius, and the cutoff condition for the first excited mode was set as $V = \pi$. However, this definition is incommensurate with original V parameter definition in waveguide theory, and in fact it corresponds to the normalized transverse attenuation constant, W parameter, as discussed in Chapter 2.

Koshiba and Saitoh [43] have proposed a V parameter definition in equation 8.3, which can provide flawless extension of waveguide properties such as cutoff wavelength, mode field diameter, and splice loss defined in conventional SI-SMFs into EIG-ASHFs. In the definition of V, they defined the effective core radius $a_{\text{eff}} = \Lambda/\sqrt{3}$ in order to keep the cutoff condition $V = 2.405$. In equation 8.3 the core refractive index is given by $n_{\text{co}} = n_{\text{SiO}_2}$, and $n_{\text{FSM}}(\lambda)$ is the corresponding effective index of the FSM. In this chapter, we will follow the definition of V parameter as given by equation 8.3, in order to keep the consistency of discussion on optical parameters in waveguides.

The starting point of waveguide analysis in EIG-ASHF is to find the effective index of the fundamental space-filling mode, n_{FSM}, where the central silica defect is removed and infinite extension is assumed. Using a vectorial finite element method (FEM), the n_{FSM} was calculated [44], which was confirmed consistent with other reports [42,43].

Effective index of the FSM is plotted in Figure 8.6, as a function of normalized wavelength, λ/Λ, for various normalized hole diameters, d/Λ. The effective index of the FSM can be made highly dispersive so that its value ranges from the maximum value of 1.45, the refractive index of silica, to very close to 1.1, which endows very flexible design of optical properties.

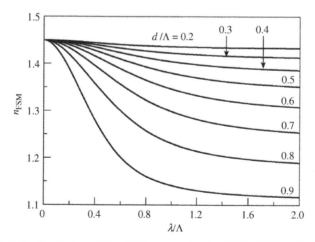

FIGURE 8.6 Effective index of the FSM, n_{FSM}, for the EIG-ASHF with different hole diameter/pitch ratios [42–44].

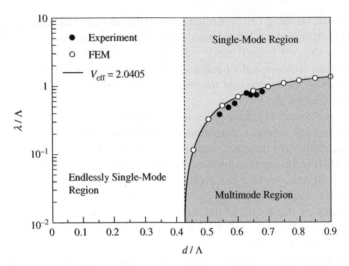

FIGURE 8.7 Relative cutoff wavelength λ/Λ as a function of relative hole diameter d/Λ [43,46].

In comparison to conventional SI-SMFs, EIG-ASHF shows unique "endless single-mode" operation where the light from UV to IR range is guided in the single mode without existence of higher order modes. Three regions of distinctive guidance properties exist in EIG-ASHFs and they are shown in Figure 8.7.

The solid line in Figure 8.7 is the trace for $V = 2.405$ that separates the single-mode region and the multimode region. For $d/\Lambda < 0.43$, as indicated by dotted vertical line, the fiber is endless single mode.

Saitoh and Koshiba [45] reported empirical expressions for normalized frequency, V parameter, and transverse attenuation constant, W parameter, which gave precise prediction of the effective index of FSM (n_{FSM}) and the effective index of the fundamental mode guided along the core (n_{eff}), and its chromatic dispersion (D).

The V and W parameters share the same functional form as given in equation 8.4:

$$V\left(\frac{\lambda}{\Lambda},\frac{d}{\Lambda}\right) = A_1 + \frac{A_2}{1 + A_3 \exp(A_4 \lambda/\Lambda)} = \frac{2\pi a_{\text{eff}}}{\lambda}\sqrt{n_{\text{co}}^2 - n_{\text{FSM}}^2(\lambda)}$$

$$W\left(\frac{\lambda}{\Lambda},\frac{d}{\Lambda}\right) = B_1 + \frac{B_2}{1 + B_3 \exp(B_4 \lambda/\Lambda)} = \frac{2\pi a_{\text{eff}}}{\lambda}\sqrt{n_{\text{eff}}^2 - n_{\text{FSM}}^2(\lambda)}$$

$$(8.4)$$

The coefficients for V parameter were found by fitting the FEM results and they are given in equation 8.5 and Table 8.1:

$$A_i = a_{i0} + a_{i1}\left(\frac{d}{\Lambda}\right)^{b_{i1}} + a_{i2}\left(\frac{d}{\Lambda}\right)^{b_{i2}} + a_{i3}\left(\frac{d}{\Lambda}\right)^{b_{i3}}$$

$$(8.5)$$

TABLE 8.1　Coefficients for V Parameter [45]

	$i=1$	$i=2$	$i=3$	$i=4$
a_{i0}	0.54808	0.71041	0.16904	−1.52736
a_{i1}	5.00401	9.73491	1.85765	1.06745
a_{i2}	−10.43248	47.41496	18.96849	1.93229
a_{i3}	8.22992	−437.50962	−42.4318	3.89
b_{i1}	5	1.8	1.7	−0.84
b_{i2}	7	7.32	10	1.02
b_{i3}	9	22.8	14	13.4

The coefficients for W parameter were found by fitting the FEM results and they are given in equation 8.6 and Table 8.2:

$$B_i = c_{i0} + c_{i1}\left(\frac{d}{\Lambda}\right)^{d_{i1}} + c_{i2}\left(\frac{d}{\Lambda}\right)^{d_{i2}} + c_{i3}\left(\frac{d}{\Lambda}\right)^{d_{i3}} \tag{8.6}$$

The comparison of V and W parameters between calculations by the above empirical fitting curves and FEM results is shown in Figure 8.8a and b, which show excellent agreements for $\lambda/\Lambda < 2.0$, $V > 0.85$, and $W > 0.1$.

From the definition of V and W parameters, we can directly compute both the effective index of fundamental space-filling mode, n_{FSM}, and the effective index of the core-guided fundamental mode, n_{eff}, using the following relations:

$$n_{FSM} = \sqrt{n_{co}^2 - \left(\frac{\lambda V}{2\pi a_{eff}}\right)^2} \tag{8.7}$$

$$n_{eff} = \sqrt{\left(\frac{\lambda}{2\pi a_{eff}}\right)^2 (W^2 - V^2) + n_{co}^2} \tag{8.8}$$

TABLE 8.2　Coefficients for W Parameter [45]

	$i=1$	$i=2$	$i=3$	$i=4$
c_{i0}	−0.0973	0.53193	0.24876	5.29801
c_{i1}	−16.70566	6.70858	2.72423	0.05142
c_{i2}	67.13845	52.04855	13.28649	−5.18302
c_{i3}	−50.25518	−540.66947	−36.80372	2.7641
d_{i1}	7	1.49	3.85	−2
d_{i2}	9	6.58	10	0.41
d_{i3}	10	24.8	15	6

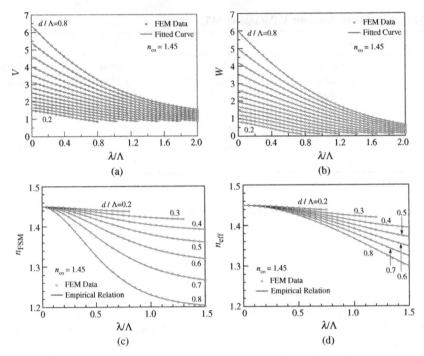

FIGURE 8.8 Effective normalized waveguide parameters: (a) V, (b) W, (c) effective index of fundamental space-filling mode, n_{FSM}, and (d) the effective index of the core-guided fundamental mode, n_{eff}, for EIG-ASHFs as a function of λ/Λ for various d/Λ [45].

These effective indices are plotted in Figure 8.8c and d, which again show excellent agreement with FEM results.

From the effective index of the core-guided mode, $n_{eff}(\lambda)$, we can calculate the chromatic dispersion by adding material dispersion contribution given by Sellmeir equations:

$$D = -\frac{\lambda}{c}\frac{d^2 n_{eff}}{d\lambda^2} \tag{8.9}$$

For EIG-ASHF, plots of chromatic dispersion as a function of wavelength are shown in Figure 8.9 for various waveguide parameters. These curves were calculated using FEM [46] and the plots very similar to these can be obtained directly from n_{eff} using empirical V and W parameters in equations 8.8 and 8.9.

In Figure 8.9, we can observe very wide ranges of dispersion curves whose values cannot be obtained in conventional SMFs as discussed in Chapters 2 and 3. Especially for short hole pitch $\Lambda = 1.0\,\mu m$ (Figure 8.9a), a very large negative dispersion could be obtained in communication bands covering O, E, S, C, and L bands, 1.2–1.6 μm. In the other extreme, for a large hole pitch $\Lambda = 3.0\,\mu m$ and small hole diameter $d/\Lambda = 0.2$ (Figure 8.9d), the dispersion is very close to that of bulk silica as expected. In the

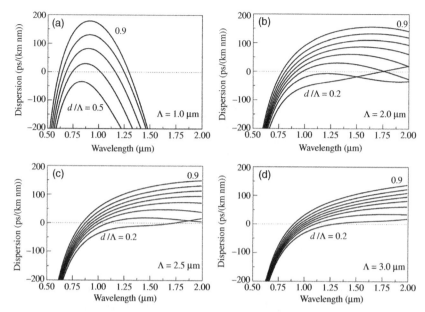

FIGURE 8.9 (a–d) Chromatic dispersion of EIG-ASHF for various waveguide parameters calculated by FEM methods [46].

intermediate regime, various dispersion parameters such as the zero-dispersion wavelength, the dispersion value, and its slope at a certain wavelength could be flexibly controlled by varying air hole diameter and/or the hole pitch. Along with endless single-mode operations and these chromatic dispersion characteristics, EIG-ASHFs have shown high potentials that still need further investigations.

Conventional SI-SMF's mode field diameter (MFD) can be expressed in terms of V parameter [47] as follows:

$$\frac{w}{a_{\text{eff}}} = 0.65 + \frac{1.619}{V_{\text{eff}}^{3/2}} + \frac{2.879}{V_{\text{eff}}^{6}} \tag{8.10}$$

where w is the half of MFD and is defined as the effective modal spot size. For the given V parameters in equations 8.4 and 8.5 and Table 8.1, MFD can be calculated and the results are summarized in Figure 8.10, as a function of normalized wavelength, λ/Λ, for various normalized hole diameters, d/Λ. The experimental results are for the commercially available fibers from NKT Photonics [48].

8.4 LARGE MODE AREA AND BENDING LOSS BASED ON EFFECTIVE INDEX GUIDING AIR–SILICA HOLEY FIBERS

There are urgent demands for large mode area optical fibers in fiber lasers and fiber amplifiers, where nonlinear effects should be sufficiently suppressed to maintain a

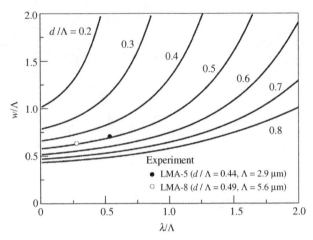

FIGURE 8.10   Mode field radius of EIG-ASHF for various waveguide parameters [46].

high level of gain at the desired spectral range. Conventional SI-SMFs have inherent limitations on the core size and the numerical aperture (NA) in the single-mode regime. Usually the MFD of conventional SI-SMF is about 9 μm at 1.55 μm wavelength. Fabrication of an SI-SMF with a large mode area would require in general a larger core diameter with a smaller refractive index difference. However, the refractive index control in conventional CVD process is within 10^{-5} or at best 10^{-6}, which directly limits the upper bound of the core radius in practice.

EIG-ASHFs have offered a very effective alternative for large mode area single-mode fibers as shown in Figure 8.4b [21]. ASHF technology can provide a very unique and precise control of the index difference between core and cladding regions by securing high precision in d/Λ control as indicated in Figure 8.8c. Conditions for large mode area or large MFD can be directly deduced from Figure 8.10, where the mode field radius w rapidly increases by several folds with decreasing d/Λ. General direction of large mode area EIG-ASHF is to either increase the air hole pitch or reduce the relative hole diameter. Single-mode HFs with mode areas as large as 680 μm^2 at 1550 nm have been reported [49] with $d/\Lambda = 0.055$ and $\Lambda = 12.8$ μm, which operates in the endless single mode. Various large mode area EIG-ASHFs have been reported for $\Lambda > 5$ μm with MFDs of 9–26 μm, where they usually show endless single-mode guidance.

Enlarging the central core by replacing three air holes with silica defects instead of one has been found to further increase the mode field diameter [50]. The size of the air holes in this case was reduced to keep the fiber in the single mode, and the hole pitch was also reduced. In Figure 8.11, the loss curves for the two large mode area fibers are shown along with the cross-sections of fibers.

The single central silica defect ASHF had $\Lambda = 10$ μm, $d/\Lambda = 0.45$, and an MFD of ~10.5 μm. The three central silica defect ASHF had $\Lambda = 6$ μm, $d/\Lambda = 0.25$, and an MFD of ~12.0 μm. Note that actual silica core sizes are similar but the MFD in the three silica defect ASHF increased by 30% without additional loss as confirmed in Figure 8.11.

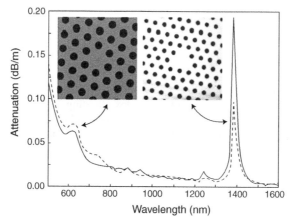

FIGURE 8.11 Attenuation spectra for two types of large mode area EIG-ASHFs. The dotted line is the spectrum for the single central silica defect ASHF, and the solid line is for the three silica defect ASHF [50].

Despite the very large MFD in EIG-ASHFs predicted in Figure 8.10, the macroscopic bend loss ultimately limits the mode area scaling that can be utilized in practical applications. In conventional SI-SMFs the bend loss increases at long wavelengths because the equivalent index profile in the bent fiber becomes leaky for the mode at a long wavelength with a lower effective index. Bending has been discussed using the equivalent index model [51], which accounts for the different path lengths seen at different transverse positions x as light travels around a bend of radius R_{bend}, which is illustrated in Figure 8.12.

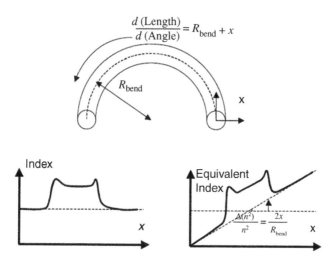

FIGURE 8.12 Different path lengths seen at different transverse positions x as light travels around a bend of radius R_{bend} [51].

Path lengths are adjusted by defining the equivalent index profile as follows:

$$n_{eq}^2(x,y) = n^2(x,y)\left(1 + \frac{2x}{R_{bend}}\right) \tag{8.11}$$

At longer wavelength, the light in the fundamental core mode of conventional SI-SMF will have more chance leak toward higher index cladding.

In contrast to SI-SMF, EIG-ASHF will, in theory, exhibit both a short and a long wavelength bend edge. EIG-ASHFs exhibit a bend loss edge at long wavelengths due to rapid expansion of the mode into the cladding, resulting in a more weakly guided mode that will suffer a greater perturbation in response to bending. They also show an additional bend loss edge at short wavelengths as a direct consequence of their unique cladding structure. In an EIG-ASHF the core–cladding index difference decreases toward short wavelengths (see Figure 8.8c and d). The mode at a shorter wavelength is more tightly confined to the core and experiences less of the air holes. This unique property gives rise to "endless single mode," and also leads to a weak guidance that could be easily perturbed by the bending. The long wavelength bend edge of EIG-ASHF occurs for $\lambda \gg \Lambda/2$, and for typical large mode area ASHF ($\Lambda > 5\,\mu$m) it is located in the nontransparent wavelength regime of silica [52]. Therefore, bending analysis in large mode area ASHFs has been confined only to a short wavelength bend edge.

The bend loss formula developed for the fundamental mode in a conventional SI-SMF [53,54] has been successfully adopted in the EIG-ASHFs' loss at short wavelength bend edges [55]. For a Gaussian mode in a conventional SI-SMF, the bending loss is given by

$$\alpha_{bend} = \frac{\sqrt{\pi}}{4}\frac{1}{A_{eff}}\frac{a_{eff}}{W}\frac{\exp[-(4/3)(R_{bend}/a_{eff})(\Delta/V^2)W^3]}{\sqrt{W(R/a_{eff}) + (V^2/2\Delta W)}}$$

$$A_{eff} = \pi w^2, \qquad a_{eff} = \frac{\Lambda}{\sqrt{3}} \tag{8.12}$$

$$\Delta = \frac{n_{co}^2 - n_{FSM}^2}{2n_{co}^2} \approx \frac{n_{co} - n_{FSM}}{n_{co}}$$

Note that these parameters are already given in equations 8.4–8.7 and 8.10, and therefore the bending loss α_{bend} at a certain bending radius R_{bend} can be readily estimated as a function of wavelength λ for the given hole parameters, d and Λ.

Nielsen et al. [55] reported both simulations and experiments on the short wavelength bending loss for various large mode area fibers and some of results are shown in Figure 8.13.

It is confirmed that the bend loss edge wavelength λ_{ble} show a specific relation as [55]

$$\lambda_{ble} \propto \sqrt{\frac{\Lambda^3}{R}} \tag{8.13}$$

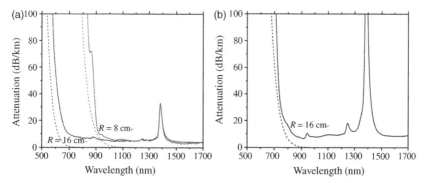

FIGURE 8.13 Loss spectra of EIG-ASHFs for various bending radius R: (a) LMA-20 fiber with the core diameter of 20.0 μm, hole pitch $\Lambda = 13.20$ μm, and $d/\Lambda = 0.485$; (b) LMA-25 fiber with the core diameter of 24.5 μm, hole pitch $\Lambda = 16.35$ μm, and $d/\Lambda = 0.500$. The dotted lines are theoretical simulations and solid lines are experimental measurements.

8.5 DISPERSION CONTROL IN EFFECTIVE INDEX GUIDING AIR–SILICA HOLEY FIBERS

As already discussed in Chapters 2 and 3, dispersion control in optical fibers is one key issue in various applications such as signal broadening control in both time domain and spectral domain for wavelength division multiplexing, optical Kerr nonlinearity control, and pulse shaping. Chromatic dispersion is directly calculated from the effective index of the fundamental core mode, n_{eff}, by the relation given as follows:

$$D = -\frac{\lambda}{c}\frac{d^2 n_{\text{eff}}}{d\lambda^2} \qquad (8.9)$$

In conventional SI-SMFs various profiles have been attempted for dispersion flattening and dispersion compensations, as discussed in Chapter 4. Typical index profiles for dispersion control in Chapter 4 are shown in Figure 8.14.

Note that the dispersion control in SI-SMFs could be achieved in a certain level of flexibility by tailoring inner cladding structures. Similar concepts have been adopted in EIG-ASHFs by either (1) controlling d and Λ as discussed in Figure 8.9 [25,26,56] or (2) introducing concentric rings of different hole diameters d_i and different pitch Λ_i near the central defect core [27,30,57–59]. Typical structures of EIG-ASHFs reported for the dispersion-flattening (DF) applications are shown in Figure 8.15.

Due to the high-index contrast between silica and air in ASHFs, and the flexibility of varying hole sizes and patterns, a much broader range of dispersion control is accessible with ASHFs than with standard fibers. For the uniform structure in Figure 8.15a, it was shown that zero-dispersion wavelengths could be varied from the infrared region far down into the visible part of the spectrum [25,26,56], simply by varying hole size and spacing, d, and Λ, maintaining the uniform distribution. However, maintaining all of the same air hole diameter in the cladding region is

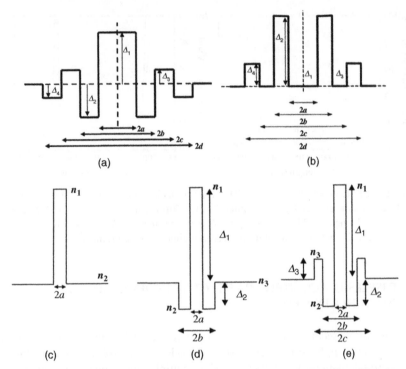

FIGURE 8.14 Dispersion control profiles in SI-SMFs: (a and b) dispersion flattening and (c–e) dispersion compensation.

not sufficient enough to control both the dispersion and the dispersion slope in a wide wavelength range in a flexible manner. In fact, EIG-ASHF does not require perfect periodicity in the cladding region to confine the guiding light to the high-index core region, and Saito et al. [27] have proposed a different approach by changing the hole diameter and arranging those holes in a ring as schematically shown in Figure 8.16.

Two different types of air hole arrangements are shown in Figure 8.16 to decrease and increase the effective index of "inner claddings" in EIG-ASHF, where d_i and Λ_i ($i = 1 - n$) are the hole diameter and hole pitch in the ith air hole ring, respectively. In Figure 8.16a the air hole diameters are $d_1 > d_2 = \Lambda = d_n$ and all the Λ_i are the same. The effective refractive index in the cladding region decreases with increasing air hole diameter to result in depressed inner cladding profile as in the bottom of Figure 8.16a. When the hole diameter is reduced in a ring as in Figure 8.16b with $d_2 < d_1 = d_3 = \ldots = d_n$, the effective refractive index increases to result in a raised inner cladding profile as in the bottom of Figure 8.16b. Note that the flexible control of the inner cladding region was the key to tailoring the dispersion slope and dispersion in the conventional SI-SMFs as shown in Figure 8.14. By optimizing the air hole diameters d_i and the hole-to-hole spacing Λ_i, both the dispersion and the dispersion slope can be controlled in a wide wavelength range. In Table 8.3 we summarize some of recent reports on dispersion-flattened EIG-ASHFs.

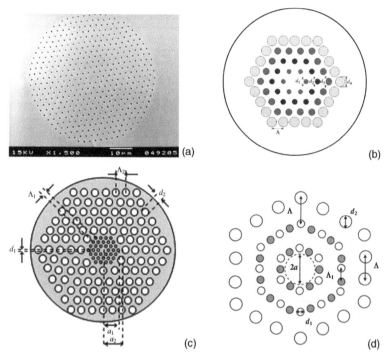

FIGURE 8.15   Dispersion-flattened EIG-ASHFs: (a) tailoring d and Λ in the uniform lattice [26], (b) changing hole sizes d_i in concentric rings keeping Λ constant [27], (c) changing hole sizes d_i and Λ_i in concentric rings [30], and (d) changing hole sizes d_i and Λ_i in concentric rings along with interstitial holes shown in gray circles [58].

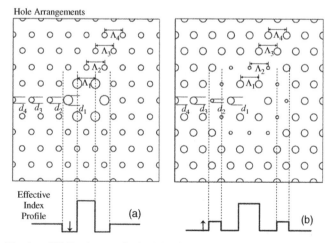

FIGURE 8.16   (a and b) Fundamental principles for refractive index tailoring in EIG-ASHF by varying hole diameter and arranging the holes in a ring.

TABLE 8.3   Comparison of the Dispersion-Flattening EIG-ASHFs

Reference	D	L_c	A_{eff}	FBW	N_r	N_Λ	N_d
Ferrando et al. 2005 [56]	0 ± 0.5	–	–	543	–	1	1
Ferrando et al. 2001 [25]	0 ± 1.5	–	–	~600	–	1	1
Reeves et al. 2002 [26]	0 ± 1.2	$>10^{-3}$	~44	600	11	1	1
Saito et al. 2003 [27]	0 ± 0.4	<0.1	~8.5	490	4	1	4
Matsui et al. 2005 [30]	4.5 ± 0.4	<0.01	>100	350	8	2	2
Wu et al. 2005 [57]	0 ± 0.25	–	13.2	430	4	1	2
Razzak et al. 2007 [58]	0 ± 0.41	10^{-4}	11.94	290	6	1	3
Kaijage et al. 2008 [59]	0 ± 0.25	10^{-3}	10.86	230	6	1	3

Note that D, chromatic dispersion, is in ps/nm·km, L_c confinement loss in dB/km. A_{eff} effective mode area in μm^2, and FBW Flattened bandwidth in nm. N_r is number of air hole rings, N_Λ number of different pitches, N_d number of different air hole diameter.

Despite the seemingly equivalent step-index profiles for dispersion-flattened optical fibers in EIG-ASHFs of Figure 8.15, and SI-SMFs of Figure 8.14, there is a fundamental difference in the chromatic dispersion curves. Typical dispersion curves of EIG-ASHFs and SI-SMFs are shown in Figure 8.17.

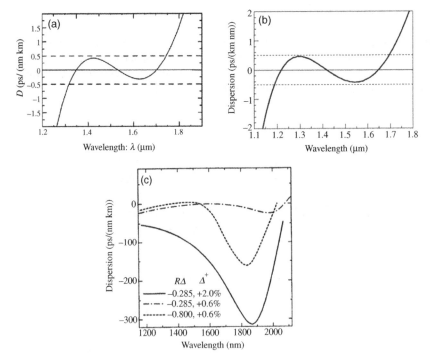

FIGURE 8.17 Typical dispersion curves of EIG-ASHFs with (a) uniform d and L [25], (b) concentric rings of different hole diameters [27], and (c) typical dispersion curves of SI_SMFs [60] (see Figure 4.27 in Chapter 4).

It is highly noteworthy to recognize the functional differences between EIG-ASHF (Figure 8.17a and b) and SI-SMFs serving the same purpose—dispersion tailoring with inner cladding structures. In the conventional SI-SMFs the dispersion curve in Figure 8.17c shows only a single extremum in the region of interest, 1.2–1.6 μm, which is achromatic. However, in the case of EIG-ASHFs, both the dispersion curves in Figure 8.17a and b show two neighboring extrema, which is apochromatic. This apochromatic behavior is attributed to unique cladding dispersions in EIG-ASHFs and this is the fundamental principle that enables these fibers to reach such a large dispersion-flattened operation bandwidth with a low third-order dispersion. EIG-ASHFs also realize a zero-dispersion wavelength of less than 1270 nm, which corresponds to the zero-dispersion wavelength of silica glass, beating the material dispersion limit in the shorter wavelengths.

Dispersion compensation fibers are indispensable for long-distance, high speed transmission based on standard single-mode fibers. For the accumulated dispersion of SMF after certain long-distance propagation, a dispersion compensation module should provide appropriate negative dispersion value and dispersion slope within the transmission band. The negative dispersion values have been achieved by dual core design as discussed in Chapter 4 [61]. The dispersion properties in conventional SMF and commercial dispersion-compensating fibers [62] at 1550 nm are summarized in Table 8.4.

Here the relative dispersion slope is defined for dispersion slope $S(\lambda)$ and dispersion $D(\lambda)$:

$$\mathrm{RDS}(\lambda) = \frac{S(\lambda)}{D(\lambda)} \tag{8.14}$$

In conventional dispersion-compensating fibers (DCFs), dual concentric ring core fibers or W-type fibers have been widely used, whose refractive index profiles are shown in Figure 8.14c–e [61,62]. In these fibers, high GeO_2-doped central core is surrounded by an outer ring core with a less GeO_2 doping concentration to obtain large negative dispersion and negative dispersion slope. These DCFs have shown about -100 ps/(nm km). To obtain a large negative dispersion coefficient, usually high GeO_2-doping is required, which induces an abrupt increase in the transmission loss, as discussed in Chapter 6 (Figure 6.13). These conventional DCFs also have limitations in the spectral range and most of them are limited in C or L bands.

TABLE 8.4 Dispersion Properties in Conventional SMF at 1550 nm [62]

Fiber Types	D (ps/(nm km))	S (ps/(nm^2 km))	RDS (m^{-1})	Effective Area (μm^2)
SMF (ITU G652)	16.5	0.058	0.0036	82
DCF-EWBDK (OFS)	-120	-0.432	-0.0036	21
DCF-LLDK (OFS)	-170	-0.612	-0.0036	21
DCF-HFDK (OFS)	-250	-0.900	-0.0036	15

EIG-ASHFs have shown a strong potential at least in optical waveguide design to overcome those limitations providing a very large negative dispersion in a wide spectral range using only silica glass. There have been two categories of dispersion-compensating EIG-ASHFs

(1) uniform air hole lattice with optimal hole diameter d and hole pitch Λ [63–66]
(2) double concentric ring structure by varying d and Λ within a few hexagonal rings [67–73]

Basic ideas of double concentric ring waveguides or W-type waveguides have been successfully adopted in dispersion-compensating ASHFs as shown in Figure 8.18a. Here the effective refractive index of the outer ring core was raised by reducing the hole diameter in a hexagonal ring, $d_2 < d_1 = d_3 = \ldots d_N$. This concept of changing the hole diameter or/and hole pitch has been widely explored by various research groups and their typical waveguide structures are shown in Figure 8.18b–f. It is noted that these types can be further categorized into two classes: (1) outer cladding modification in a uniform lattice and (2) inner cladding modification. The waveguides (b) and (c) belong to the outer cladding modification type, where the ASHFs form a

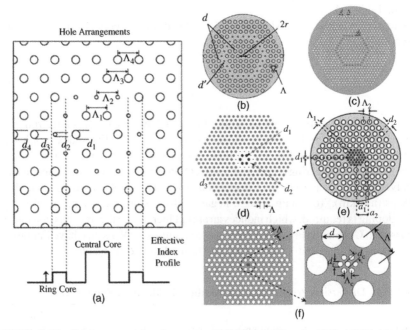

FIGURE 8.18 Dual concentric ring core (DCRC) profiles in EIG-ASHF structure: (a) schematic diagram of DCRC with the central core and the ring core, by varying the hole size in a ring, (b) DCRC for a large negative dispersion in a narrow band [67], (c) for a large negative dispersion in C-band [70], (d) for S + C + L bands [71], (e) for O + E + S + C + L bands [72], and (f) for S + C + L + U bands [73].

uniform lattice with a constant d and Λ along with the central SiO_2 defect core except the holes with a reduced diameter, d', in the outer ring core region. In contrast, the inner cladding modification types modify the holes in the nearest neighbor rings from the central core as in Figure 8.18d–f.

It is well known that ASHF is very useful for chromatic dispersion management and it is possible to obtain even $-10,000 \, ps/(nm \, km)$ dispersion coefficient without Ge doping [65]. However, the available wavelength bandwidth for dispersion compensation of SMF is very narrow because of the concave dispersion profile of dual concentric core fibers. Therefore, it is necessary to compensate for the dispersion slope of SMF and for broadband operation. Although dual concentric core fibers with carefully chosen structural parameters have been proposed for broadband dispersion compensation of SMF [61,62], the problem due to Ge doping still remains. Huttunen and Torma reported a DCASHF with high-index core and large air holes that has dispersion values as large as $-59,000 \, ps/(km \, nm)$ over a broad wavelength range of 100 nm [69]; however, the central core of the proposed DCASHF is not truly single mode and the relative dispersion slope has not been optimized to match with that of a standard SMF in wide wavelength range (Table 8.5).

Despite various novel structures for dispersion-controlled ASHFs, only a few experimental reports are available due to fundamental difficulties in ASHF fabrication process such as hole diameter uniformity along the fiber axis and consistency of the hole arrangements on the cross-sectional plane. Labonté et al. [74] reported very high sensitivity of dispersion to the variation of the geometrical parameters of ASHF based on both experiments and theory. They found a hole size variation as small as 10 nm can induce a significant modification in chromatic dispersion in terms of dispersion value and slope over a wide wavelength range. Such a high sensitivity was mainly attributed to the strong interaction between the guided field and the first ring of holes whose diameters or position can change along the fiber. Labonté et al. [74] also predicted that this geometrical sensitivity in the ASHFs can

TABLE 8.5 Comparison of the Dispersion-Compensating EIG-ASHFs

Reference	D	CBW	A_{eff}	N_r	N_Λ	N_d
Birks et al. 1999 [63]	−680	100	<3	–	1	1
Shen et al. 2003 [64]	−474.5	236	1.6	–	1	1
Poli et al. 2003 [65]	−1,700	100	<2	–	1	1
Mangan et al. 2004 [66]	−1,080	30	–	–	1	1
Gérôme et al. 2004 [67]	−2,200	<30	~10	7	1	2
Huttunen et al. 2005 [69]	−59,000	>100	30	–	1	2
Fujisawa et al. 2006 [70]	−538	40	23.6	14	1	2
Varshney et al. 2007 [71]	−98.3	150	21.6	11	1	3
Matsui et al. 2007 [72]	−75	365	17.2	7	2	2
Franco et al. 2008 [73]	−179	195	~6	9	2	3

Note that D, chromatic dispersion, is in ps/nm·km, CBW Compensating bandwidth in nm and A_{eff} effective mode area in μm^2. N_r is number of air hole rings, N_Λ number of different pitches, N_d number of different air hole diameter.

be reduced to a certain extent by increasing the core size. Up to now, from a technological point of view, bringing d and Λ under control over a long length of fiber with an accuracy better than a few tens of nanometers seems to be very challenging. Consequently, the chromatic dispersion cannot be considered as constant along the fiber in real ASHFs.

8.6   OPTICAL LOSS IN EFFECTIVE INDEX GUIDING AIR–SILICA HOLEY FIBERS

ASHFs have a potential to achieve lower loss because they do not have the concentration fluctuation of index controlling dopants such as GeO_2, P_2O_5, or F, which induces significant scattering loss [75]. Since the first fabrication report of ASHFs [4], various efforts to reduce their optical loss have been reported and significant improvements in loss have been achieved in recent years [76–82]. The optical loss reduction trend is summarized in Figure 8.19. The lowest loss EIG-ASHF has been reported by Tajima et al. [82,83] and the progress on fiber manufacturing process especially low-loss fabrication has been reviewed recently by Zhou et al. [83] and Fini and Bise [84].

The wavelength-dependent optical loss in an EIG-ASHF can be expressed by addition of several contributions as follows [83]:

$$\alpha(\lambda) = \frac{A_R(\lambda)}{\lambda^4} + B_i(\lambda) + C_{UV} \exp\left(\frac{C_2}{\lambda}\right) + D_{IR} \exp\left(\frac{-D_2}{\lambda}\right) + E_{OH}(\lambda) + L_c(\lambda)$$

$$(8.15)$$

The right-hand side of the above equation consists of Rayleigh scattering loss (A_R), imperfection loss (B_i), UV absorption (C_{UV}), IR absorption (D_{IR}), OH loss (E_{OH}), and confinement loss (L_c).

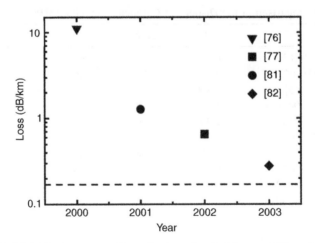

FIGURE 8.19   Recent trend in reduction of optical loss at 1550 nm [84].

TABLE 8.6   Attenuation and Contributing Coefficients in EIG-ASHFs [84]

Loss (dB/km) at 1550 nm	A_R (dB/(km μm^4))	B_i (dB/km)	OH Loss (dB/km) at 1380 nm	Reference
1	2.0	0.6	30	[76]
1.3	1.9	1	>20[a]	[77]
2.0	4.2	1	7	[78]
2.6	N.A.	N.A.	50	[80]
0.58	1.0	0.27	18[a]	[81]
0.28	1.0	0.08	0.8	[83]

N.A.: not available.
[a] Estimated values.

In optical communication window near 1 μm wavelength region, the effect of UV absorption is negligible. The confinement loss can be reduced to a negligible value with proper arrangements of air holes in several rings [85], which will be discussed in the latter part of this section. Therefore, the optical loss of an EIG-ASHF in the 1 μm wavelength region is mainly affected by four contributions as follows:

$$\alpha(\lambda) = \frac{A_R(\lambda)}{\lambda^4} + B_i(\lambda) + E_{OH}(\lambda) + D_{IR} \exp\left(\frac{-D_2}{\lambda}\right) \qquad (8.16)$$

In Table 8.6, these coefficients are summarized for the reported EIG-ASHFs [76–78,80,81,83].

In the calculation of each contribution in equation 8.16, we have to consider the modal power distribution $P_\lambda(r, \theta)$ for the fundamental mode. In comparison to conventional SI-SMF, EIG-ASH provides highly wavelength-dependent $P_\lambda(r, \theta)$, which extends over a wide region in the cladding. Rayleigh scattering coefficient remains constant, $A_{core} = 0.75$ dB/(km μm^4) [83], and scattering is divided into the core contribution and cladding contribution as follows:

$$\alpha_R(\lambda) = A_R(\lambda)\frac{1}{\lambda_4} = \frac{1}{\lambda_4}\frac{\int A_R(r,\theta)P_\lambda(r,\theta)r\,d\theta\,dr}{\int P_\lambda(r,\theta)r\,d\theta\,dr}$$

$$= A_{core}\frac{1}{\lambda_4} + \frac{\int_{clad}A_R(r,\theta)P_\lambda(r,\theta)r\,d\theta\,dr}{\int P_\lambda(r,\theta)r\,d\theta\,dr}\frac{1}{\lambda_4} \qquad (8.17)$$

Note that in EIG-ASHFs, the interior surface roughness of holes distributed over the cladding could be large and contributes significantly in the Rayleigh scattering loss. Experimentally Zhou et al. [83] found that despite the low power distribution in the cladding area, the scattering loss contribution in the cladding became a dominant factor in the total loss of EIG-ASHF at longer wavelengths.

In Table 8.7, the most recent achievement of low loss in ASHFs [82,83] is compared with conventional SMF. The fabricated ASHF was EIG-PCF, where the

TABLE 8.7   Optical Loss Comparison Between Low-Loss EIG-ASHF and Conventional SMF [83]

	EIG-ASHF	Conventional SMF
Loss at 1310 nm (dB/km)	0.44	0.35
Loss at 1550 nm (dB/km)	0.28	0.20
OH loss at 1380 nm (dB/km)	0.4	<0.2
Imperfection loss (dB/km)	0.08	<0.01
Rayleigh scattering coefficient (dB/(km μm^4))	1.0	1.0

hole diameter d and hole pitch Λ were 4 and 8 μm, respectively. The total number of holes was 90 arranged in five hexagonal rings. The variation in the outside diameter of the fiber was less than 1% throughout the fabrication process. The optical loss spectrum of the fabricated low-loss EIG-ASHF is shown in Figure 8.20.

In the lowest loss EIG-ASHF, the Rayleigh scattering loss was reduced to the same level of conventional SMF by improving the roughness of the hole interior surface. The OH loss at 1380 nm was also reduced to 0.4 dB/km by series of dehydration process and using silica tubes made by VAD process.

The imperfection loss in EIG-ASHF was still significantly higher than that of conventional SMF. Fiber diameter fluctuation during fiber drawing process can cause an additional imperfection loss if the hole size and hole pitch change along the fiber length. Pressure variation within the holes could result in diameter and pitch variation to add more imperfection loss. Controlling hole size or preventing hole closure through hole pressurization may be accomplished either by directly applying an

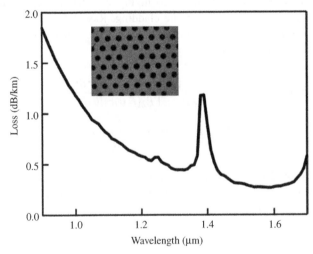

FIGURE 8.20   Optical loss spectrum of the low-loss EIG-ASHF. The inset is the cross-section of the fabricated fiber [83].

overpressure or by sealing one end of the preform before drawing. The sealing method is often preferred since this technique resulted in an unexpected self-feedback mechanism to prevent hole closure and maintains hole sizes, whereas the direct application of an overpressure was found to be very sensitive to the pressure applied. Recently comparison of numerical analysis and experiments [85–88] is being conducted to understand the hole diameter/pitch control process, which will eventually reduce the imperfection loss.

The confinement loss is a unique loss mechanism that occurs in single-material holey fibers such as ASHFs. In ASHFs the core has the same refractive index as the exterior cladding beyond the finite holey region, so propagating modes are intrinsically leaky [89]. It has been both experimentally and theoretically confirmed that addition of rings of air holes that surround the solid core, thereby increasing the physical separation between the solid core and the exterior cladding, reduces the confinement loss effectively [89–92]. The waveguide structures such as the hole diameter, hole pitch, and the number of air hole rings affect and control the confinement loss.

For a leaky mode the effective index can be expressed in terms of a complex number:

$$n_{eff} = n'_{eff} + i n''_{eff} \qquad (8.18)$$

where n'_{eff} and n''_{eff} are the real and imaginary parts of the effective index, respectively. For a hexagonal EIG-ASHF n'_{eff} and n''_{eff} were computed by White et al. [89] and results are shown in Figure 8.21. Note that the real part of the effective index, n'_{eff}, decreases with the wavelength in contrast to imaginary part, n''_{eff}, that increases with the wavelength.

For each leaky mode, both the chromatic dispersion D (ps/(nm km)) and the confinement loss (dB/m) can be calculated from n'_{eff} and n''_{eff}, respectively, as follows:

$$D(\lambda) = -\frac{\lambda}{c} \frac{d^2 n'_{eff}(\lambda)}{d\lambda^2} \qquad (8.19)$$

$$L_c(\lambda) = \frac{20}{\ln 10} \frac{2\pi}{\lambda} n''_{eff}(\lambda) \qquad (8.20)$$

where λ is the vacuum wavelength.

For a hexagonal EIG-ASHF, the confinement loss was calculated by Ferrarini et al. [90] using equation 8.20 for various waveguide parameters and the results are summarized in Figure 8.22.

It is noted that the losses quickly decrease by increasing the numbers of rings and the hole diameter as shown in Figure 8.22a. By increasing the hole diameter to $d = 0.92\,\mu m = 0.4\Lambda$, the upper value of the range goes to 0.5×10^5 dB/m, whereas the lower value goes to 10^{-5} that is negligible with respect to all the other causes of losses. Figure 8.22b shows that a loss reduction is also obtained by fixing the ratio d/Λ and by increasing the pitch Λ. The pitch and the hole diameter are changed by the same scale

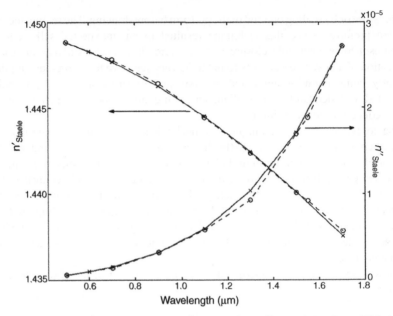

FIGURE 8.21   Real (n'_{eff}) and imaginary (n''_{eff}) parts of the effective index for an EIG-ASHF with the six holes in a hexagonal ring, $d = 2.26\,\mu m$ and $\Lambda = 4\,\mu m$. The crosses are from the multipole method and circles are from the vector beam propagation method [89].

factor; consequently, greater pitches correspond to greater core size and thus more confined fields.

Despite significant progress, the optical loss of ASHF is still high in comparison to that of conventional SMF. It is important to understand detailed optical loss mechanisms in order to further reduce the optical loss for highly efficient nonlinear fiber devices and practical transmission system applications.

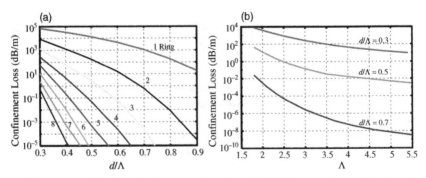

FIGURE 8.22   (a) Confinement loss as a function of hole diameter d normalized to pitch $\Lambda = 2.3\,\mu m$ for different numbers of rings. (b) Confinement loss as a function of pitch Λ for different ratios d/Λ. In both cases a wavelength $\lambda = 1.55\,\mu m$ is assumed [90].

8.7   KERR NONLINEARITY IN EFFECTIVE INDEX GUIDING AIR–SILICA HOLEY FIBERS

Nonlinearity and optical loss should be considered at the same time for practical nonlinear fiber optic device applications and the figure of merit in a nonlinear fiber is given as

$$\text{Figure of merit (FOM)} = \frac{\gamma(\lambda)}{\alpha(\lambda)} \tag{8.21}$$

$$\gamma = \frac{2\pi}{\lambda} \frac{n_2}{A_{\text{eff}}} \tag{8.22}$$

where γ is the effective nonlinearity in $(\text{W km})^{-1}$ and α is the attenuation in dB/km. n_2 is the nonlinear coefficient of the material, λ is the vacuum wavelength, and A_{eff} is the effective mode area. The effective nonlinearity γ therefore can be controlled by either material characteristics (n_2) or waveguide property (A_{eff}) (Figure 8.23).

Despite the relative low nonlinear coefficient in silica ($n_2 \approx 2.2 \times 10^{-20}\,\text{m}^2/\text{W}$), silica optical fibers can be utilized by utilizing the high confinement (small A_{eff}) and long interaction length (low optical loss α). Conventional SMF such as Corning SMF28 has an A_{eff} of $\sim 90\,\mu\text{m}^2$ and the n_2 of low-concentration GeO_2-doped silica core is $\sim 2.2 \times 10^{-20}\,\text{m}^2/\text{W}$, and then we have γ of the order of 1 $(\text{W km})^{-1}$ at 1550 nm. By reducing the core diameter and subsequently reducing $A_{\text{eff}} \sim 11\,\mu\text{m}^2$,

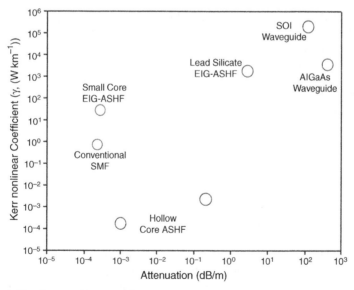

FIGURE 8.23   Kerr nonlinear coefficient as a function of optical attenuation for standard optical fiber [92], silica-based small core EIG-ASHF [93,94], silicon-on-insulator waveguide [95], AlGaAs waveguide [96], a highly nonlinear lead silicate glass effective index guiding holey fiber EIG-HF from Ref. [97], and a silica-based hollow core ASHF [98].

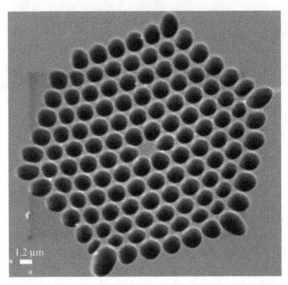

FIGURE 8.24   Cross-section of highly nonlinear EIG-ASHF [93,100].

and doping high GeO_2 concentration within the core to enhance n_2 to $\sim 5.4 \times 10^{-20}$ m^2/W, the value of γ as high as ~ 20 (W km)$^{-1}$ has been reported in conventional solid waveguide structure [99].

EIG-ASHF fibers can provide a significantly enhanced light confinement than conventional solid fibers because the cladding can be mostly comprised of air when the air-filling fraction is large ($d/\Lambda \sim 0.9$). For an EIG-ASHF with a large d/Λ value and a small silica defect core diameter, a very small value of A_{eff} can be achieved to result in a γ an order of magnitude higher than conventional SMFs.

One example of such a highly nonlinear EIG-ASHF is shown in Figure 8.24. The silica defect core size, d, was as small as 1.2 μm and the average air-filling factor d/Λ of 0.9, which results in a very small $A_{eff} \sim 1.5$ μm^2. The fiber showed a normal dispersion about -30 ps/(nm km) at 1550 nm and the effective fiber nonlinearities of more than $\gamma \approx 60$ (W km)$^{-1}$ at 1550 nm have been measured in pure silica holey fibers [93,100]. Hence, it is possible to achieve fiber nonlinearities more than 50 times higher than in standard telecommunications fiber by reducing the core diameter d, or equivalently A_{eff}.

However, reducing the core diameter to dimensions comparable to the wavelength of light no longer provides a tight confinement. This limit in the small core can be attributed to two physical routes:

(1) In small core fibers the light is under the influence of the diffraction and the evanescent wave interaction at the air/glass boundaries near the core becomes dominant, so the effect of surface roughness at the boundaries becomes very significant [101].

(2) Unless many rings of holes are used, the core mode can interact with the pure silica exterior cladding region to result in significant increase of confinement loss [94].

Lower confinement loss and tighter mode confinement have been achieved by either using larger air-filling fractions or floating the core region with very fine silica struts. However, even using a large air-filling fraction, in the limit that the core dimensions are smaller than the wavelength of light guided by the fiber, more than six rings of air holes are required to ensure low-loss operation [94].

The physical limit of small core in EIG-ASHFs for nonlinear enhancement has been theoretically analyzed by Zheltikov [102]. The product $I \cdot l_{\text{eff}}$, where I is the light field intensity in an optical fiber and l_{eff} is the effective interaction length, gives a figure of merit for the efficiency of a broad class of nonlinear optical interactions. The waveguide enhancement of a nonlinear optical process in a fiber where a low loss α supports a long fiber interaction length l to satisfy $\alpha \cdot l \gg 1$, with respect to a tightly focused Gaussian beam, is given by [102,103]

$$\xi = \frac{I_{\text{f}} l_{\text{eff}}^{\text{f}}}{I_{\text{t}} l_{\text{eff}}^{\text{t}}} \approx \frac{\lambda \eta}{\pi a^2 \alpha} \qquad (8.23)$$

Note that the waveguide nonlinear enhancement factor ξ is directly dependent on the waveguide parameters. In equation 8.23, a is the fiber core radius and η is the ratio of optical power confined to the fiber core to the total power guided by the fiber.

For a step-index optical fiber, the power ratio in the core is known from the early optical fiber theories such as by Snyder and Love [104] and is described as follows [102]:

$$\eta = \frac{P_{\text{core}}}{P_{\text{core}} + P_{\text{clad}}}$$

$$P_{\text{core}} = \frac{kn_{\text{core}}^2}{\beta J_1^2(u)} \left\{ a_1 a_3 \left[J_0^2(u) + J_1^2(u) \right] + a_2 a_4 \left[J_2^2(u) + J_1(u) J_3(u) \right] \right\}$$

$$P_{\text{core}} = \frac{kn_{\text{core}}^2}{\beta K_1^2(u)} \frac{u^2}{w^2} \left\{ a_1 a_5 \left[K_0^2(w) - K_1^2(w) \right] + a_2 a_6 \left[K_2^2(w) + K_1(w) K_3(w) \right] \right\}$$

$$(8.24)$$

Here the constants are defined as follows:

$$a_1 = \frac{F_2 - 1}{2}, \qquad a_2 = \frac{F_2 + 1}{2}, \qquad a_3 = \frac{F_1 - 1}{2}, \qquad a_4 = \frac{F_1 + 1}{2}$$

$$a_5 = \frac{F_1 - 1 + 2\Delta}{2}, \qquad a_6 = \frac{F_1 + 1 - 2\Delta}{2}$$

$$F_1 = \frac{(uw)^2}{V^2} [b_1 + (1 - 2\Delta) b_2], \qquad F_2 = \frac{V^2}{(uw)^2 (b_1 + b_2)} \qquad (8.25)$$

$$b_1 = \frac{1}{2u} \left(\frac{J_0(u)}{J_1(u)} - \frac{J_2(u)}{J_1(u)} \right), \qquad b_2 = \frac{1}{2W} \left(\frac{K_0(W)}{K_1(W)} + \frac{K_2(W)}{K_1(W)} \right)$$

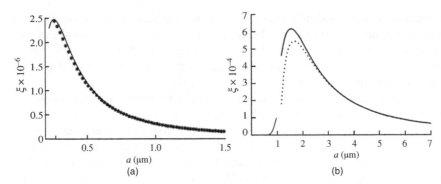

FIGURE 8.25   The waveguide nonlinear enhancement factor ξ as a function of core radius at the wavelength of $1\,\mu m$ in silica step-index optical fibers [102,103] for: (a) a silica core suspended in the air and (b) conventional step-index fiber.

It is noted that the above formulae for a conventional step-index SMF can be readily applicable to EIG-ASHF using the empirical expression in equations   to find the nonlinear enhancement effects in EIG-ASHF with a small core and holey cladding.

Zheltikov [102,103] analyzed two extreme cases: (1) small silica fiber suspended in the air ($n_{core} \approx 1.45, n_{clad} \approx 1, \Delta \approx 0.26$) and (2) conventional SMF with a small refractive index difference ($n_{core} \approx 1.45, \Delta \approx 0.005$) for the propagating light at $\lambda = 1.0\,\mu m$. The simulation results for these two cases are summarized in Figure 8.25a and b, respectively.

Comparing the two graphs we can confirm that the nonlinear enhancement factor ξ is about 50–60 times higher in the silica–air structure than conventional solid silica fiber, which is consistent with the experimental reports. Furthermore, it is also observed that ξ increases nonlinearly with decreasing core radius a, showing a maximum peak an optimal value a_{opt}. Further reduction of the core radius smaller than a_{opt} decreased ξ. In the silica core/air cladding case (Figure 8.25a), we have $a_{opt} \approx 0.27\,\mu m$ and in the conventional SMF case, we have $a_{opt} \approx 1.5\,\mu m$.

Nonlinear effects in EIG-ASHF can be used for a wide range of optical processing applications including supercontinuum generation, optical data regeneration, wavelength conversion, optical demultiplexing, and Raman amplification [92]. Among these nonlinear optical properties, supercontinuum generation [105] has been most widely cited. Recent review articles fully describe this unique phenomenon [106,107]. In these nonlinear optical applications, it is important to note that three parameters should be considered to optimize specific performances: (1) the effective area, (2) dispersion, and (3) material nonlinearity. As we have seen in the previous sections, the central defect diameter (d) and the air-filling factor (d/Λ) can control the effective area and dispersion but these two are not independent, which requires optimal parameters.

The material nonlinearity could be further enhanced by replacing silica with a high refractive index glass. For example, the chalcogenide glass As_2S_3 is 100 times more nonlinear than silica ($n \approx 2.4$, $n_2 \approx \times 10^{-18}\,m^2/W$ at $\lambda = 1550\,nm$) [108] and the Schott lead glass SF57 is 20 times more nonlinear than silica ($n \approx 1.8$, $n_2 \approx 4 \times 10^{-19}\,m^2/W$ at $\lambda = 1550\,nm$) [109]. Although these high refractive index glasses are clearly attractive for nonlinear devices, their application has been very limited in conventional solid fiber structure because the core material index should be raised with a low imperfection population at the core–cladding interface. In the holey fiber technology, these requirements were obviated and waveguide structure can be designed and fabricated with a single glass material. The effective nonlinearity of SF57 EIG-holey fiber was measured to be $\gamma \approx 550\,(W\,km)^{-1}$ at 1550 nm [110], more than 500 times higher than standard optical fiber. Further development toward a new highly nonlinear material and highly confined waveguide structure in EIG-holey fiber would open a new avenue toward nonlinear fiber devices with unprecedented low operating powers (10–100 mW) and remarkably short device lengths (0.1–1 m).

8.8 BIREFRINGENCE CONTROL IN EFFECTIVE INDEX GUIDING AIR–SILICA HOLEY FIBERS

One of salient features of EIG-ASHFs is their strong birefringence, which is an order of magnitude larger than in conventional birefringent fibers that have been discussed in Chapter 7. In comparison to conventional solid core/cladding fibers, EIG-ASHFs can realize high birefringence because the index contrast between the silica core and holey cladding is higher than conventional fibers and their fabrication process allows very convenient allocation of asymmetric microstructure around the core. Utilizing high birefringence and cutoff control, EIG-ASHF structure also permits single-polarization single-mode fiber, which guides only one polarization state of the fundamental mode.

In general, the modal birefringence B comprises three anisotropy contributions [111]:

$$B = B_G + B_S^i + B_S^e \qquad (7.6)$$

Here B_G is the geometrical component induced by the core shape anisotropy. B_S^i is the self-stress component induced by the thermal expansion difference of the asymmetric core and the outer stress component. B_S^e is induced by the stress-applying parts imbedded in the cladding.

Birefringence of conventional single mode is in the range of $B = 10^{-6}$ to 10^{-5}. In conventional high-birefringence fibers the nominal value of modal birefringence B exceeds 10^{-4}. In the case of low-birefringence fibers, B is in the range of 10^{-9} [112].

In recent EIG-ASHFs, these birefringence mechanisms have been successfully adopted and provided high B over 10^{-3} [113–121], which is an order of magnitude larger than conventional high-birefringence fibers. Various types of high-birefringent

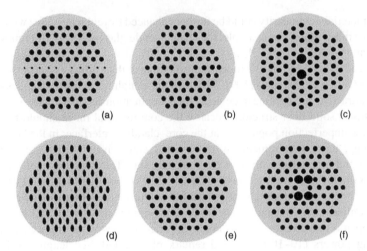

FIGURE 8.26 Types of birefringent EIG-ASHFs using defect engineering: (a) small holes in a line [113], (b) two silica core defects [114], (c) two large holes near the core [115], (d) elliptic hole [116], (e) three silica core defects [117], and (f) four large holes near the core [118]. Here the horizontal and vertical axes correspond to slow and fast axes, respectively.

EIG-ASHFs have been reported and schematic waveguide structures are as shown in Figure 8.26. The waveguide structures can be categorized into

(1) dissimilar hole size defects, which correspond to Figure 8.26a, c, and f;
(2) multiple silica defects, which correspond to Figure 8.26b and e;
(3) elliptic holes, which correspond to Figure 8.26d; and
(4) externally applied stress.

The configuration of Figure 8.26a [113] has a horizontal array of smaller diameter air holes, which contributes to higher effective index along the horizontal axis. The waveguide structures of Figure 8.26b and e [114,117] are based on removal of air holes replacing them with silica defects near the core, which also increases the effective index along the horizontal axes. In Figure 8.26c and f [115,118] larger holes are located near the core region to reduce the local cladding index and applying stresses.

In Figure 8.26d [116], the regular homogeneous hole arrangement is maintained but the holes are elliptic, which makes the cladding index highly anisotropic to generate birefringence.

The birefringent EIG-ASHFs in Figure 8.26 showed birefringence over 10^{-3} and one of experimental reports and corresponding theoretical analysis results are shown in Figure 8.27. Figure 8.27a is the actual photograph of the fabricated EIG-ASHF, which corresponds to Figure 8.26a [113]. The lowest guided modes in the slow and fast axes are shown in Figure 8.27c and d, respectively. In the calculation of the effective index of guided modes and birefringence, full-vector FEM has been employed [46].

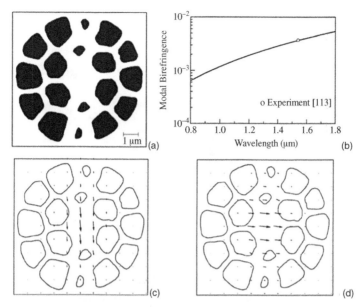

FIGURE 8.27   (a) Cross-section of fabricated EIG-ASHF [113], (b) birefringence calculated by FEM [46], (c) fundamental mode in the slow axis [46], and (d) fundamental mode in the fast axis [46].

The effective index of the slow-axis mode is larger than that of the fast-axis mode and Figure 8.27b shows the birefringence as a function of wavelength. The birefringence obtained by FEM simulation is 3.66×10^{-3} at $\lambda = 1.54\,\mu m$ [46] and is in good agreement with the measured value of 3.7×10^{-3} [113].

Besides these high-birefringent EIG-ASHFs utilizing defect structures, there have been various attempts to obtain high birefringence by including stress-applying parts (SAPs) as in PANDA fibers and elliptic core/cladding fibers described in Chapter 7. The waveguide structures of these types are shown in Figure 8.28.

Folkenberg et al. [119] included the SAP made of B_2O_3-doped SiO_2 in the exterior cladding as in Figure 8.28a, which is very similar in structure to that of PANDA fiber in Chapter 7. Instead of using borosilicate glass, Issa et al. [120] utilized two large air holes in the exterior cladding as SAPs as in Figure 8.28b. They further examined the fiber structure and found that the air holes were elliptic. Schreiber et al. [121] introduced borosilicate defects in the holey cladding to make them SAPs as in Figure 8.29c. These fibers showed birefringence of 10^{-4} similar to those of conventional birefringent fibers. However, in the case of Figure 8.28a, a PANDA-like EIG-ASF, the fiber showed a high birefringence over a very wide range as shown in Figure 8.29a due to "endless single-mode" nature of the EIG-ASHF [119]. In the case of birefringent fiber based on borosilicate defect SAPs as in Figure 8.28c, the fiber showed a very wide single-mode single-polarization region from 880 to 1600 nm at the bending radius of 25 cm [121].

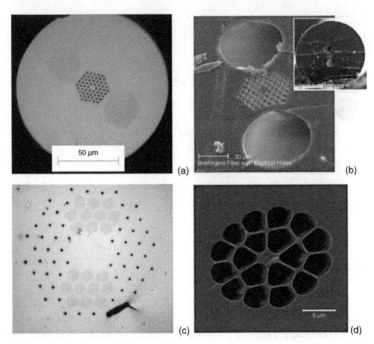

FIGURE 8.28   Birefringent EIG-ASHFs based on stress-applying parts: (a) borosilicate SAP in exterior cladding [119], (b) elliptic hollow SAP [120], (c) borosilicate SAP inside holey cladding [121], and (d) elliptic core/cladding [122].

For the fiber in Figure 8.28d, regular microstructure capillaries were stacked and the ellipticity of the core and holes surrounding the core was introduced by controlling the lateral tension between the outer jacket tube and the capillary stack [122,123]. This elliptic core/holey-clad EIG-ASHF showed a large birefringence of 7.5×10^{-3} and a short beat length of 0.3 mm at $\lambda = 1550$ nm [122].

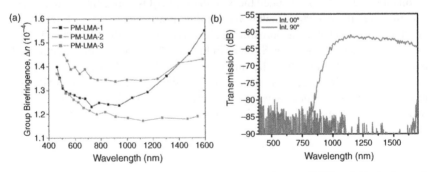

FIGURE 8.29   (a) High birefringence over a wide spectral range [119] and (b) single-mode single-polarization over a wide spectral range [121].

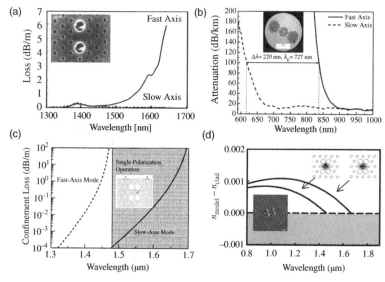

FIGURE 8.30   Single-mode single-polarization EIG-ASHFs: (a) two large air holes near the silica core [124], (b) borosilicate stress-applying parts (SAPs) [126], (c) four large air holes near the silica core [118] (theoretical simulation), and (d) four large air holes near the silica core (experiment and theoretical simulation) [128].

Single-mode single-polarization fibers (SMSPFs) have been extensively studied in conventional birefringent fibers as discussed in Chapter 7 in order to remove polarization dispersion, and polarization coupling in high-birefringent fibers. Flexible design capability of EIG-ASHFs has resulted in various types of SMSPFs [118,124–128] and the single-polarization characteristics are shown in Figure 8.30.

As in the case of solid core/cladding SMSPFs, one of two fundamental polarization modes on the fast axis is designed to be cutoff at a shorter wavelength and the remnant mode polarized along the slow axis became the single-guided mode. This differential cutoff can be achieved by either waveguide design itself or macrobending.

8.9   HOLLOW OPTICAL FIBER AND ITS APPLICATIONS

Hollow ring core fiber in Figure 8.1 (d), is also called as hollow optical fibers (HOFs). HOFs, composed of a central air hole, GeO_2–SiO_2 ring core, and SiO_2 cladding, have been recently proposed to take advantage of optical guiding along the ring core in an annulus mode, and adiabatic mode transformation to conventional solid core fibers [129]. The schematic waveguide structure and the photograph of the fabricated fiber cross-section are shown in Figure 8.31.

Similar glass capillary structures have been used in the early stage of optical fiber development such as liquid core silica capillary fibers [130] or dye-filled capillaries [131,132]. However, these prior capillary optical guiding structures used the high refractive index material filled in the central hole as the core and the modal distribution was in essence very similar to conventional solid core glass fibers.

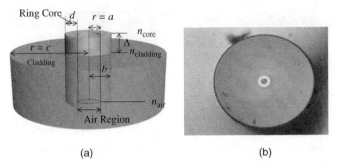

FIGURE 8.31   (a) Waveguide structure and (b) cross-section of a hollow optical fiber [129].

In other avenue of atomic physics, hollow fibers have been used to guide atoms as well [133,134].

HOF shown in Figure 8.31 provides efficient optical guidance along the germanosilicate ring core to result in an annulus mode distribution and these annulus modes can be efficiently converted to the conventional linear polarized modes in an adiabatic mode transformation with minimal loss [135]. Based on HOF, varieties of new devices have been proposed and experimentally demonstrated such as a launching conditioner for multimode fiber bandwidth control in gigabit Ethernet [135], a mode converter for high-order mode dispersion compensation [136], tunable wavelength selective devices [137–142], and birefringence control [143–145]. Furthermore, the HOF structure has been applied to EIG-ASHFs to control birefringence and chromatic dispersion [146–148].

For a given waveguide structure as in Figure 8.31a, the longitudinal component of the electric field in a core mode is analytically given as

$$E_z(r,\theta) = \begin{cases} AI_\nu(vr)\sin(\nu\theta + \phi), & r < a \\ (BJ_\nu(ur) + CN_\nu(ur))\sin(\nu\theta + \phi), & a \le r \le b \\ DK_\nu(wr)\sin(\nu\theta + \phi), & r > b \end{cases} \qquad (8.26)$$

for

$$\begin{bmatrix} v = \sqrt{\beta^2 - k^2} \\ u = \sqrt{k^2 n_1^2 - \beta^2} \\ w = \sqrt{\beta^2 - k^2 n_2^2} \end{bmatrix}$$

Here $J_\nu, N_\nu, I_\nu,$ and K_ν are Bessel functions of integer order ν. Characteristic equation can be derived for the given boundary conditions to satisfy continuity requirements for the fields and their derivatives. The propagation constant β and subsequent propagation properties can be obtained from the characteristic equation.

The fundamental mode and the first five excited modes guided along the ring core were identified as in Figure 8.32 at the wavelength of 800 nm. HOF parameters were:

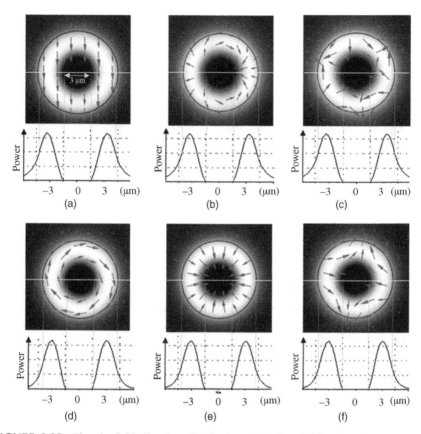

FIGURE 8.32 Electric field direction distributions in HOF cladding modes: (a) HE_{11}, (b) EH_{11}, (c) HE_{31}, (d) TE_{01}, (e) TM_{01}, and (f) HE_{21}. Here (a–c) are symmetric and (d–f) are antisymmetric core modes of HOF. The wavelength of the light is 800 nm. The intensity distributions are also shown below the electric field direction distribution as a function of radial position.

the air hole radius, $a = 1.5\,\mu m$; thickness of the GeO_2-doped SiO_2 ring cores, $d = 2.6\,\mu m$; and relative refractive index difference between core and cladding, $\Delta = 0.45\%$.

The direction distribution of the electric fields in the core modes was similar to that of conventional SMFs. It is noteworthy to find that the fundamental HE_{11} mode is linearly polarized across the ring core. It is also highly notable that the excited modes TE_{01}, TM_{01}, and HE_{21} are nondegenerate, as described in Section 8.9.1, in contrast to the fourfold degenerate LP_{11} mode in conventional SMFs. Therefore, the guided modes in HOF were assigned following the hybrid mode nomenclature rather than weakly guiding LP mode assignment in conventional SMFs.

Three symmetric modes, HE_{11}, HE_{31}, and EH_{11}, have the twofold degeneracy for two orthogonal polarization directions. The antisymmetric modes TE_{01} and TM_{01} in

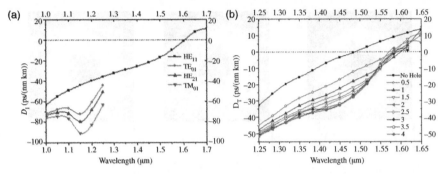

FIGURE 8.33 Chromatic dispersion of the core-guided modes: (a) fundamental mode and first excited modes and (b) impacts of the hole diameter over the chromatic dispersion.

Figure 8.32d and e are nondegenerate and the HE_{21} mode has the twofold degeneracy. The intensity distributions of these core modes were very similar to one another in contrast to conventional SMFs, which can make the overlap integral between the modes in HOF significantly larger than conventional SMFs.

For the fundamental core mode HE_{11} and first excited modes, the chromatic dispersion calculation results are shown in Figure 8.33a. HOF parameters were: the air hole radius, $a = 1.5\,\mu m$; thickness of the GeO_2-doped SiO_2 ring cores, $d = 2.6\,\mu m$; and relative refractive index difference between core and cladding, $\Delta = 0.45\%$. The HE_{11} mode showed the zero dispersion near 1600 nm, which is quite different from 1310 nm of conventional SMFs. Anomalous dispersion range is significantly reduced and in the U band the dispersion is less than 10 ps/(nm km). The dispersion values in 1310 and 1550 nm are -31 and -9.35 ps/(nm km), respectively. The sign of dispersion slope near 1550 nm is also opposite to conventional single-mode fiber. The excited modes have relatively large negative dispersion near the cutoff as expected. The impact of the air hole radius on chromatic dispersion was investigated and compared with the conventional single-mode fiber, where the hole radius is equal to zero as in Figure 8.33b. The air hole generally decreased the dispersion in O, E, S, and C bands.

First few cladding modes are categorized into symmetric and antisymmetric modes as in Figure 8.34. Antisymmetric cladding modes are (a)–(f), and (g)–(i) are symmetric modes. Note that TE_{0x} and TM_{0x} modes are nondegenerate, and HE and EH are twofold degenerate modes. In the calculation we assumed the cladding diameter of 125 μm, surrounded by air of refractive index 1. Due to presence of the central air hole, the cladding modes of HOF also showed significantly different intensity distribution compared with conventional SMFs, where the symmetric cladding modes have intensity peaks at the center while the antisymmetric modes have nodes. HOFs, both symmetric and antisymmetric modes, have the annular structures and modal overlap integrals between the fundamental core mode HE_{11} would be significantly different from those in conventional fibers, which could provide a new opportunity in periodically coupled optical fiber device design.

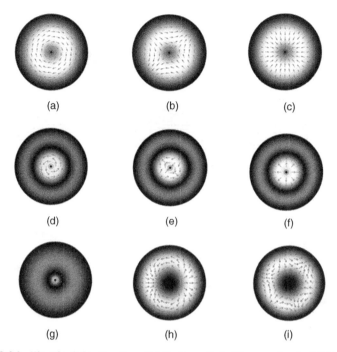

FIGURE 8.34 Electric field direction distribution of cladding modes in HOF: (a) TE_{01}, (b) HE_{21}, (c) TM_{01}, (d) TE_{02}, (e) HE_{22}, (f) TM_{02}, (g) HE_{12}, (h) EH_{11}, and (i) HE_{31}. (a–f) are antisymmetric and (g–i) are symmetric cladding modes.

HOF has the exactly same glass composition for the cladding, pure SiO_2, and the core, GeO_2-doped SiO_2, as the conventional SMFs, which makes HOF the most compatible to SMFs compared to other types of silica index guiding holey fibers. When an HOF is locally heated, for example, by applying arc in a fusion splicer, the fiber collapsed to a solid core fiber that could be made identical to conventional SMF. When the width of the ring core and the index difference between the ring core and silica cladding are designed to match the core parameters of conventional SMF, a very-low-loss conversion of the Gaussian-like LP_{01} mode in the conventional SMF into the annulus mode in HOF can be expected as illustrated in Figure 8.35. All fiber novel mode converters can be, therefore, fabricated utilizing local adiabatic collapse along HOF [129,135].

In experiments, the lowest insertion loss from the adiabatic mode conversion was measured as 0.3 dB for a taper length of 10 mm. The loss is mainly attributed to the mismatch in mode field area between the collapsed core of HOF and the SMF core as well as imperfect tapering structure [129].

8.9.1   Mode Converter for Gigabit LAN

In this section, practical application of HOF in short-haul gigabit Ethernet will be described based on the results of Choi et al. [135]. As discussed in Chapter 5, it is

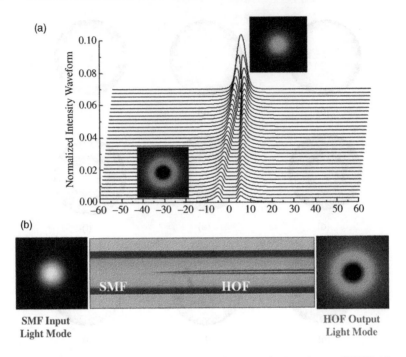

FIGURE 8.35 Unique adiabatic mode conversion along the SMF and tapered HOF: (a) mode field evolution along a tapered hollow optical fiber and (b) photograph of a mode converter that converts the fundamental mode of SMF to an annulus mode of HOF [129].

required to reduce differential modal delay (DMD) that is caused by a central refractive index dip of the conventional low-bandwidth MMFs. It is necessary to establish a well-defined launching condition, especially in the case of laser-based MMF systems.

The HOFs with the air hole/outer diameter of 7–8/125 μm and with the length of 100–150 mm were used for a mode converter that transforms the LP_{01} mode of SMF into an annular mode to avoid the central defect in MMF. The insertion loss of the mode converter was less than 0.6 dB at the operating wavelength of 1.31 μm. Schematic of the SMF–HOF–MMF mode converter unit is shown in Figure 8.36.

The mode converter, designated by "SMF–HOF–MMF" in Figure 8.36, was coupled to SMF pigtailed light source and its output was connected to 500 m of MMF. An uncooled Fabry–Perot (FP) laser with a central wavelength of 1.31 μm was directly modulated with $2^{23} - 1$ pseudorandom bit sequence (PRBS) nonreturn to zero (NRZ) format at 2.5 Gb/s. The extinction ratio of the data signal was 10.95 dB and the eye diagram of directly modulated FP laser is shown in inset of Figure 8.37a. The graded index MMF of 50/125 μm was used as a transmission medium in this experiment. The bit error rate (BER) versus the received power was compared with and without the mode converter as shown in Figure 8.37b.

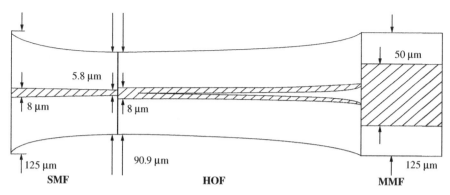

FIGURE 8.36 Structure of SMF–HOF–MMF mode converter for ring mode launching [135].

The receiver sensitivity of 500 m MMF links was improved over 0.72 dB at the BER of 10^{-9}, using the proposed mode converter, which is comparable to prior offset launching technique [150]. The use of mode converter based on the annulus mode of HOF does have a strong potential to reduce the modal dispersion by filtering modes of the conventional MMFs and subsequently to expand the maximum transmission distance.

8.9.2 Polarization Control

HOF can provide a high birefringence by inducing geometrical asymmetry using a method that has been widely accepted in elliptic core fiber as discussed in Chapter 7. There have been continuous research efforts in polarization control based on HOF such as simulation [143] and experiments [144,145].

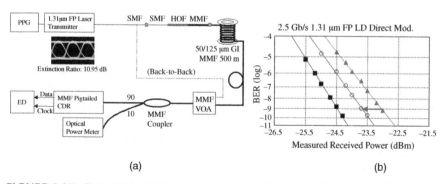

FIGURE 8.37 SMF–HOF–MMF mode converter application in high speed MMF optical link [135]. (a) Transmission experiment setup (PPG, pulse pattern generator; VOA, variable optical attenuator; CDR, clock/data receiver; ED, error detector). (b) BER versus the received power. Here data with solid rectangles are for back-to-back measurement, hollow circles are for the results with the mode converter, and solid triangles are without mode converter [135].

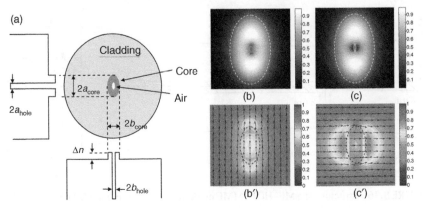

FIGURE 8.38 (a) Schematic diagram of elliptic HOF, (b) intensity distribution of fundamental mode in the slow axis (b′) and its field and intensity distribution near the air hole, and (c) intensity distribution of fundamental mode in the fast axis (c′) and its field and intensity distribution near the air hole [143].

Elliptic HOF is schematically shown in Figure 8.38a and the fundamental modes along the slow axis (Figure 8.38b) and fast axis (Figure 8.38c) are illustrated.

In the cross-section of the elliptic HOF in Figure 8.38a, the elliptic air hole is concentric to the elliptic ring core. Key waveguide parameters are: the major axes of the air hole and the ring core, a_{hole} and a_{core}; the minor axes, b_{hole} and b_{core}; and the refractive index difference of core and cladding, Δn. In Figure 8.38, the electric field profiles of (b) y-polarization and (c) x-polarization modes are shown. The white dashed curve indicates the core–cladding boundary. The core–cladding index difference was set to be 0.02. The central air hole region is magnified in (b′) and (c′) to show the details. The black dashed curve indicates the elliptic air–core interface. Note that electric field amplitude in the x-polarization is more significantly suppressed at the interface than that in the y-polarization.

The birefringence can be flexibly controlled by varying waveguide parameters of elliptic HOF as listed in Figure 8.38a. The maximum birefringence is plotted as a function of index difference between the core and cladding, Δn, in Figure 8.39a and it is compared with the elliptic core fiber without air hole. It is noted that by inclusion of central air hole birefringence increases more than 250%. The impact of ellipticity over the birefringence is also analyzed in Figure 8.39b, for a given ellipticity high Δn resulted in a higher birefringence.

The elliptic HOF has been fabricated and its birefringence characteristics have been reported by Jung et al. [144]. Schematic fabrication process is shown in Figure 8.40a, where conventional flat-sided preform technique was applied. Fabricated elliptic HOF's cross-section is shown in Figure 8.40b. Cross-section of the fabricated elliptic HOF with air hole collapsed is shown in Figure 8.40c. It is noted that the cladding is slightly elliptic because the fiber was drawn at a lower temperature.

The fabricated fiber showed a very high birefringence over 2×10^{-3} that is larger than conventional high-birefringent fibers such as PANDA fiber by several factors as

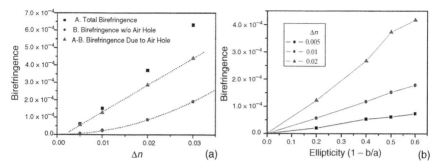

FIGURE 8.39 (a) The birefringence maxima achievable with each Δn: A, the maxima of the total birefringence; B, the birefringence at $a_{hole} = 0$; $A–B$, the birefringence due to the core–air boundary effect. The data in diamond and triangle symbols are fitted with a parabolic and a linear function, respectively. (b) Birefringence as a function of ellipticity. The ellipticity of the hole and the core was varied simultaneously. Each curve corresponds to each Δn [143].

shown in Figure 8.41. It is also noted that the hole diameter, subsequently the ring core thickness, did significantly affect the birefringence such that a larger hole diameter and thin ring core showed a higher birefringence.

8.9.3   Application to Air–Silica Hollow Fiber for Defect Engineering

Unique guidance of HOF has been also adopted in air–silica holey fiber that has opened a new avenue of EIG-ASHF design [146–148]. The concept of combining HOF with EIG-ASHF is schematically illustrated in Figure 8.42.

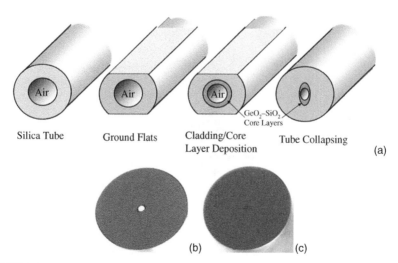

FIGURE 8.40 (a) Schematic diagram for fabrication process of elliptic HOF, (b) cross-section of fabricated HOF, and (c) cross-section of fiber with air hole collapsed [144].

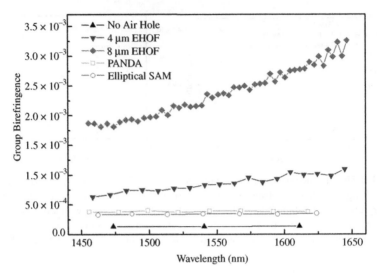

FIGURE 8.41 Comparison of birefringence of elliptic HOF with conventional birefringent fibers [144].

In conventional EIG-ASHFs the defects were usually silica or doped silica and the only free parameter was its size. When the defect is replaced by HOF structure as in Figure 8.42, various waveguide parameters such as the ring index Δn (or Δ equivalently), ring thickness W_{ring}, and hole diameter d can be flexibly controlled for novel EIG-ASHF design (Figure 8.43). The hollow defect can further improve the detection sensitivity of gas and liquid by increasing the evanescent wave interaction at the core in comparison to conventional silica defect PCF [149].

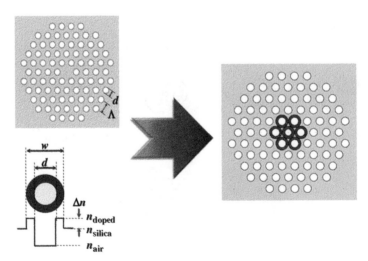

FIGURE 8.42 The central silica defect in conventional EIG-ASHF is replaced with HOF structure that can be flexibly designed with various waveguide parameters of HOF.

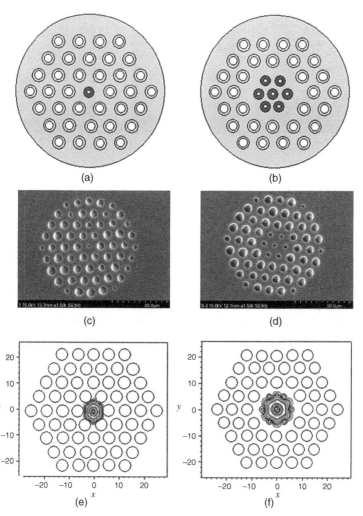

FIGURE 8.43   Schematic of EIG-ASHF with HOF defects: (a) one-layer defects, (b) two-layer defects, (c and d) actual cross-section of fabricated fibers, and (e and f) fundamental modes guided by the fibers [146,147].

The EIG-ASHF with HOF defects showed a good guidance along the core and the modal intensity profiles of the fundamental mode and two excited modes are shown in Figure 8.44. Due to the nature of HOF, the intensity profile of the fundamental mode showed annular property, which is not observed in conventional EIG-ASHFs with silica defects.

Another notable feature of EIG-ASHF is flexible control of effective area, which is highly dependent on wavelength as shown in Figure 8.45.

The fiber also showed flexible control of dispersion value and dispersion slope independently.

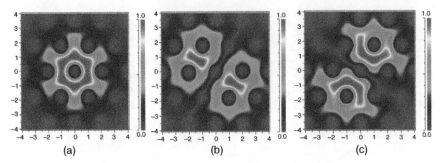

FIGURE 8.44 Annulus mode intensity profiles $|H|^2$ for two-layer germanosilicate ring core IGHF at $\lambda = 600$ nm: (a) fundamental, (b) second-order, and (c) third-order modes. $\Lambda = 2$ μm, $d/\Lambda = 0.5$, $W_{ring} = 1.6$ μm, and $\Delta = 0.013$. Here the cutoff wavelength $\lambda_{cutoff} = 900$ nm [146].

As shown in Figure 8.46a, the dispersion slope can be changed by changing the hole diameter d in the fiber. In this case only the d is changed with all the other waveguide parameters fixed. It is also noted that very flat dispersion could be achieved at optimal parameters as shown in the inset of Figure 8.46a.

In Figure 8.46b, when both d and W_{ring} were changed keeping a certain ratio of $W_{ring}/d = 1.6$, then the zero-dispersion wavelength could be shifted from 1150 to 1250 nm.

Another salient feature of ASHF with HOF defects is its low splice loss with conventional SMF. There have been various methods to reduce the splice loss between

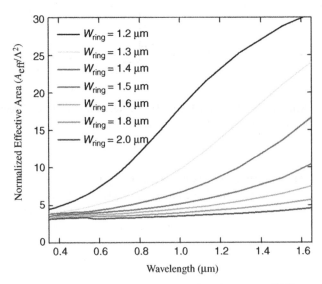

FIGURE 8.45 Effective area of two-layer GeO₂–SiO₂ ring core IGHFs as a function of wavelength for different W_{ring}. $d/\Lambda = 0.5$, $\Lambda = 2$ μm, $\Delta = 0.013$, and $W_{ring} = 1.2$–2.0 μm [146].

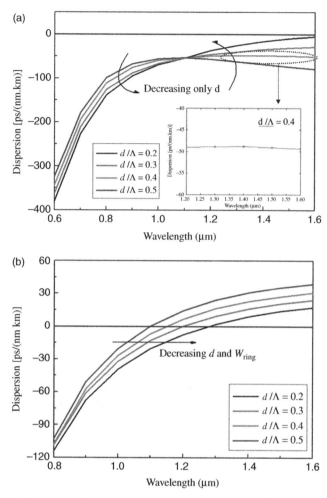

FIGURE 8.46 The chromatic dispersion of the two-layer germanosilicate ring core IGHF for the different d/Λ in the range of 0.2–0.5 (a) as decreasing air hole diameter d with $\Lambda = 2\,\mu m$, $W_{ring} = 1.6\,\mu m$, and $\Delta = 0.013$ and (b) as decreasing d and W_{ring} with $\Lambda = 4\,\mu m$, $W_{ring}/d = 1.6$, and $\Delta = 0.013$ [146].

ASHFs and SMFs [151–153], which requires intermediate fibers or special device. In the ASHF with HOF defects, a very low loss less than 0.65 dB was achieved inherently, which is attributed to adiabatic mode transformation nature of HOF. The mechanism of low splice loss is shown in Figure 8.47. In conventional ASHFs the holes are collapsed and pure silica segment will be spliced to SMF to result in a large mode mismatch and subsequent optical loss. In contrast, the rings will collapse to a high-index core in ASHF with HOF defects to provide a nearly symmetric mode that can be matched to LP_{01} mode distribution as shown in Figure 8.47.

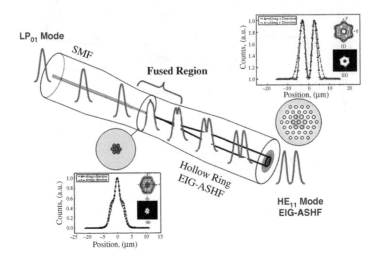

FIGURE 8.47   Adiabatic mode transformation along the SMF–AShF splice. The ASHF with hollow ring defects will provide a nearly symmetric intensity profile in the fused region to enable a low-loss splice [147].

REFERENCES

[1] P. V. Kaiser and H. W. Astle, "Low-loss single material fibers made from pure fused silica," Bell System Technical Journal, vol. 53, no. 6, pp. 1021–1039, 1974.

[2] T. Okoshi, K. Oyamada, M. Nishimura, and H. Yokota, "Side tunnel fiber: an approach to polarization-maintaining optical waveguiding scheme," Electronics Letters, vol. 18, no. 19, pp. 824–826, 1982.

[3] L. Li, G. Wylangowski, D. N. Payne, and R. D. Birch, "Broad-band metal glass single-mode fiber polarizers," Electronics Letters, vol. 22, no. 19, pp. 1020–1022, 1986.

[4] J. C. Knight, T. A. Birks, P. St. J. Russell, and D. M. Atkin, "All-silica single-mode optical fiber with photonic cladding," Optics Letters, vol. 21, no. 19, pp. 1547–1549, 1996.

[5] P. St. J. Russell, "Photonic crystal fibers," Journal of Lightwave Technology, vol. 24, no. 12, pp. 4729–4749, 2006.

[6] J. C. Knight, "Photonic crystal fibres," Nature, vol. 424, no. 6950, pp. 847–851, 2003.

[7] P. St. J. Russell, "Photonic crystal fibers," Science, vol. 299, no. 5605, pp. 358–362, 2003.

[8] T. M. Monro and D. Richardson, "Holey optical fibres: fundamental properties and device applications," Comptes Rendus Physique, vol. 4, pp. 175–186, 2003.

[9] J. A. West, "Photonic crystal fibers," The 14th Annual Meeting of the IEEE Lasers and Electro-Optics Society (LEOS2001), San Diego, USA, vol. 1, pp. 14–15, 2001.

[10] B. J. Eggleton, C. Kerbage, P. S. Westbrook, R. S. Windeler, and A. Hale, "Microstructured optical fiber devices," Optics Express, vol. 9, no. 13, pp. 698–713, 2001.

[11] J. Broeng, D. Mogilevstev, S. E. Barkou, and A. Bjarklev, "Photonic crystal fibers: a new class of optical waveguides," Optical Fiber Technology, vol. 5, no. 3, pp. 305–330, 1990.

[12] A. M. Zheltikov, "Microstructure optical fibers for a new generation of fiber-optic sources and converters of light pulses," Physics-Uspekhi, vol. 50, no. 7, pp. 705–727, 2007.

[13] A. Bjaklev, J. Broeng, and A. S. Bjarklev, "Photonic crystal fibres," Springer Science, New York, 2003.

[14] F. Zolla, G. Renversez, A. Nicolet, B. Kuhlmey, S. Guenneau, and D. Felbacq, "Foundations of photonic crystal fibres," Imperial College Press, London, 2005.

[15] J. D. Joannopoulos, S. G. Johnson, J. N. Winn, and R. D. Meade, "Photonic crystals: molding the flow of light," 2nd edition, Princeton University Press, New Jersey, 2008.

[16] V. V. Ravi Kanth Kumar, A. K. George, W. H. Reeves, J. C. Knight, P. St. J. Russell, F. G. Omenetto, and A. J. Taylor, "Extruded soft glass photonic crystal fiber for ultrabroad supercontinuum generation," Optics Express, vol. 10, no. 25, pp. 1520–1525, 2002.

[17] K. M. Kiang, K. Frampton, T. M. Monro, R. Moore, J. Tucknott, D. W. Hewak, D. J. Richardson, and H. N. Rutt, "Extruded single-mode non-silica glass holey optical fibers," Electronics Letters, vol. 38, no. 12, pp. 546–547, 2002.

[18] R. T. Bise and D. Trevor, "Sol–gel-derived microstructured fibers: fabrication and characterization," Optical Fiber Communication Conference and Exposition and the National Fiber Optic Engineers Conference, Technical Digest (CD), Optical Society of America, 2005, paper OWL6.

[19] T. A. Birks, J. C. Knight, and P. St. J. Russell, "Endlessly single mode photonic crystal fiber," Optics Letters, vol. 22, no. 13, pp. 961–963, 1997.

[20] K. Saitoh, Y. Tsuchida, M. Koshiba, and N. A. Mortensen, "Endlessly single-mode holey fibers: the influence of core design," Optics Express, vol. 13, no. 26, pp. 10833–10839, 2005.

[21] J. C. Knight, T. A. Birks, R. F. Cregan, P. St. J. Russell, and P. D. de Sandro, "Large mode area photonic crystal fibre," Electronics Letters, vol. 34, no. 13, pp. 1347–1348, 1998.

[22] N. A. Mortensen, M. D. Nielsen, J. R. Folkenberg, A. Petersson, and H. R. Simonsen, "Improved large-mode-area endlessly single-mode photonic crystal fibers," Optics Letters, vol. 28, no. 6, pp. 393–395, 2003.

[23] Y. Tsuchida, K. Saitoh, and M. Koshiba, "Design of single-mode holey fibers with large-mode-area and low bending losses: the significance and the ring-core region," Optics Express, vol. 15, no. 4, pp. 1794–1803, 2007.

[24] J. C. Knight, J. Arriaga, T. A. Birks, A. Ortigosa-Blanch, W. J. Wadsworth, and P. St. J. Russell, "Anomalous dispersion in photonic crystal fiber," IEEE Photonics Technology Letters, vol. 12, pp. 807–809, 2000.

[25] A. Ferrando, E. Silvestre, P. Andrés, J. J. Miret, and M. V. Andrés, "Designing the properties of dispersion-flattened photonic crystal fibers," Optics Express, vol. 9, no. 13, pp. 687–697, 2001.

[26] W. H. Reeves, J. C. Knight, P. St. J. Russell, and P. J. Roberts, "Demonstration of ultra-flattened dispersion in photonic crystal fibers," Optics Express, vol. 10, no. 14, pp. 609–613, 2002.

[27] K. Saito, M. Koshiba, T. Hasegawa, and E. Sasaoka, "Chromatic dispersion control in photonic crystal fibers: application to ultra-flattened dispersion," Optics Express, vol. 11, no. 8, pp. 843–852, 2003.

[28] W. H. Reeves, D. V. Skryabin, F. Biancalana, J. C. Knight, P. St. J. Russell, F. G. Omenetto, A. Efimov, and A. J. Taylor, "Transformation and control of ultrashort pulses in dispersion-engineered photonic crystal fibres," Nature, vol. 24, no. 31, pp. 511–515, 2003.

[29] K. Hansen, "Dispersion flattened hybrid-core nonlinear photonic crystal fiber," Optics Express, vol. 11, no. 13, pp. 1503–1509, 2003.

[30] T. Matsui, J. Zhou, K. Nakajima, and I. Sankawa, "Dispersion-flattened photonic crystal fiber with large effective area and low confinement loss," Journal of Lightwave Technology, vol. 23, no. 12, pp. 4178–4183, 2005.

[31] N. H. Hai, Y. Namihira, S. Kaijage, F. Begum, S. M. Abdur Razzak, and K. Miyagi, "Broadband nearly-zero ultra-flattened dispersion single mode index guiding holey fiber," Optical Review, vol. 16, no. 3, pp. 351–354, 2009.

[32] W. J. Wadsworth, A. Ortigosa-Blanch, J. C. Knight, T. A. Birks, T.-P. Martin Man, and P. St. J. Russell, "Supercontinuum generation in photonic crystal fibers and optical fiber tapers: a novel light source," Journal of Optical Society of America B, vol. 19, no. 9, pp. 2148–2155, 2002.

[33] W. J. Wadsworth, N. Joly, J. C. Knight, T. A. Birks, F. Biancalana, and P. St. J. Russell, "Supercontinuum and four-wave mixing with Q-switched pulses in endlessly single-mode photonic crystal fibres," Optics Express, vol. 12, no. 2, pp. 299–309, 2004.

[34] D. A. Sidorov-Biryukov, A. Fernandez, L. Zhu, A. Verhoef, P. Dombi, A. Pugzlys, E. E. Serebryannikov, A. M. Zheltikov, J. C. Knight, and A. Baltuška, "Solitonic dynamics of ultrashort pulses in a highly nonlinear photonic-crystal fiber visualized by spectral interferometry," Optics Letters, vol. 33, no. 5, pp. 446–448, 2008.

[35] A. Ortigosa-Blanch, J. C. Knight, W. J. Wadsworth, J. Arriaga, B. J. Mangan, T. A. Birks, and P. St. J. Russell, "Highly Birefringent Photonic Crystal Fibers," Optics and Photonics News, vol. 12, no. 12, pp. 17–17, 2001.

[36] K. Suzuki, H. Kubota, S. Kawanish, M. Tanaka, and M. Fujita, "Optical properties of a low-loss polarization-maintaining photonic crystal fiber," Optics Express, vol. 9, no. 13, pp. 676–680, 2001.

[37] K. Saitoh and M. Koshiba, "Single polarization single mode photonic crystal fiber," Photonics Technology Letters, vol. 15, no. 10, pp. 1384–1386, 2003.

[38] P. R. Chaudhuri, V. Paulose, C. Zhao, and C. Lu, "Near elliptic core polarization-maintaining photonic crystal fiber: modeling birefringence characteristics and realization," Photonics Technology Letters, vol. 16, no. 5, pp. 1301–1303, 2004.

[39] D. Chen and L. Shen, "Highly birefringent elliptical-hole photonic crystal fibers with double defect," Journal of Lightwave Technology, vol. 25, no. 9, pp. 2700–2705, 2007.

[40] B. T. Kuhlmey, R. C. McPhedran, and C. M. de Sterke, "Modal cutoff in microstructured optical fibers," Optics Letters, vol. 27, no. 19, pp. 1684–1686, 2002.

[41] M. D. Nielsen and N. A. Mortensen, "Photonic crystal fiber design based on the V-parameter," Optics Express, vol. 11, no. 21, pp. 2762–2768, 2003.

[42] N. A. Mortensen, J. R. Folenberg, M. D. Nielsen, and K. P. Hansen, "Modal cutoff and the V parameter in photonic crystal fibers," Optics Letters, vol. 28, no. 20, pp. 1879–1881, 2003.

[43] M. Koshiba and K. Saitoh, "Applicability of classical optical fiber theories to holey fibers," Optics Letters, vol. 29, no. 15, pp. 1739–1741, 2004.

[44] F. Brechet, J. Marcou, D. Pagnoux, and P. Roy, "Complete analysis of the characteristics of propagation into photonic crystal fibers, by the finite element method," Optical Fiber Technology, vol. 6, no. 2, pp. 181–191, 2000.

[45] K. Saitoh and M. Koshiba, "Empirical relations for simple design of photonic crystal fibers," Optics Express, vol. 13, no. 1, pp. 267–274, 2005.

[46] K. Saitoh and M. Koshiba, "Numerical modeling of photonic crystal fibers," Journal of Lightwave Technology, vol. 23, no. 11, pp. 3580–3590, 2005.

[47] D. Marcuse, "Loss analysis of single-mode fiber splices," Bell System Technology Journal, vol. 56, no. 5, pp. 703–718, 1977.

[48] http://www.nktphotonics.com/.

[49] J. C. Baggett, T. M. Monro, K. Furusawa, and D. J. Richardson, "Comparative study of large-mode holey and conventional fibers," Optics Letters, vol. 26, no. 14, pp. 1045–1047, 2001.

[50] N. Mortensen, M. Nielsen, J. Folkenberg, A. Petersson, and H. Simonsen, "Improved large-mode area endlessly single-mode photonic crystal fibers," Optics Letters, vol. 28, no. 6, pp. 393–395, 2003.

[51] D. Marcuse, "Influence of curvature on the losses of doubly clad fibers," Applied Optics, vol. 21, no. 23, pp. 4208–4213, 1982.

[52] T. Sørensen, J. Broeng, A. Bjarklev, E. Knudsen, and S. E. B. Libori, "Macro-bending loss properties of photonic crystal fibre," Electronics Letters, vol. 37, no. 5, pp. 287–289, 2001.

[53] J. Sakai and T. Kimura, "Bending loss of propagation modes in arbitrary-index profile optical fibers," Applied Optics, vol. 17, no. 10, pp. 1499–1506, 1978.

[54] J. Sakai, "Simplified bending loss formula for single mode optical fiber," Applied Optics, vol. 18, no. 7, pp. 951–952, 1979.

[55] M. D. Nielsen, N. A. Mortensen, M. Albertsen, J. R. Folkenberg, A. Bjarklev, and D. Bonacinni, "Predicting macro-bending loss for large-mode area photonic crystal fibers," Optics Express, vol. 12, no. 8, pp. 1775–1779, 2004.

[56] A. Ferrando, E. Silvestre, P. Andres, J. J. Miret, and M. Andres, "Nearly zero ultraflattened dispersion in photonic crystal fibers," Optics Letters, vol. 25, no. 11, pp. 790–792, 2000.

[57] T. L. Wu and C. H. Chao, "A novel ultraflattened dispersion photonic crystal fiber," IEEE Photonics Technology Letters, vol. 17, no. 1, pp. 67–69, 2005.

[58] S. M. A. Razzak, Y. Namihira, K. Miyagi, F. Begum, S. Kaijage, N. H. Hai, T. Kinjo, and N. Zou, "Dispersion and confinement loss control in modified hexagonal photonic crystal fibers," Optical Review, vol. 14, no. 1, pp. 14–16, 2007.

[59] S. F. Kaijage, Y. Namihira, N. H. Hai, F. Begum, S. M. A. Razzak, T. Kinjo, H. Higa, and N. Zou, "Multiple defect-core hexagonal photonic crystal fiber with flattened dispersion and polarization maintaining properties," Optical Review, vol. 15, no. 1, pp. 31–37, 2008.

[60] Y. Akasaka, R. Sugizaki, S. Arai, Y. Suzuki, and T. Kamiya, "Dispersion flat compensation fiber for dispersion shifted fiber," 22nd European Conference on Optical Communications (ECOC 96), Technical Digest, pp. 2.221–2.224, 1996, paper TuP.01.

[61] B. P. Pal and K. Pande, "Optimization of a dual-core dispersion slope compensating fiber for DWDM transmission in the 1480–1610 nm band through G. 652 single-mode fibers," Optics Communications, vol. 201, no. 4–6, pp. 335–344, 2002.

[62] L. Grüner-Nielsen, M. Wandel, P. Kristensen, C. Jørgensen, L. V. Jørgensen, B. Edvold, B. Pálsdóttir, and D. Jakobsen, "Dispersion-compensating fibers," Journal of Lightwave Technology, vol. 23, no. 11, pp. 3566–3579, 2005.

[63] T. A. Birks, D. Mogilevtsev, J. C. Knight, and P. St. J. Russell, "Dispersion compensation using single-material fibers," IEEE Photonics Technology Letters, vol. 11, no. 6, pp. 674–676, 1999.

[64] L. P. Shen, W.-P. Huang, G. X. Chen, and S. S. Jian, "Design and optimization of photonic crystal fibers for broad-band dispersion compensation," IEEE Photonics Technology Letters, vol. 15, no. 4, pp. 540–542, 2003.

[65] F. Poli, A. Cucinotta, M. Fuochi, S. Selleri, and L. Vincetti, "Characterization of microstructured optical fibers for wideband dispersion compensation," Journal of Optical Society of America A, vol. 20, no. 10, pp. 1958–1962, 2003.

[66] B. J. Mangan, F. Couny, L. Farr, A. Langford, P. J. Roberts, D. P. Williams, M. Banham, M. W. Mason, D. F. Murphy, E. A. M. Brown, H. Sabert, T. A. Birks, J. C. Knight, and P. St. J. Russell, "Slope-matched dispersion compensating photonic crystal fibre," Proceedings of Conference on Lasers and Electro-Optics (CLEO 2004), San Francisco, CA, 2004, paper CPDD3.

[67] F. Gérôme, J.-L. Auguste, and J.-M. Blondy, "Design of dispersion-compensating fibers based on a dual concentric-core photonic crystal fiber," Optics Letters, vol. 29, no. 23, pp. 2725–2727, 2004.

[68] Y. Ni, L. Zhang, L. An, J. Peng, and C. Fan, "Dual-core photonic crystal fiber for dispersion compensation," IEEE Photonics Technology Letters, vol. 16, no. 6, pp. 1516–1518, 2004.

[69] A. Huttunen and P. Torma, "Optimization of dual-core and microstructure fiber geometries for dispersion compensation and large mode area," Optics Express, vol. 13, no. 2, pp. 627–635, 2005.

[70] T. Fujisawa, K. Saitoh, K. Wada, and M. Koshiba, "Chromatic dispersion profile optimization of dual-concentric-core photonic crystal fibers for broadband dispersion compensation," Optics Express, vol. 14, no. 2, pp. 893–900, 2006.

[71] S. K. Varshney, N. J. Florous, K. Saito, M. Koshiba, and T. Fujisawa, "Numerical investigation and optimization of a photonic crystal fiber for simultaneous dispersion compensation over $S + C + L$ wavelength bands," Optics Communications, vol. 274, pp. 74–79, 2007.

[72] T. Matsui, K. Nakajima, and I. Sankawa, "Dispersion compensation over all the telecommunication bands with double-cladding photonic-crystal fiber," Journal of Lightwave Technology, vol. 25, no. 3, pp. 757–762, 2007.

[73] M. A. R. Franco, V. A. Serrão, and F. Sircilli, "Microstructured optical fiber for residual dispersion compensation over $S + C + L + U$ wavelength bands," IEEE Photonics Technology Letters, vol. 20, no. 9, pp. 751–753, 2008.

[74] L. Labonté, P. Roy, D. Pagnoux, F. Louradour, C. Restoin, G. Melin, and E. Burov, "Experimental and numerical analysis of the chromatic dispersion dependence upon the actual profile of small core microstructured fibres," Journal of Optics A: Pure and Applied Optics, vol. 8, pp. 933–938, 2006.

[75] K. Tsujikawa, M. Ohashi, K. Shiraki, M. Tateda, and S. Shibata, "Scattering property of F and GeO_2 codoped silica glasses," Electronics Letters, vol. 30, no. 4, pp. 351–352, 1994.

[76] J. Canning, M. A. van Eijkelenborg, T. Ryan, M. Kristensen, and K. Lyytikainen, "Complex mode coupling within air–silica structured optical fibres and applications," Optics Communications, vol. 185, no. 4–6, pp. 321–324, 2000.

[77] K. Suzuki, H. Kubota, S. Kawanishi, M. Tanaka, and M. Fujita, "Optical properties of polarization maintaining photonic crystal fiber," Optics Express, vol. 9, no. 13, pp. 676–680, 2001.

[78] R. Bise, E. Monberg, F. Dimarcello, and D. Trevor, "Impact of preform fabrication and fiber draw on the optical properties of microstructured optical fibers," Proceedings of International Wire and Cable Symposium (IWCS), Orlando, p. 339, 2002.

[79] K. Tajima, K. Nakajima, K. Kurokawa, N. Yoshizawa, and M. Ohashi, "Low-loss photonic crystal fibers," Optical Fiber Communications Conference, A. Sawchuk (ed), OSA Trends in Optics and Photonics, Vol. 70, Optical Society of America, 2002, paper ThS3.

[80] J. A. West, N. Venkataramam, C. M. Smith, and M. T. Gallagher, "Photonic crystal fiber," 27th European Conference on Optical Communications (ECOC'01) Amsterdam, Technical Digest, vol. 4, pp. 582–585, 2001.

[81] L. Farr, J. C. Knight, B. J. Mangan, and P. J. Roberts, "Low loss photonic crystal fibre," 28th European Conference on Optical Communications (ECOC'02), Copenhagen, 2002, paper PD1.3.

[82] K. Tajima, J. Zhou, K. Nakajima, and K. Sato, "Ultra low loss and long length photonic crystal fiber," Optical Fiber Communications Conference Digest, 2003, paper PD1.

[83] J. Zhou, K. Tajima, K. Nakajima, K. Kurokawa, C. Fukai, T. Matsui, and I. Sankawa, "Progress on low loss photonic crystal fibers," Optical Fiber Technology, vol. 11, no. 2, pp. 101–110, 2005.

[84] J. Fini and R. Bise, "Progress in fabrication and modeling of microstructured optical fibers," Japanese Journal of Applied Physics, vol. 43, no. 8B, pp. 5717–5730, 2004.

[85] C. J. Voyce, A. D. Fitt, J. R. Hayes, and T. M. Monro, "Mathematical modeling of the self-pressurizing mechanism for microstructured fiber drawing," Journal of Lightwave Technology, vol. 27, no. 7, pp. 871–878, 2009.

[86] C. J. Voyce, A. D. Fitt, and T. M. Monro, "The mathematical modeling of rotating capillary tubes for holey-fibre manufacture," Journal of Engineering Mathematics, vol. 60, no. 1, pp. 69–87, 2008.

[87] C. J. Voyce, A. D. Fitt, and T. M. Monro, "Mathematical model of the spinning of microstructured fibres," Optics Express, vol. 12, no. 23, pp. 5810–5820, 2004.

[88] A. D. Fitt, K. Furusawa, T. M. Monro, and C. P. Please, "Modeling the fabrication of hollow fibers: capillary drawing," Journal of Lightwave Technology, vol. 19, no. 12, pp. 1924–1931, 2001.

[89] T. P. White, R. C. McPhedran, and C. M. de Sterke, "Confinement losses in micro-structured optical fibers," Optics Letters, vol. 26, no. 21, pp. 1660–1662, 2001.

[90] D. Ferrarini, L. Vincentti, and M. Zoboli, "Leakage properties of photonic crystal fibers," Optics Express, vol. 10, no. 23, pp. 1314–1319, 2002.

[91] T. Matsui, Z. Jian, K. Nakajima, I. Sankawa, "Dispersion-flattened photonic crystal fiber with large effective area and low confinement loss," Journal of Lightwave Technology, vol. 23, no. 12, pp. 4178–4183, 2005.

[92] J. C. Knight and D. V. Skryabin, "Nonlinear waveguide optics and photonic crystal fiber," Optics Express, vol. 15, no. 23, pp. 15365–15376, 2007.

[93] W. Belardi, J. H. Lee, K. Furusawa, Z. Yusoff, P. Petropoulos, M. Ibsen, T. M. Monro, and D. J. Richardson, "A 10 Gbit/s tunable wavelength converter based on four-wave mixing in highly nonlinear holey fibre," 28th European Conference on Optical Communications (ECOC'02), Copenhagen, 2002, paper PD1.2.

[94] V. Finazzi, T. M. Monro, and D. J. Richardson, "Small core silica holey fibers: nonlinearity and confinement loss trades-off," Journal of Optical Society of America B, vol. 20, no. 7, pp. 1427–1436, 2003.

[95] J. Zhang, Q. Lin, G. Piredda, R. W. Boyd, G. P. Agrawal, and P. M. Fauchet, "Optical solitons in a silicon waveguide," Optics Express, vol. 15, no. 12, pp. 7682–7688, 2007.

[96] U. Peschel, R. Morandotti, J. M. Arnold, J. S. Aitchison, H. S. Eisenberg, Y. Silberberg, T. Pertsch, and F. Lederer, "Optical discrete solitons in waveguide arrays. 2. Dynamic properties," Journal of Optical Society of America B, vol. 19, no. 11, pp. 2637–2644, 2002.

[97] J. Y. Y. Leong, P. Petropoulos, J. H. V. Price, H. Ebendorff-Heidepriem, S. Akimakis, R. C. Moore, K. E. Frampton, V. Finazzi, X. Feng, T. M. Monro, and D. J. Richardson, "High-nonlinearity dispersion-shifted lead-silicate holey fibers for efficient 1-µm pumped supercontinuum generation," Journal of Lightwave Technology, vol. 24, no. 1, pp. 183–190, 2006.

[98] G. Humbert, J. C. Knight, G. Bouwmans, P. St. Russell, D. P. Williams, P. J. Roberts, and B. J. Mangan, "Hollow core photonic crystal fibers for beam delivery," Optics Express, vol. 12, no. 8, pp. 1477–1484, 2004.

[99] T. Okuno, M. Onishi, T. Kashiwada, S. Ishikawa, and M. Nishimura, "Silica-based functional fibers with enhanced nonlinearity and their applications," IEEE Journal of Selected Topics in Quantum Electronics, vol. 5, no. 5, pp. 1385–1391, 1999.

[100] W. Belardi, "Holey optical fibres for high nonlinearity devices," Ph.D. Thesis, University of Southampton, 2003.

[101] K. Tajima, Z. Jian, K. Nakajima, and K. Sato, "Ultralow loss and long length photonic crystal fiber," Journal of Lightwave Technology, vol. 22, no. 1, pp. 7–10, 2004.

[102] A. M. Zheltikov, "The physical limit for the waveguide enhancement of nonlinear-optical processes," Optics and Spectroscopy, vol. 95, no. 3, pp. 410–415, 2003.

[103] A. M. Zheltikov, "Nonlinear optics of microstructured fibers," Physics-Uspekhi, vol. 47, no. 1, pp. 69–98, 2004.

[104] A. W. Snyder and J. D. Love, "Optical waveguide theory," Chapman and Hall, 1983.

[105] J. K. Ranka, R. S. Windeler, and A. J. Stentz, "Visible continuum generation in air–silica microstructure optical fibers with anomalous dispersion at 800 nm," Optic Letters, vol. 25, no. 1, pp. 25–27, 2000.

[106] J. M. Dudley, G. Genty, and S. Coen, "Supercontinuum generation in photonic crystal fiber," Review of Modern Physics, vol. 78, no. 4, pp. 1135–1184, 2006.

[107] J. M. Dudley and J. R. Taylor, "Ten years of nonlinear optics in photonic crystal fibre," Nature Photonics, vol. 3, pp. 85–90, 2009.

[108] M. Asobe, "Nonlinear optical properties of chalcogenide glass fibers and their application to all-optical switching," Optical Fiber Technology, vol. 3, no. 2, pp. 142–148, 1997.

[109] P. Petropoulos, H. Ebendorff-Heidepriem, V. Finazzi, R. Moore, K. Frampton, D. Richardson, and T. Monro, "Highly nonlinear and anomalously dispersive lead silicate glass holey fibers," Optics Express, vol. 11, no. 26, pp. 3568–3573, 2003.

[110] K. M. Kiang, K. Frampton, T. M. Monro, R. Moore, J. Tucknott, D. W. Hewak, and D. J. Richardson, "Extruded single-mode nonsilica glass holey optical fibres," Electronics Letters, vol. 38, no. 12, pp. 546–547, 2002.

[111] S. C. Rashleigh, "Origins and control of polarization effects in single-mode fibers," Journal of Lightwave Technology, vol. LT-1, no. 2, pp. 312–331, 1983.

[112] D. N. Payne, A. J. Barlow, and J. J. R. Hansen, "Development of low- and high-birefringence optical fibers," IEEE Journal of Quantum Electronics, vol. QE-18, no. 4, pp. 477–488, 1982.

[113] A. Ortigosa-Blanch, J. C. Knight, W. J. Wadsworth, J. Arriaga, B. J. Mangan, T. A. Birks, and P. St. J. Russell, "Highly birefringent photonic crystal fibers," Optics Letters, vol. 25, no. 18, pp. 1325–1327, 2000.

[114] T. P. Hansen, J. Broeng, S. E. B. Libori, E. Knudsen, A. Bjarklev, J. R. Jensen, and H. Simonsen, "Highly birefringent index-guiding photonic crystal fibers," IEEE Photonics Technology Letters, vol. 13, no. 6, pp. 588–590, 2001.

[115] K. Suzuki, H. Kubota, S. Kawanishi, M. Tanaka, M. Fujita, "Optical properties of polarization maintaining photonic crystal fiber and its application," IEIC Technical Report, vol. 102, no. 135, pp. 23–26, 2002.

[116] M. J. Steel and R. M. Osgood, Jr., "Polarization and dispersive properties of elliptical-hole photonic crystal fibers," Journal of Lightwave Technology, vol. 19, no. 4, pp. 495–503, 2001.

[117] M. Szpulak, G. Statkiewicz, J. Olszewski, T. Martynkien, W. Urbańczyk, J. Wójcik, M. Makara, J. Klimek, T. Nasilowski, F. Berghmans, and H. Thienpont, "Experimental and theoretical investigations of birefringent holey fibers with a triple defect," Applied Optics, vol. 44, no. 13, pp. 2652–2658, 2005.

[118] K. Saitoh and M. Koshiba, "Single-polarization single-mode photonic crystal fibers," IEEE Photonics Technology Letters, vol. 15, no. 10, pp. 1384–1386, 2003.

[119] J. R. Folkenberg, M. D. Nielsen, N. A. Mortensen, C. Jakobsen, and H. R. Simonsen, "Polarization maintaining large mode area photonic crystal fiber," Optics Express, vol. 12, no. 5, pp. 956–960, 2004.

[120] N. A. Issa, M. A. van Eijkelenborg, M. Fellew, F. Cox, G. Henry, and M. C. J. Large, "Fabrication and study of microstructured optical fibers with elliptical holes," Optics Letters, vol. 29, no. 12, pp. 1336–1338, 2004.

[121] T. Schreiber, H. Schultz, O. Schmidt, F. Röser, J. Limpert, and A. Tünnermann, "Stress-induced birefringence in large-mode-area micro-structured optical fibers," Optics Express, vol. 13, no. 10, pp. 3637–3646, 2005.

[122] A. Ortigosa-Blanch, A. Diez, M. Delgado-Pinar, J. L. Cruz, and M. V. Andres, "Ultrahigh birefringent nonlinear microstructured fiber," IEEE Photonics Technology Letters, vol. 16, no. 7, pp. 1667–1669, 2004.

[123] W. Belardi, G. Bouwmans, L. Provino, and M. Douay, "Form-induced birefringence in elliptical hollow photonic crystal fiber with large mode area," IEEE Journal of Quantum Electronics, vol. 41, no. 12, pp. 1558–1564, 2005.

[124] H. Kubota, S. Kawanishi, S. Koyanagi, M. Tanaka, and S. Yamaguchi, "Absolutely single polarization photonic crystal fiber," IEEE Photonics Technology Letters, vol. 16, no. 1, pp. 182–184, 2004.

[125] J. Ju, W. Jin, and M. S. Demokan, "Design of single-polarization single-mode photonic crystal fiber at 1.30 and 1.55 μm," Journal of Lightwave Technology, vol. 24, no. 2, pp. 825–830, 2006.

[126] J. R. Folkenberg, M. D. Nielsen, and C. Jakobsen, "Broadband single-polarization photonic crystal fiber," Optics Letters, vol. 30, no. 12, pp. 1446–1448, 2005.

[127] M. Delgado-Pinar, A. Diez, J. L. Cruz, and M. V. Andres, "High extinction-ratio polarizing endlessly single-mode photonic crystal fiber," IEEE Photonics Technology Letters, vol. 19, no. 8, pp. 562–564, 2007.

[128] M. Delgado-Pinar, A. Díez, S. Torres-Peiró, M. V. Andrés, T. Pinheiro-Ortega, and E. Silvestre, "Waveguiding properties of a photonic crystal fiber with a solid core surrounded by four large air holes," Optics Express, vol. 17, no. 9, pp. 6931–6938, 2009.

[129] K. Oh, S. Choi, Y. Jung, and J. W. Lee, "Novel hollow optical fibers and their applications in photonic devices for optical communications," Journal of Lightwave Technology, vol. 23, no. 2, pp. 524–532, 2005.

[130] J. Stone, "Optical transmission loss in liquid-core hollow fibers," Journal of Quantum Electronics, vol. QE-8, no. 3, pp. 386–388, 1972.

[131] T. Kobayashi and W. J. Blau, "Laser emission from conjugated polymer in fibre waveguide structure," Electronics Letters, vol. 38, no. 2, pp. 67–68, 2002.

[132] M. Saito, A. Honda, and K. Uchida, "Photochromic liquid-core fibers with nonlinear input–output characteristics," Journal of Lightwave Technology, vol. 21, no. 10, pp. 2255–2261, 2003.

[133] S. Hayashi, A. Ishimizu, T. Tohei, and M. Tachikawa, "Parametric excitation of laser-guided Cs atoms in a hollow-core optical fiber," Physical Review A, vol. 68, p. 053408, 2003.

[134] E. A. Nersesov, S. V. Popruzhenko, D. F. Zaretsky, and W. Becker, "The gain of high harmonics in an atomic jet and in a hollow-core fiber," Optics Communications, vol. 183, no. 1–4, pp. 289–297, 2000.

[135] S. Choi, K. Oh, W. Shin, C. S. Park, U. C. Paek, K. J. Park, Y. C. Chung, G. Y. Kim, and Y. G. Lee, "Novel mode converter based on hollow optical fiber for gigabit LAN communication," IEEE Photonics Technology Letters, vol. 14, no. 2, pp. 248–250, 2002.

[136] S. Choi and K. Oh, "A new LP_{02} mode dispersion compensation scheme based on mode converter using hollow optical fiber," Optics Communications, vol. 221, no. 4–6, pp. 307–312, 2003.

[137] S. Choi, T. J. Eom, J. Yu, B. H. Lee, and K. Oh, "Novel all-fiber bandpass filter based on hollow optical fiber," IEEE Photonics Technology Letters, vol. 14, no. 12, pp. 1701–1703, 2002.

[138] S. Choi, Y. Jung, T. J. Eom, B. H. Lee, and K. Oh, "Broadband tunable all-fiber bandpass filter based on hollow optical fiber," IEEE Photonics Technology Letters, vol. 17, no. 1, pp. 115–117, 2005.

[139] Y. Jung, S. B. Lee, Jhang, W. Lee, and K. Oh, "Bandwidth tuning in a novel hybrid fiber acousto-optic filter," Optics Letters, vol. 30, no. 1, pp. 84–86, 2005.

[140] Y. Jung, S. B. Lee, B. H. Lee, and K. Oh, "Acousto-optic tunable gap type bandpass filter with a broad stopband," IEEE Photonics Technology Letters, vol. 19, no. 17, pp. 1331–1333, 2007.

[141] Y. Jung, S. Lee, B. H. Lee, J. Kobelke, and K. Oh, "Observation of tunable band-pass characteristics in a hollow optical fiber (HOF)-micro structured fiber (MSF) composite structure using bend loss edge shift effects," Optics Letters, vol. 33, no. 24, pp. 2946–2948, 2008.

[142]   Y. Jung, S. Lee, B. H. Lee, and K. Oh, "Ultra-compact in-line broadband Mach–Zehnder interferometer using a composite leaky hollow optical fiber waveguide," Optics Letters, vol. 33, no. 24, pp. 2934–2936, 2008.

[143]   I.-K. Hwang, Y.-H. Lee, K. Oh, and D. N. Payne, "High birefringence in elliptical hollow optical fiber," Optics Express, vol. 12, no. 9, pp. 1916–1923, 2004.

[144]   Y. Jung, S. R. Han, Soan Kim, U. C. Paek, and K. Oh, "Versatile control of geometric birefringence in elliptic hollow optical fiber," Optics Letters, vol. 31, no. 18, pp. 2681–2683, 2006.

[145]   Y. Jung, B. H. Lee, J. W. Yu, and K. Oh, "Acousto-optic polarization dependent mode coupling in a dual mode hollow optical fiber," Optics Letters, vol. 32, no. 10, pp. 1305–1307, 2007.

[146]   S. Kim, U. C. Paek, and K. Oh, "New defect design in index guiding holey fiber for uniform birefringence and negative flat dispersion over a wide spectral range," Optics Express, vol. 13, no. 16, pp. 6039–6050, 2005.

[147]   S. Kim, Y. Jung, K. Oh, J. Kobelke, K. Schuster, and J. Kirchhof, "New defect and lattice structure for air silica index guiding holey fiber," Optics Letters, vol. 31, no. 2, pp. 164–166, 2006.

[148]   S. Kim and K. Oh, "Large-mode-area index-guiding holey fibers with ultra-low ultra-flattened dispersion using a novel hollow-ring defect structure," Journal of the Korean Physical Society, vol. 48, no. 5, pp. 897–901, 2006.

[149]   Z. Zhi-guo, Z. Fang-di, and Z. Min, and Y. Pei-da, "Gas sensing properties of index-guided PCF with air–core," Optics Laser Technology, vol. 40, pp. 167–174, 2008.

[150]   L. Raddatz, et al., "An experimental and theoretical study of the offset launching technique for the enhancement of the bandwidth of multimode fiber links," Journal of Lightwave Technology, vol. 16, no. 3, pp. 324–331, 1998.

[151]   B. Bourliaguet, C. Paré, F. Émond, A. Croteau, A. Proulx and R. Vallée, "Micro-structured fiber splicing," Optics Express, vol. 11, no. 25, pp. 3412–3417, 2003.

[152]   J. H. Chong, M. K. Rao, Y. Zhu, P. Shum, "An effective splicing method on photonic crystal fiber using CO_2 laser," IEEE Photonics Technology Letters, vol. 15, pp. 942–944, 2003.

[153]   L. Xiao, M. S. Demenko, W. Jin, Y. Wang, and C-L Zhao, "Fusion Splicing Photonic Crystal Fibers and Conventional Single-Mode Fibers: Microhole Collapse Effect," Journal of Lightwave Technology, vol. 25, no. 11, pp. 3563–3574, 2007.

CHAPTER NINE

Fiber Mode Analysis Using OFACAD

9.1 THEORETICAL REVIEW ON CASCADED BOUNDARY MATRIX METHOD

In Chapter 2, we have reviewed the basis of the waveguide properties in a step-index cylindrical fiber with a single core–cladding boundary. As explained in Chapter 4, we need more than simple step refractive index profile to cope with various optical specifications for current optical communication networks. There have been extensive researches on numerical modeling of an optical fiber with an arbitrary structure to predict and analyze the guiding properties [1–4]. Recently fiber design tools [5,6] are also commercially available based on various numerical techniques.

In this book, we will provide a fiber analysis software, optical fiber analysis computer-aided design (OFACAD) [7]. OFACAD is an efficient Maxwell's equation vector solver including effects of material dispersion. The fundamental approach in OFACAD is to divide a given arbitrary refractive index profile of an optical fiber into meshes within which the index profiles are approximated as a step index. The boundary conditions at the interface between the neighboring meshes will then be the same as in equations 2.59–2.64. The vector components of the electric field and magnetic field are related as in equation 2.73 in terms of Bessel functions. The boundary conditions in each mesh will add up to form a cascaded matrix whose determinant will be used to find the effective index at a given frequency. We named this efficient full-vector analysis of dielectric waveguide as cascaded boundary matrix (CBM) method. In this section, theoretical basis of CBM will be discussed and then OFACAD examples will be explained.

Silica Optical Fiber Technology for Devices and Components: Design, Fabrication, and International Standards, First Edition. By Kyunghwan Oh and Un-Chul Paek.
© 2012 John Wiley & Sons, Inc. Published 2012 by John Wiley & Sons, Inc.

9.1.1   Theoretical Basis of Cascaded Boundary Matrix Method

For a step index with a single core–cladding interface, we obtained a matrix equation for the boundary conditions:

$$\overleftrightarrow{\Lambda} \cdot \vec{X} = 0 \qquad (2.73)$$

where

$$\overleftrightarrow{\Lambda} = \begin{bmatrix} J_m(ua) & 0 & -K_m(wa) & 0 \\ 0 & J_m(ua) & 0 & -K_m(wa) \\ \dfrac{m\beta}{u^2 a^2}J_m(ua) & \dfrac{\omega\mu_0}{ua}J'_m(ua) & \dfrac{m\beta}{w^2 a^2}K_m(wa) & \dfrac{\omega\mu_0}{wa}K'_m(wa) \\ \dfrac{\omega\varepsilon_0 n_1^2}{ua}J'_m(ua) & \dfrac{m\beta}{u^2 a^2}J_m(ua) & \dfrac{\omega\varepsilon_0 n_2^2}{wa}K_m(wa) & \dfrac{m\beta}{w^2 a^2}K'_m(wa) \end{bmatrix} \quad \text{and} \quad \vec{X} = \begin{bmatrix} A \\ B \\ C \\ D \end{bmatrix}$$

The above matrix, $\overleftrightarrow{\Lambda}$, can be further decomposed into two 4×2 matrix blocks as in Figure 9.1.

In Figure 9.1 $\alpha = (m\beta/\beta_t^2 a^2)$, $\delta = (\omega\mu_0/\beta_t a)$, and $\gamma = (\omega\varepsilon_0 n^2/\beta_t a)$. General Bessel function of order m, denoted as ζ_m, will correspond to J_m and K_m for the core and the cladding, respectively. In the same way the transverse propagation constant, $\beta_t = \sqrt{|n^2 k^2 - \beta^2|}$ in equations 2.38–2.41, does depend on the refractive index, n, and β_t in Figure 9.1 will be equal to u and w in the region of core and cladding, respectively. Therefore, the boundary conditions for electromagnetic fields can be represented by a 4×2 matrix, shown in a bold line box in Figure 9.1, and we will denote the matrix as "boundary matrix" hereafter. In the above example, the effective index of the step-index fiber with a single core–cladding boundary can be found by the determinant of a matrix concatenated by two boundary matrices. This idea can be

	$n = n_1$		$n = n_2$		
$\zeta_m(\beta_t a)$	0	$\zeta_m(\beta_t a)$	0	A	
0	$\zeta_m(\beta_t a)$	0	$\zeta_m(\beta_t a)$	B	$= 0$
$\alpha\zeta_m(\beta_t a)$	$\delta\zeta_m(\beta_t a)$	$\alpha\zeta_m(\beta_t a)$	$\delta\zeta_m(\beta_t a)$	C	
$\gamma\zeta_m(\beta_t a)$	$\alpha\zeta_m(\beta_t a)$	$\gamma\zeta_m(\beta_t a)$	$\alpha\zeta_m(\beta_t a)$	D	
Core		Cladding			

FIGURE 9.1   Decomposition of the $\overleftrightarrow{\Lambda}$ into 4×2 matrix blocks, "boundary matrix."

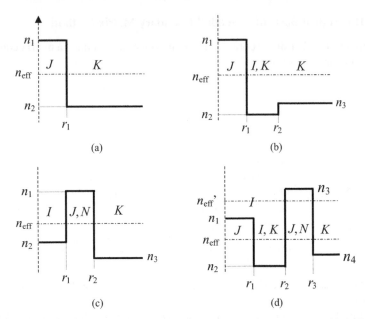

FIGURE 9.2 Step-index profiles with multiple interfaces: (a) a step-index profile with a single boundary at $r = r_1$ ($i = 1$), (b) a depressed inner cladding W-type fiber with two boundaries at $r = r_1$ and r_2 ($i = 2$), (c) a ring core fiber with two boundaries at $r = r_1$ and r_2 ($i = 2$), (d) concentric double ring core with three boundaries, $r = r_1, r_2$, and r_3 ($i = 3$). I, J, K, and N are the Bessel functions. i represents the number of boundaries.

further generalized for an arbitrary number of boundaries and equivalently arbitrary index profile. We have developed "cascaded boundary matrix (CBM)" technique that can provide a full-vector solution of Maxwell's equation in cylindrical optical fiber, which will give the effective index, and vector components of the fields as function of wavelength.

As an introduction to the concept of CBM method, let us consider the step-index profiles with multiple boundaries as shown in Figure 9.2. Vertical axis and horizontal axis in each figure are the refractive index and the radial position in optical fiber, respectively.

In the case of simple step index (a) in Figure 9.2, we have a single interface where the boundary matrix is constructed by the Bessel functions $J_m(r)$, $I_m(r)$, and $K_m(r)$. For other cases we have two interfaces in (b) W-type fiber and (c) ring core fiber, and three interfaces for (d) concentric double ring core fiber.

Before we proceed to the discussion of CBM, it is worthwhile to note how Bessel functions are chosen in reference to the location of effective index, n_{eff}:

(1) If the n_{eff} is lower than the given index of the waveguide in a region, for example, $r < r_1$ in Figure 9.2a, b, and d; $r_1 < r < r_2$ in Figure 9.2c; and $r_2 < r < r_3$ in Figure 9.2d; the electric field, E_z; and magnetic field, H_z; are in general given in a linear combination of $J_m(ua)$ and $N_m(ua)$.

(2) When the region furthermore includes the origin, $r = 0$, as in the region $r < r_1$ in Figure 9.2a, b, and d, then the only solution is $J_m(ua)$ due to the singularity of $N_m(ua)$ at the origin.

(3) On the contrary, if the n_{eff} is larger than the given index of the waveguide in a region, for example, $r > r_1$ in Figure 9.2a, $r_1 < r < r_2$ and $r > r_3$ in Figure 9.2b, $r < r_1$ and $r > r_2$ in Figure 9.2c, and $r_1 < r < r_2$ and $r > r_3$ in Figure 9.2d, the longitudinal electromagnetic fields are given in a linear combination of $K_m(ua)$ and $I_m(ua)$.

(4) When the region includes $r \to \infty$, for example, $r > r_1$ in Figure 9.2a, $r > r_2$ in Figure 9.2b, $r > r_2$ in Figure 9.2c, and $r > r_3$ in Figure 9.2d, then the only solution is $K_m(ua)$ since $I_m(ua)$ diverges at $r \to \infty$.

(5) If the region includes $r = 0$ as in $r < r_1$ in Figure 9.2c, the solution is $I_m(ua)$ due to the singularity of $K_m(ua)$ at the origin.

In the case of Figure 9.2d, if the effective index is larger than n_1 in the inner core, $r < r_1$, as $n'_{eff} > n_1$, then the longitudinal electromagnetic field components are given in $I_m(wa)$, not in $J_m(ua)$.

9.1.2 Application of CBM in Optical Fibers with Step-Index Interfaces

Now consider W-type fiber as in Figure 9.2b. This waveguide structure has been extensively applied in dispersion-compensating fibers (DCFs), which have been discussed in Chapter 4. We will discuss its dispersion properties in detail utilizing CBM method in this chapter. Note that in this case we have two interfaces, $r = r_1$ and $r = r_2$. The matrix representation of the boundary conditions at $r = r_1$ is given by Figure 9.3. Here the arguments of Bessel functions are $\beta_t r_1$. Note that we have two independent blocks, J and T_1.

At $r = r_2$, we have another matrix shown in Figure 9.4.

In Figure 9.3, we notice that the matrix for boundary conditions is decomposed into two blocks, a 4×2 boundary matrix J at $r = r_1$ for $n = n_1$ and a 4×4 boundary matrix

$n = n_1$		$n = n_2$					
J_m	0	I_m	0	K_m	0	B	
0	J_m	0	I_m	0	K_m	C	
αJ_m	$\delta J_m{'}$	αI_m	$\delta I_m{'}$	αK_m	$\delta K_m{'}$	D	
$\gamma J_m{'}$	αJ_m	$\gamma I_m{'}$	αI_m	$\gamma K_m{'}$	αK_m	E	
J		T_1				F	

(Column headers at far right: A, X_I; X_T. Rows top-to-bottom labelled A, B, C, D, E, F.)

FIGURE 9.3 Matrix at the core–inner cladding boundary in W-type fiber at $r = r_1$. The arguments of Bessel functions are $\beta_t r_1$.

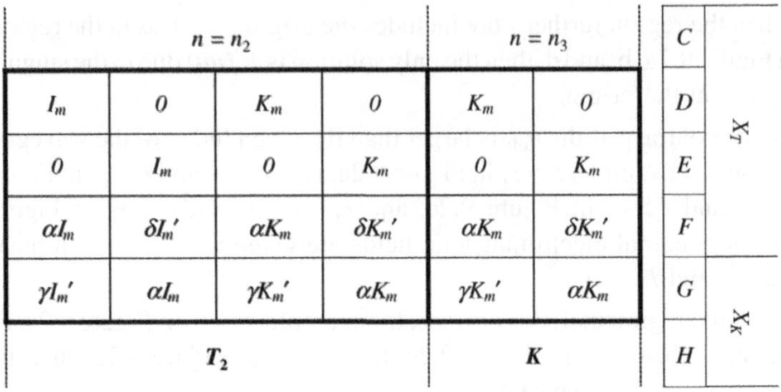

$n = n_2$				$n = n_3$			
I_m	0	K_m	0	K_m	0	C	
0	I_m	0	K_m	0	K_m	D	X_T
αI_m	$\delta I_m{}'$	αK_m	$\delta K_m{}'$	αK_m	$\delta K_m{}'$	E	
$\gamma I_m{}'$	αI_m	$\gamma K_m{}'$	αK_m	$\gamma K_m{}'$	αK_m	F	
		T_2			K	G	X_K
						H	

FIGURE 9.4 Matrix at the inner cladding–outer cladding boundary in W-type fiber at $r = r_2$. The arguments of Bessel functions are $\beta_t r_2$.

T_1 at $r = r_1$ for $n = n_2$. Similarly we can decompose the boundary matrix in Figure 9.4 into two blocks, a 4×4 boundary matrix T_2 at $r = r_2$ for $n = n_2$ and a 4×2 boundary matrix K at $r = r_2$ for $n = n_3$. We will also denote the column vectors, $[A, B] = X_J$, $[C, D, E, F] = X_T$, and $[G, F] = X_K$. For these block matrices and column vectors, we have the following equations that represent the boundary conditions in W-type fibers:

$$J \cdot X_J + T_1 \cdot X_T = 0 \tag{9.1}$$

$$T_2 \cdot X_T + K \cdot X_K = 0 \tag{9.2}$$

By combining matrices in the above two equations, we obtain an 8×8 matrix equation for the boundary conditions at two interfaces at both $r = r_1$ and $r = r_2$ (Figure 9.5).

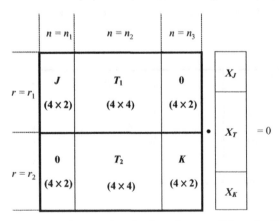

FIGURE 9.5 An 8×8 matrix equation for boundary conditions in W-type fiber. The arguments of Bessel functions are $\beta_t r_i$ and $i = 1$ and 2.

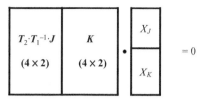

FIGURE 9.6 Matrix representation of the 4×4 matrix equation for boundary conditions in W-type fiber.

We also note that equations 9.1 and 9.2 can be further reduced to a single equation:

$$T_2 \cdot \{(T_1)^{-1} \cdot J \cdot (-X_J)\} + K \cdot X_K = 0 \qquad (9.3)$$

And this is represented by a 4×4 matrix equation as shown in Figure 9.6.

The effective index can, therefore, be computed by the determinant of the 4×4 cascaded boundary matrix in Figure 9.6. Note that minus sign in front of X_J in equation 9.3 can be neglected because it does not affect the effective index value and we will keep only positive sign in the matrix equation hereafter.

In order to further generalize CBM methods, we will consider the concentric double ring core fiber in Figure 9.2d. In this case we have three interfaces, $r = r_1, r_2,$ and r_3, and therefore we will have a 12×12 matrix equation to represent the boundary conditions at the interfaces. These boundary conditions can be arranged into subgroups of J, T, and K as represented in Figure 9.7.

Similar to the case of W-fiber, the matrix equation in Figure 9.7 can be further reduced to a single equation:

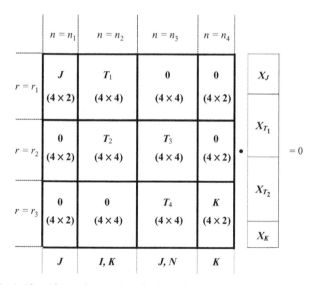

FIGURE 9.7 A 12×12 matrix equation for boundary conditions in concentric double core fiber. The arguments of Bessel functions are $\beta_t r_i$ and $i = 1 - 3$.

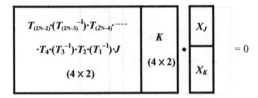

FIGURE 9.8 A 4×4 matrix equation for boundary conditions for N interfaces.

$$T_4 \cdot [(T_3)^{-1}] \cdot T_2 \cdot [(T_1)^{-1}] \cdot J \cdot (-X_J) + K \cdot X_K = 0 \qquad (9.4)$$

From equations 9.3 and 9.4, we can generalize the cascaded boundary matrix equation for step-index fibers with N interfaces as follows (Figure 9.8):

$$T_{2N-2} \cdot [(T_{2N-3})^{-1}] \cdot T_{2N-4} \cdots \cdots T_4 \cdot [(T_3)^{-1}] \cdot T_2 \cdot [(T_1)^{-1}] \cdot J(-X_J) + K \cdot X_K = 0 \qquad (9.5)$$

Therefore, we can evaluate the effective index by finding the root for the 4×4 cascaded boundary matrix in Figure 9.8, from which all the vector components of electromagnetic fields are obtained. Note that the matrix does have the inherent dependence on the azimuthal number m, or equivalently azimuthal index l as defined in equation 2.87, which determines the mode order and corresponding Bessel functions, J_m, N_m, I_m, and K_m.

9.2 ALGORITHM FOR CBM TO FIND OPTICAL PROPERTIES OF GUIDED MODES

The algorithmic routine for CBM method is discussed below. First, for a given refractive index profile, the profile is divided into stepwise meshes to set up the number of interfaces, N. The azimuthal number m of a guided mode will be also assumed. In each mesh bounded by the given interfaces, the stepwise refractive index value, n_i, will be defined along with the radial boundary, $r_{i-1} < r < r_i$. The effective index, n_{eff}, of a guided mode should lie between the cladding index and the largest index in the core region, and we pick a trial n_{eff} in this range to construct a 4×4 cascaded boundary matrix as schematically shown in Figure 9.8. The determinant of the boundary matrix is evaluated for the trial n_{eff} until it makes the determinant zero. The schematic of the program is shown in Figure 9.9. We used bisection method [8] to find the root of the determinant, which will provide evaluation of the effective index within 10^{-20} error.

When we vary the wavelength, the effective index also changes and when $n_{eff}(\lambda)$ has the same value as the refractive index of outer cladding at $\lambda = \lambda_{cutoff}$, the mode will leak out into the cladding and the mode will be cutoff. This way we can numerically find the cutoff wavelengths of guided modes.

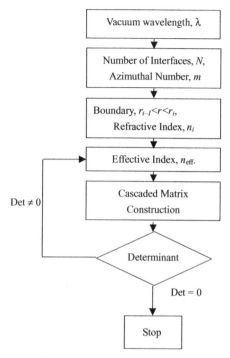

FIGURE 9.9 The block diagram of cascaded boundary matrix method to find the effective index.

Group delays and chromatic dispersion can also be obtained for $n_{\text{eff}}(\lambda)$ using

$$\tau_g = \frac{L}{v_g} = \frac{Ln_g}{c} = \frac{L}{c}\frac{d\beta}{dk} = \frac{L}{c}\left[n_{\text{eff}} - \lambda\frac{dn_{\text{eff}}}{d\lambda}\right] \qquad (2.105)$$

$$D = \frac{1}{L}\frac{d\tau_g}{d\lambda} = -\frac{\lambda}{c}\frac{d^2 n_{\text{eff}}}{d\lambda^2} \qquad (2.116)$$

Once the effective index, $n_{\text{eff}}(\lambda)$, is known, the electromagnetic fields are obtained by solving the matrix equation 9.5, evaluating the cascaded boundary matrix for n_{eff}. The relative ratios of the linear coefficients, $A, B, C, \ldots$, are obtained and by setting one of the coefficients, for example, A, equal to 1, we can find the electromagnetic field distribution. The relative intensity profile of a mode can be also obtained using

$$\left\langle \vec{S} \right\rangle = \frac{1}{2}\text{Re}(\vec{E} \times \vec{H}^*) \quad [\text{W/m}^2] \qquad (2.96)$$

The mode field can be furthermore evaluated using the electromagnetic fields.

In the program, the contribution of material dispersion in the effective index is included assuming that index raising is achieved by GeO_2 doping and index depression is by fluorine doping. The Sellmeir coefficients for the corresponding glass are used as listed in Table 2.5.

9.3   MODE ANALYSIS EXAMPLE USING OFACAD

In the following, we show some of the mode analysis examples using OFACAD. The figures in this section are actual outputs from OFACAD.

OFACAD was initially developed by Dr. Hong Seok Seo and Dr. Woojin Shin during their Ph.D. programs in Gwangju Institute of Science and Technology (GIST) and they were supervised for their degrees by authors. Dr. Seo and Dr. Shin are senior research scientists in Electronic Telecommunication Research Institute (ETRI) and Advanced Photonics Research Institute (APRI) in Korea, respectively.

OFACAD is based on numerical root-finding techniques for determinant equation of a CBM for a given stepwise refractive index profile, which was discussed in Section 9.2. In the OFACAD, various optical properties can be calculated such as effective index, group delay, chromatic dispersion, field distribution, intensity profile, mode field diameter, and effective area. Compared with the publicly known optical properties of commercial optical fibers, the OFACAD calculation results show excellent agreements, which assure practical application of the program in optical fiber waveguide designs.

In this book, we provide educational version of the program OFACAD_EDU in a separate CD. Detailed procedures for data inputs and menu selections are given in Appendix A. In this section, we will briefly summarize the mode analysis processes using OFACAD for various optical fibers commercially available, whose refractive index profiles are known.

9.3.1   Simulation Examples of a Conventional Single-Mode Fiber

As a starting point of discussion, we will describe mode analysis process for a conventional single-mode fiber (SMF) such as Corning SMF28, whose refractive index and chromatic dispersion spectrum are shown in Figure 9.10.

Step 1: Refractive Index Setup   SM-28 is a step-index fiber and its refractive index profile can be programmed by simply selecting the radial position (mm) and corresponding refractive index in the *Refractive Index Setup* window. As a reference, the refractive index of SiO_2 glass is shown in the window, which is calculated by Sellmeir equation using coefficients in Table 2.5. For the given input data (radial position, refractive index), the index difference Δn and relative index difference $\Delta(\%)$ referenced to SiO_2 are automatically calculated along with the refractive index profile graph on the top left in Figure 9.11. For SMF-28, we used the following input data available from Corning SMF28 datasheet: core radius of $4\,\mu m$ and Δ of $\sim 0.34\%$. Actual program screen shot is shown in Figure 9.11.

Step 2: Mode Selection and Cutoff Wavelength Calculation   In the *Mode Selection* window, we can select one of guided modes, HE_{lm}, TE_{lm}, TM_{lm}, and

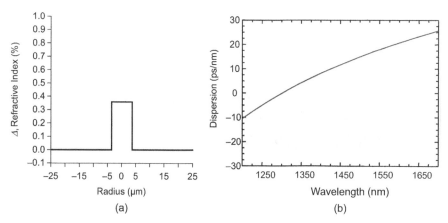

FIGURE 9.10   (a) Refractive index profile and (b) chromatic dispersion of SMF-28 fiber.

EH_{lm} modes, which are explained in Chapter 2. In this case we chose the fundamental mode, HE_{11}. After selecting the mode, we can find the cutoff wavelengths of the higher order modes in *Cutoff Wavelength* window on the upper right side in Figure 9.12. In this window, we can select a mode among HE_{lm}, TE_{lm}, TM_{lm}, and EH_{lm} except the HE_{11} mode because in conventional step-index profiles, the fundamental HE_{11} mode does not have the cutoff. After selecting mode, the wavelength of interests is specified by *Start Wave (μm)* and *Stop Wave (μm)*. By clicking OK button in the window, cutoff wavelength of the selected mode is calculated within the chosen wavelength range. For

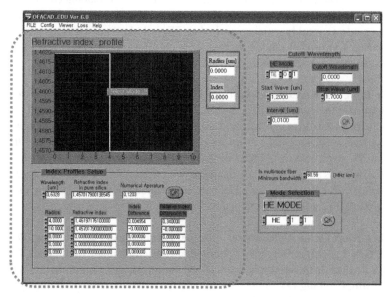

FIGURE 9.11   *Step 1* for SMF-28 analysis. Refractive index profile data are defined in the *Index Profile Setup* window on the lower left side.

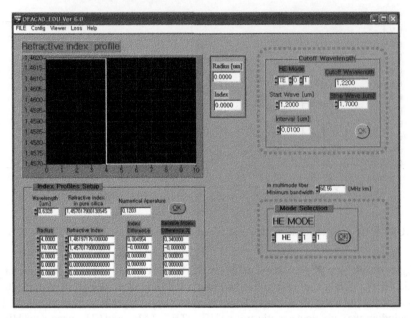

FIGURE 9.12 *Step 2* for SMF-28 analysis. The HE_{11} mode was selected in *Mode Selection*. The cutoff wavelength of an excited mode, TE_{01} mode, is calculated within the chosen wavelength range defined on the *Cutoff Wavelength* window.

SMF-28, we selected the first excited mode TE_{01} and we chose the spectral range from 1.2 to 1.7 μm. The cutoff wavelength of the TE_{01} mode was calculated as 1220 nm, which is a typical value for SMF-28. Note that in the given refractive index profile, the waveguide is weakly guiding such that TE_{01}, TM_{01}, and HE_{21} modes are all degenerate to form an LP_{11} mode with the same cutoff wavelength. The results of Step 2 in the program are shown in Figure 9.12.

Step 3: Effective Index Calculation   Various optical properties of the selected guide mode, in this example HE_{01} mode, can be calculated using OFACAD. In the *Viewer* menu on the top menu bar, there are six optical characteristics analysis options: *Effective Index, Group Delay, Dispersion, Field, Intensity,* and *MFD and Effective Area*. In this step, we select *Effective Index* in *Viewer* menu and the *EFFECTIVE INDEX* window is activated. In this example, we selected the wavelength range *Start*: 1.2 μm and *End*: 1.7 μm. Clicking *OK* button the calculation results for the effective index of the HE_{11} mode are obtained as in Figure 9.13.

Step 4: Group Delay Calculation   In the Viewer menu, select *Group delay* to activate *GROUP DELAY* window. Specify the wavelength range of interests that you want to calculate. In this example, we selected the wavelength range *Start*: 1.2 μm and *End*: 1.7 μm. Clicking *OK* button the calculation results for the group delay of the HE_{11} mode are obtained as in Figure 9.14.

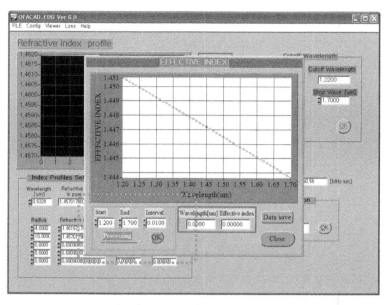

FIGURE 9.13 *Step 3* for SMF-28 analysis. The effective index of the HE_{11} mode is calculated within the chosen wavelength range defined on the *EFFECTIVE INDEX* window of the *Viewer* menu.

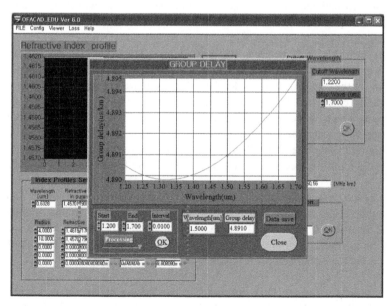

FIGURE 9.14 *Step 4* for SMF-28 analysis. The group delay of the HE_{11} mode is calculated within the chosen wavelength range defined on the *GROUP DELAY* window of the *Viewer* menu.

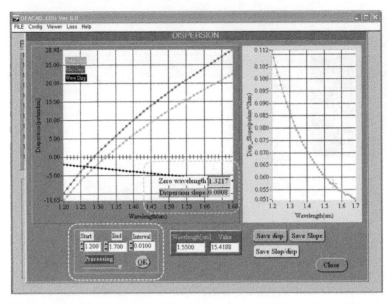

FIGURE 9.15 *Step 5* for SMF-28 analysis. Chromatic of the HE_{11} mode is calculated within the chosen wavelength range defined on the *DISPERSION* window of the *Viewer* menu.

Step 5: Chromatic Dispersion Calculation In the Viewer menu, select *Dispersion* to activate *DISPERSION* window. Specify the wavelength range of interests. Clicking *OK* button the calculation results for the dispersion of the HE_{11} mode are obtained as in Figure 9.15.

In the DISPERSION window, the contributions from material and waveguide dispersions are specified in separate curves along with the total chromatic dispersion spectrum for the selected HE_{11} mode. Zero-dispersion wavelength and its dispersion slope are also calculated. Dispersion slope in the specified wavelength range is plotted in a separate graph on the right-hand side of Figure 9.15. In this step-index profile, the zero dispersion occurs near 1.32 µm along with a slope of ~0.08 ps/(nm^2 km). The dispersion and its slope at 1550 nm are ~15 ps/(nm km) and ~0.06 ps/(nm^2 km), which are within the specification of ITU-T G.652 single-mode fiber standard.

Step 6: Electromagnetic Field Calculation In the Viewer menu, select *Field* to activate the *FIELD* window. Specify the wavelength of interests. In this example we selected the *WAVELENGTH* at 1.55 µm range. Clicking *OK* button the calculation results for the radial, azimuthal, and longitudinal components of electric field and magnetic field are obtained as in Figure 9.16.

Step 7: Modal Intensity Distribution In the Viewer menu, select *Intensity* to activate the *INTENSITY* window. Specify the wavelength of interests. In this example we selected the *WAVELENGTH* at 1.55 µm range. Clicking *Plot* button the calculation results for the modal intensity on x–y plane are obtained as in Figure 9.17. By clicking

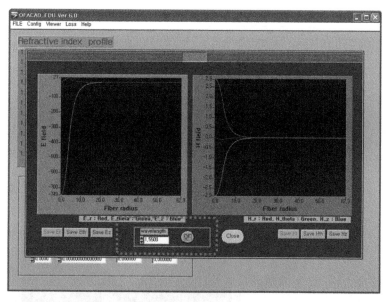

FIGURE 9.16 *Step 6* for SMF-28 analysis. Distribution of electric and magnetic field components of the HE_{11} mode is calculated at the chosen wavelength defined on the *FIELD* window of the *Viewer* menu.

3D Plot, we can also view the intensity distribution in *3 DIMENSIONAL MODE FIELD PATTERN* window as in Figure 9.17.

Step 8: Mode Field Diameter and Effective Area Calculation   In the Viewer menu, select *Mode field and Effective Area* to activate *Mode field/Effective area* window. Specify the wavelength range of interests. In this example we selected the wavelength range *Start*: 1.2 µm and *End*: 1.7 µm. Clicking *Calculate* button the mode field diameter and effective area are obtained in the specified spectral range as in Figure 9.18.

Besides these fundamental properties, bending loss and splicing loss can be also estimated by OFACAD.

9.3.2 Simulation Examples for a Dispersion-Flattened Fiber

In the following, we will show the simulation results for a dispersion-flattened fiber (DFF), which is discussed in Section 4.3, whose refractive index profile is shown in Figure 9.19.

Modal Analysis of DFF   As shown in Figure 9.19, the refractive index profile is composed of rectangular step-index segments. Therefore, we run *OFACAD_EDU* program as in the case of SM-28 described in the previous section.

The first step of modal analysis is to set up the refractive index profile. As described in Figure 9.11, refractive index profile data are defined in the *Index Profile Setup*

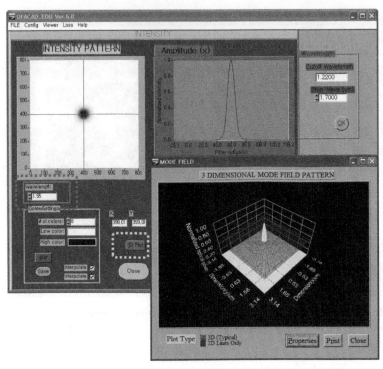

FIGURE 9.17 *Step 7* for SMF-28 analysis. Modal intensity pattern of the HE_{11} mode is calculated at the chosen wavelength defined on the *INTENSITY* window of the *Viewer* menu.

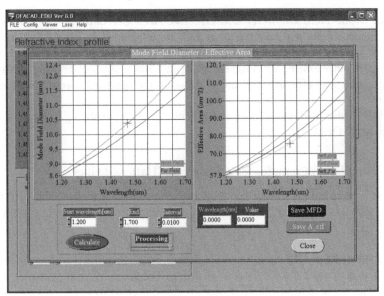

FIGURE 9.18 *Step 8* for SMF-28 analysis. Mode field diameter and effective area of the HE_{11} mode are calculated within the chosen wavelength range defined on the *MODE FIELD DIAMETER AND EFFECTIVE AREA* window of the *Viewer* menu.

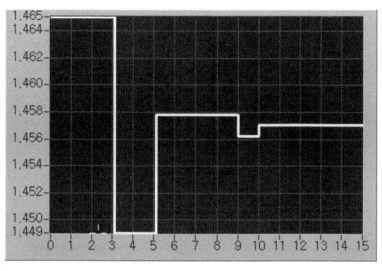

FIGURE 9.19   Refractive index profiles of a DFF. The x and y axes are the radial positions in micrometers and the refractive index is evaluated at $\lambda = 632.8$ nm.

window. The TE_{01} mode cutoff wavelength was calculated to be $1.200\,\mu m$, which ensures the single-mode operation in O, E, S, C, and L bands.

For the refractive index profile in Figure 9.20, the group delay and chromatic dispersion are calculated and the results are shown in Figure 9.21a and b. We have zero chromatic dispersion near $1.58\,\mu m$ and its slope is $\sim 0.06\,ps/(nm^2\,km)$. The dispersion

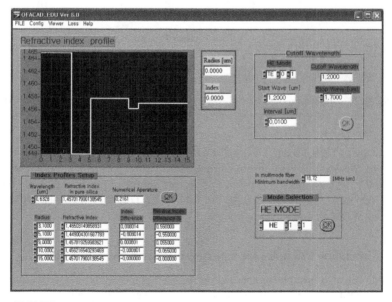

FIGURE 9.20   Setting up the refractive index profile of a DFF in OFACAD.

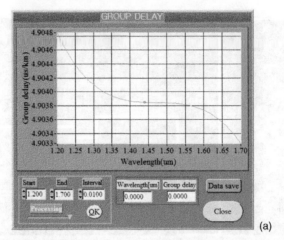

(a)

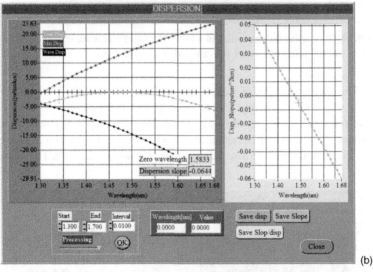

(b)

FIGURE 9.21 Simulation results for the rectangular index NZ-DSF: (a) group delay and (b) chromatic dispersion, dispersion slope.

and its slope at 1550 nm are ~ -1 ps/(nm km) and ~ 0.02 ps/(nm^2 km), respectively. Therefore, this specific fiber design will provide a small chromatic dispersion whose magnitude is less than ~ 5 ps/(nm km) for entire $O + S + C + L$ bands (1260–1625 nm), which are defined in Table 2.8. This DFF can be utilized in wideband WDM systems as described in Chapter 4.

9.3.3 Simulation Examples for a Nonzero Dispersion-Shifted Fiber

OFACAD can also simulate nonzero DSFs with a high accuracy, and we will discuss step-index DSF, whose refractive index profile is given in Figure 9.22.

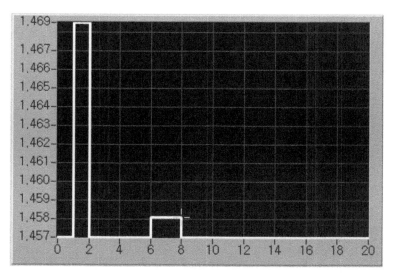

FIGURE 9.22   Refractive index profiles of a nonzero dispersion-shifted fiber. The x and y axes are the radial positions in micrometers and the refractive index is evaluated at $\lambda = 632.8$ nm.

The refractive index profile of nonzero dispersion-shifted fiber (NZ-DSF) is composed of rectangular step-index segments (Figure 9.23). We run *OFACAD_EDU* program as in the case of SM-28 described in the previous section. The first step of modal analysis is to set up the refractive index profile. As described in Figure 9.11, refractive index profile data are defined in the *Index Profile Setup* window. The cutoff

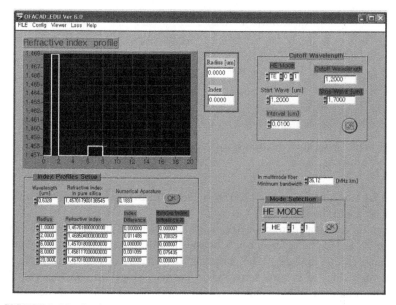

FIGURE 9.23   Setting up the refractive index profile of an NZ-DFS in OFACAD.

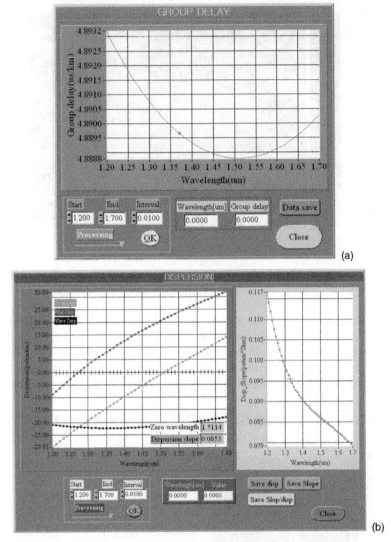

FIGURE 9.24 Simulation results for the NZ-DSF: (a) group delay and (b) chromatic dispersion, dispersion slope.

wavelength was calculated to be 1200 nm, which ensures the single-mode operation in O, E, S, C, and L bands.

The group delay and chromatic dispersion are calculated and the results are shown in Figure 9.24a and b. The zero-dispersion wavelength is near 1.51 μm located outside the C band (1530–1655 nm). The chromatic dispersion has a nonzero finite value in the range of 3–8 ps/(nm km) and a positive slope of $\sim$0.08 ps/(nm^2 km), which can provide a WDM window covering both C and L bands, as described in Section 4.3.

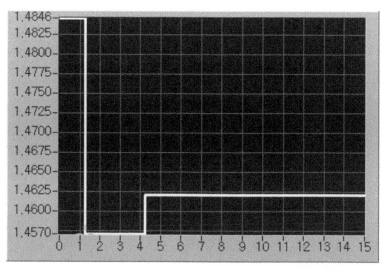

FIGURE 9.25 Refractive index profiles of a dispersion-compensating fiber. The *x* and *y* axes are the radial positions in micrometers and the refractive index is evaluated at $\lambda = 632.8$ nm.

9.3.4 Simulation Examples for a Dispersion-Compensating Fiber

OFACAD can also simulate DCF that is described in Section 4.4. We will discuss a step-index DCF, whose refractive index profile is given in Figure 9.25.

The refractive index profile of DCF is composed of rectangular step-index segments (Figure 9.26). We run *OFACAD_EDU* program as in the case of SM-28

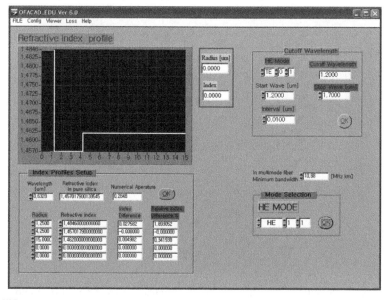

FIGURE 9.26 Setting up the refractive index profile of a rectangular index DCF in OFACAD.

described in the previous section. The first step of modal analysis is to set up the refractive index profile. As described in Figure 9.11, refractive index profile data are defined in the *Index Profile Setup* window. The cutoff wavelength was calculated to be 1200 nm, which ensures the single-mode operation in O, E, S, C, and L bands.

The group delay and chromatic dispersion are calculated and the results are shown in Figure 9.27a and b. In C band (1530–1655 nm) the chromatic dispersion has a large negative value of -120 to -100 ps/(nm km) and a negative slope of -0.37 to

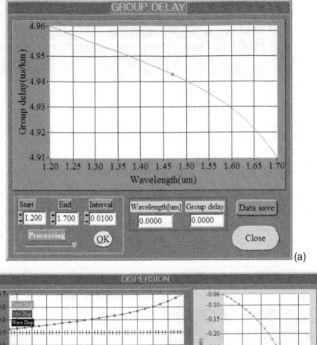

(a)

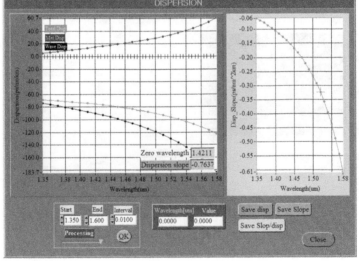

(b)

FIGURE 9.27 Simulation results for the rectangular index DCF: (a) group delay and (b) chromatic dispersion, dispersion slope.

$- 0.52$ ps/(nm^2 km), which can compensate dispersion and dispersion slope of SMFs, as described in Section 4.4.

REFERENCES

[1] D. Marcuse, "Theory of dielectric optical waveguides," 2nd edition, Academic Press, 1991.
[2] A. W. Snyder and J. D. Love, "Optical waveguide theory," Kluwer Academic Publishers, 1983.
[3] K. Okamoto, "Fundamentals of optical waveguides," Academic Press, 2006.
[4] M. L. Calvo and V. Lakshminarayanan, "Optical waveguides," CRC Press, 2007.
[5] OptiFiber™, Optiwave, http://www.optiwave.com/products/fiber.html.
[6] Fiber Optic Mode Solver TM, Apollo Photonics, http://www.apollophoton.com/apollo/page.php?id=27
[7] H. S. Seo, "Parametric studies on optical fibers for Raman amplifiers and double clad fiber lasers and their applications in optical communications," Ph.D. Dissertation, Gwangju Institute of Science and Technology, Gwangju, Republic of Korea, 2001.
[8] R. Kress, "Numerical analysis," Springer, 1998.

OFACAD Installation/ Operation Manual

A.1   INSTALLATION

The simulation program referenced in this book may be found at the following ftp site: ftp://ftp.wiley.com/public/sci_tech_med/optical_fiber. On the site, you will find files shown in Figure A.1.

Click the "setup" icon to install OFACAD in your computer. Then the installation window will pop up as shown in Figure A.2.

Select an appropriate directory for the program and click "Finish." Then you will have the message shown in Figure A.3, which indicates the OFACAD program is successfully installed. Your computer should be rebooted to complete the installation.

Along with the message in Figure A.3, the OFACAD directory will show in your PC screen as in Figure A.4.

A.2   OPERATION OF OFACAD

After rebooting the computer, click the OFACAD_EDU icon in the designated directory as in Figure A.4. Then the program will start with the monitor optimization notice as shown in Figure A.5. Click "OK" to proceed.

A.2.1   Menu Bar

On the top of OFACAD, there is a menu bar where operation menus such as "FILE," "Config," "Viewer," "Loss," and "Help" are listed (Figure A.6).

Silica Optical Fiber Technology for Devices and Components: Design, Fabrication, and International Standards, First Edition. By Kyunghwan Oh and Un-Chul Paek.
© 2012 John Wiley & Sons, Inc. Published 2012 by John Wiley & Sons, Inc.

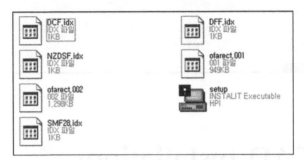

FIGURE A.1   Files in the OFACAD program CD.

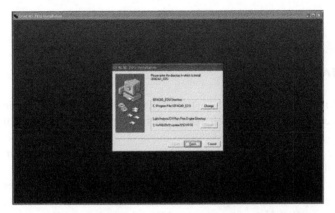

FIGURE A.2   Installation window for OFACAD where the program directories are defined.

FIGURE A.3   Message for successful installation of OFACAD.

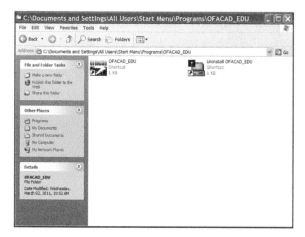

FIGURE A.4 OFACAD program in the designated directory.

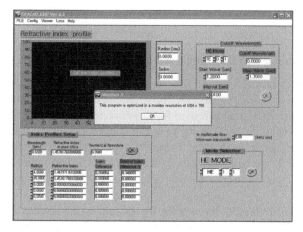

FIGURE A.5 Monitor resolution notice.

FIGURE A.6 Menu bar in the OFACAD.

In the FILE menu, you will find the following commands: "New," "Load index profile," "Save index profile," and "Quit." The command "New" will clear index profile parameters. The command "Load index profile" will load a refractive index profile and the "Save index profile" command will save the index profile into a file in the format *.idx.

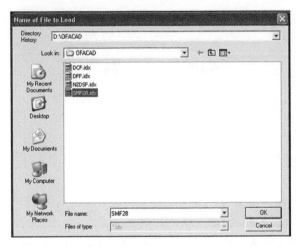

FIGURE A.7   Loading a refractive index profile file from CD. Here we selected SMF28—a conventional single-mode optical fiber.

A.2.2   File Loading and Defining a Refractive Index Profile

Click "Load index profile," and set the directory to the CD drive. In the CD you will find four refractive index profiles: DCF.idx, DFF.idx, NZDSF.idx, and SMF28.idx (Figure A.7). Here DCF is an acronym for dispersion-compensating fiber, DFF for a dispersion-flattened fiber, NZDSF for a nonzero dispersion-shifted fiber, and SMF28 for a conventional single-mode fiber. These fibers are discussed in detail in Chapters 2 and 4. In Chapter 9, you will find more explanation of DCF, DFF, and NZ-DSF. In this appendix, we will confine our discussion only to SMF28.idx—a conventional single-mode fiber.

Choose the "SMF28.idx" file among the sample index profiles and then click "OK." The file is loaded into program and you will find the detailed parameters in the "Index Profile Setup" on the main page of OFACAD as shown in Figure A.8. Note that radial coordinate, "Radius," is given in micrometers. The refractive index at each radial segment is referenced to that of pure silica glass calculated by Sellmeir equation, which is described in Table 2.5. In this program, we will assume that the higher refractive index is achieved by doping GeO_2 in silica glass, and the lower refractive index by adding F. In this assumption, we will use the refractive index formulae for a binary silica glass, similar to equation NaN for GeO_2–SiO_2 and F–SiO_2 glasses. The refractive indices are evaluated at $\lambda = 0.6328\,\mu m$, the HeNe laser wavelength.

When you click OK in the "Index Profile Setup" menu as in Fig. A8, the index profile is plotted as in Fig.A9.

Now you are ready to start mode analysis, and the OFACAD program asks you to select the mode of interests as shown in the notice "Select Mode" in Figure A.9.

Index Profiles Setup

Wavelength [um]	Refractive index in pure silica	Numerical Aperature	
0.6328	1.457017900138545	0.1203	OK

Radius	Refractive index	Index Difference	Relative Index Difference %
4.0000	1.46197200000000	0.004954	0.340016
62.5000	1.457018000000000	0.000000	0.000007
0.0000	0.000000000000000	0.000000	0.000000
0.0000	0.000000000000000	0.000000	0.000000
0.0000	0.00000000000000	0.000000	0.000000

FIGURE A.8 "Index Profile Setup" screen, where the radial positions and corresponding refractive indices are defined. In this example, it is the refractive index profile of SMF28—conventional single-mode fiber.

A.2.3 Mode Selection and Cutoff Wavelength Calculation

You can select one of HE_{lm}, TE_{lm}, TM_{lm}, and EH_{lm} modes to proceed further in the modal analysis using OFACAD. Here, as in Figure A.10, we will select the fundamental mode "HE_{11}," which is the same as the LP_{01} mode in the conventional single-mode fiber.

After selecting the mode of interest, we need to check the cutoff wavelengths of higher order modes. In the "Cutoff Wavelength" window, select one of higher order modes from HE_{lm}, TE_{lm}, TM_{lm}, and EH_{lm} (Figure A.11). Note that in weakly guiding approximation, the HE_{11} mode does not have the cutoff. In this example, we will choose the "TE_{01}" mode to find its cutoff wavelength. Note that TE_{01}, TM_{01}, and HE_{21} modes consist of the LP_{11} mode in a weakly guiding approximation and the cutoff wavelength of TE_{01} mode in this case is equivalent to that of LP_{11} mode. Define the

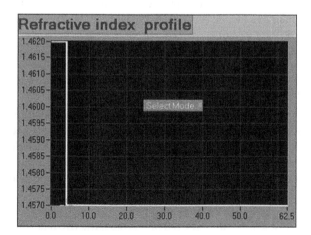

FIGURE A.9 Refractive index profile SMF.idx.

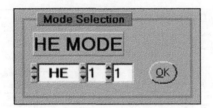

FIGURE A.10 "Mode Selection" window to select a mode of interest that is guided by the optical fiber in the given refractive index profile.

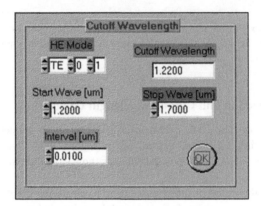

FIGURE A.11 "Cutoff Wavelength" window.

wavelength range to locate the cutoff wavelength by choosing "Start Wave" and "Stop Wave," which define the lower and upper limits of wavelength range, respectively.

When you click "OK," the cutoff wavelength will appear in the unit of micrometers. For the given refractive index profile, SMF28.idx, shown in Figure A.9, the cutoff wavelength of TE_{01} mode is calculated to be 1.22 µm. This ensures the single-mode guidance at $\lambda > 1.22$ µm.

A.2.4 Plotting the Effective Index of the Selected Mode Over the Refractive Index Profile

After selecting the mode of interest and confirming the cutoff wavelength of the higher order modes, we can preview the effective index of the selected mode in comparison to the given refractive index profile in OFACAD. This could be achieved by the "Material Property" command in "Config" menu. Choose a wavelength in the selection box, shown in red dotted line. Click "OK," and then three colored plots are generated as in Figure A.12.

In Figure A.12, three colored lines are plotted at a selected wavelength. In this example, we chose $\lambda = 1.310$ µm. The red plot is the refractive index profile, the blue line is the effective index of the HE_{11} mode, and the green line is the refractive index of

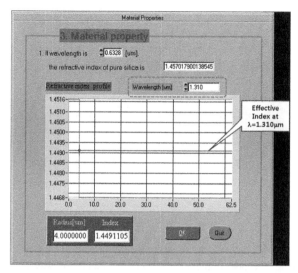

FIGURE A.12 Comparison of the refractive index profile and the effective index of the selected mode at a given wavelength.

pure silica cladding. Note that all the values in these plots are evaluated at the given wavelength. Note that the effective index lies between the refractive index of core and that of cladding.

A.2.5 Calculating the Effective Index as a Function of Wavelength

The core of the modal analysis in optical fibers is to find the effective index of a mode as a function of wavelength, $n_{eff}(\lambda)$, from which various propagation parameters can be derived. OFACAD provides a window to plot the calculated effective index. Choose "Effective Index" command from "Viewer" menu. Then the "EFECTIVE INDEX" window pops out as in Figure A.13.

Define the spectral range to calculate the effective index by specifying "Start," "End," and "Interval." In this example, we chose the range from 1.2 to 1.7 μm with the calculation interval of 0.01 μm. Click "OK" and the bar in "Processing" will move from left to right indicating the program is calculating the effective index. Then the plot of the effective index as a function of wavelength is generated as in Figure A.13. You can store the data as a text file by clicking "Data save."

A.2.6 Calculating the Group Delay as a Function of Wavelength

The group delay of a mode is calculated from the effective index as given by equations 2.104 and 2.105:

$$n_g = n_{eff} + k \frac{dn_{eff}}{dk} = n_{eff} - \lambda \frac{dn_{eff}}{d\lambda} \tag{2.104}$$

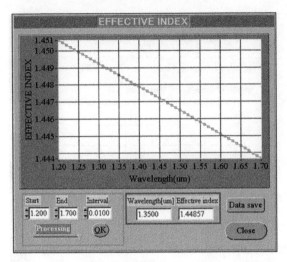

FIGURE A.13   Effective index of the selected mode as a function of wavelength in a specified spectral range.

$$\tau_g = \frac{L}{v_g} = \frac{Ln_g}{c} = \frac{L\,d\beta}{c\,dk} = \frac{L}{c}\left[n_{\text{eff}} - \lambda\frac{dn_{\text{eff}}}{d\lambda}\right] \tag{2.105}$$

Conventionally group delays are evaluated for $L = 1\,\text{km}$ and OFACAD is also calculated as group delay per unit length (μs/km).

Choose "Group Delay" command from "Viewer" menu. Then the "GROUP DELAY" window pops out as in Figure A.14.

Define the spectral range to calculate the group delay by specifying "Start," "End," and "Interval." In this example, we chose the range from 1.2 to 1.7 μm with the calculation interval of 0.01 μm. Click "OK" and the bar in "Processing" will move from left to right indicating the program is calculating the group delay. Then the plot of the group delay as a function of wavelength is generated as in Figure A.14. You can store the data as a text file by clicking "Data save."

A.2.7   Calculating Chromatic Dispersion

Chromatic dispersion, $D(\lambda)$, is one of the most important optical parameters for applications in optical communications and nonlinear optics, and it is defined as in equation 2.116:

$$D(\lambda) = \frac{1}{L}\frac{d\tau_g}{d\lambda} = -\frac{\lambda}{c}\frac{d^2 n_{\text{eff}}}{d\lambda^2} \tag{2.116}$$

The dispersion slope is defined as follows:

$$\text{Dispersion slope} = \frac{dD(\lambda)}{d\lambda}$$

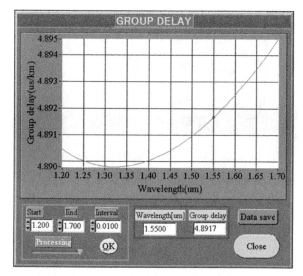

FIGURE A.14   Group delay of the selected mode as a function of wavelength in a specified spectral range.

Dispersion and dispersion slope are given in the unit of ps/(nm km) and ps/(nm^2 km), respectively.

Choose "Dispersion" command from "Viewer" menu. Then the "DISPERSION" window pops out as in Figure A.15.

Define the spectral range to calculate the dispersion and its slope by specifying "Start," "End," and "Interval." In this example, we chose the range from 1.2 to 1.7 μm with the calculation interval of 0.01 μm. Click "OK" and the bar in "Processing" will move from left to right indicating the program is calculating the dispersion and its slope. Then the plot of the dispersion and its slope as a function of wavelength are

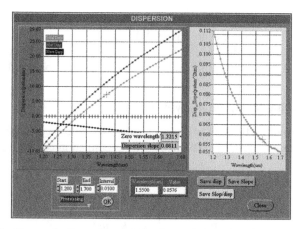

FIGURE A.15   Chromatic dispersion and its slope of the selected mode as a function of wavelength in a specified spectral range.

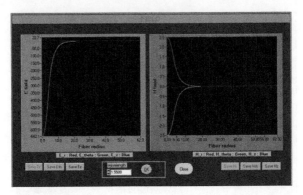

FIGURE A.16 Calculation of electromagnetic field components of the selected mode at a specified wavelength.

generated as in Figure A.15. You can store the data as a text file by clicking "Save disp" or "Save Slope" for chromatic dispersion and its slope, respectively. In the dispersion, there are three colored plots: black for waveguide dispersion, blue for material dispersion, and red for total chromatic dispersion, which is explained in Section 2.1.

A.2.8 Calculating Field Distribution at a Given Wavelength

The electric and magnetic components of the selected mode can be calculated using OFACAD. Choose "Field" command from "Viewer" menu. Then the "FIELD" window pops out as in Figure A.16. These fields are calculated based on Table 2.3 for the given propagation parameters in equations 2.49, 2.53, and 2.101, as follows:

$$\beta = n_{\text{eff}}k = \frac{2\pi n_{\text{eff}}}{\lambda}, \qquad u^2 = n_1^2 k^2 - \beta^2, \qquad w^2 = \beta^2 - n_2^2 k^2$$

When you select a certain wavelength and click "OK," the fields are plotted as a function of radial coordinate as in Figure A.16.

In this example, we set the wavelength at 1.55 μm. Each component of electric and magnetic fields is plotted in a different color, and the data can be saved individually.

A.2.9 Calculating Modal Intensity Profile

From the electromagnetic field, we can calculate the intensity distribution of the selected mode using OFACAD. The intensity I is defined in equation 2.97:

$$\left\langle \left| \vec{S} \right| \right\rangle = \langle S\hat{z} \rangle = I = \frac{1}{2\eta} \left| E_x^0 \right|^2 = \begin{cases} \dfrac{n_1}{2\eta_0} \dfrac{J_0^2(ur)}{J_0^2(ua)} |E_0|^2 & (r \le a) \\[3mm] \dfrac{n_2}{2\eta_0} \dfrac{K_0^2(ur)}{K_0^2(ua)} |E_0|^2 & (r \ge a) \end{cases} \qquad (2.97)$$

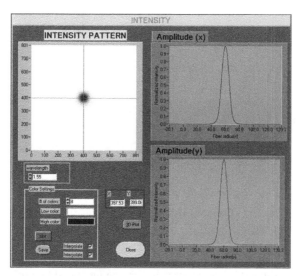

FIGURE A.17   Calculation and 2D plotting of the modal intensity of the selected mode at a specified wavelength.

Choose "Intensity" command from "Viewer" menu. Then the "INTENSITY" window pops out as in Figure A.17. Set the "wavelength" and click "plot"; then the INTENSITY PATTERN window will show a two-dimensional plot of the modal intensity distribution. In this example, we chose $\lambda = 1.55\,\mu$m and the intensity of the HE_{11} mode is presented in Figure A.17. Two windows on the right-hand side are line-intensity profiles along x and y axes. When you click "3D Plot," OFACAD generates a three-dimensional graph of the modal intensity distribution as in Figure A.18.

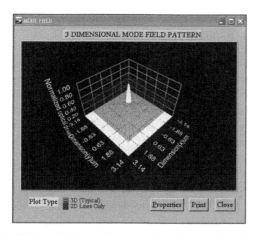

FIGURE A.18   3D plotting of the modal intensity of the selected mode at a specified wavelength.

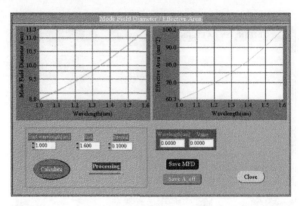

FIGURE A.19   Calculation of mode field diameter and effective area of the selected mode at a specified wavelength range.

3D plot can be rotated by the mouse, and can be increased by pushing down Ctrl key on the keyboard.

A.2.10   Calculating Mode Field Diameter and Effective Area

Mode field diameter is an important parameter in fiber optics and it is related to bending loss and splice loss. Mode field diameter and effective area definitions are explained in Section 2.3.

Choose "MFD and Effective Area" command from "Viewer" menu. Then the "Mode Field Diameter/Effective Area" window pops out as in Figure A.19. Set the "Start wavelength," "End," and "Interval." Click "Calculate"; then the window will show mode field diameter and effective area of the HE_{11} mode in the specified wavelength range.

A.2.11   Calculating Bending Loss

When an optical fiber is bent in a loop, the lateral refractive index profile is deformed to effectively raise the refractive index of cladding outward, and the light guided along the core leaks out to cladding to result in a bending loss. There are two types of bending loss: macrobending and microbending loss. In this example of OFACAD, we will focus only on macrobending loss whose bending radius is of order of millimeters and larger. Discussion on bending loss is given in Section 8.4. See Refs [53,54] of Chapter 8.

Choose "Bending" command from "Loss" menu. Then the "BENDING LOSS" window pops out as in Figure A.20. In the "Macro Bending Loss" box, set the "Bending Radius." Set the "Start wavelength," "End," and "Interval." Click "Calculate"; then the window will show macrobending loss as a function of wavelength for the HE_{11} mode. In this example, the bending loss abruptly increases near $1.5\,\mu m$.

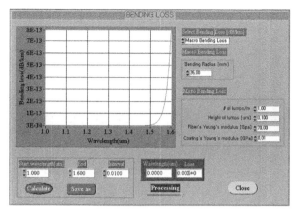

FIGURE A.20 Calculation of macrobending loss of the selected mode at a specified wavelength range for a given bending radius.

A.2.12   Calculating Splicing Loss

When two different optical fibers are spliced together, splicing loss can be generated due to mismatch in the mode field diameter, or offset between the fiber axes, angular misalignment, and the index matching oil. Using OFACAD we can estimate the splice loss between the designed fiber and an arbitrary fiber at different conditions.

Choose "Splicing" command from "Loss" menu. Then the "SPLICING LOSS" window pops out as in Figure A.21. In the "Splicing Loss" box, set the "MFD" of a fiber that is to be spliced with the designed fiber. Set the "Matching Oil Index." You

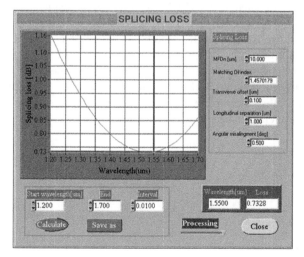

FIGURE A.21 Calculation of splicing loss of the designed fiber with a fiber of an arbitrary mode field diameter (MFD) for various offset conditions.

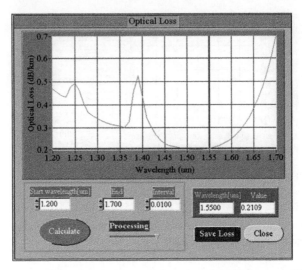

FIGURE A.22 Calculation of optical loss of the selected mode at a specified wavelength range.

can put the silica refractive index in the case of fusion splicing. Set the "Transverse offset," "Longitudinal separation," and "Angular misalignment." Set the "Start wavelength," "End," and "Interval." Click "Calculate"; then the window will show the splice loss of the HE_{11} mode in the specified wavelength range.

A.2.13   Calculating Optical Loss

Fiber loss can be estimated in OFACAD. The origin of optical loss in optical fiber is explained in detail in Section 2.3. The loss consists of contributions from UV absorption, Rayleigh scattering, IR absorption, and OH radical absorption (see equations 2.139–2.141).

Choose "Optical Loss" command from "Loss" menu. Then the "Optical Loss" window pops out as in Figure A.22. Set the "Start wavelength," "End," and "Interval." Click "Calculate"; then the window will show the loss spectrum of the HE_{11} mode in the specified wavelength range.

Operation Manual of OFACAD

B.1   CONVENTIONAL STEP INDEX SINGLE MODE FIBER SIMULATION

As shown in Figure B.1, choose "SMF28.idx" from the default index profiles and then click "OK." In OFACAD_EDU main page, click "OK" from "Index Profiles Setup" that is shown in Figure B.2.

Related refractive index profile will appear as shown in Figure B.3.

As indicated in Figure B.3, mode should be selected now. This program gives the facility to choose HE, TE, TM, and EH modes. Let us choose HE_11, which is the fundamental mode, as shown in Figure B.4, and then click "OK."

Now it is the appropriate time to check for the cutoff wavelength. The wavelength at which a mode ceases to propagate is called the cutoff wavelength for that mode. Usually just TE_01 mode is checked for cutoff wavelength. Click "OK" button (Figure B.5).

B.1.1   Material Property

Choose "Material property" from "Config" in the menu bar. Press "OK" button. The plot is demonstrated in Figure B.6. By putting wavelength in the upper part, refractive index profile of pure silica will be defined. Here for wavelength equal to 0.6328 µm, refractive index of pure silica is 1.4570179.

B.1.2   Effective Index versus Wavelength (µm)

Choose "Effective Index" from "Viewer" in menu tab. Then press "OK." And then curve of "Effective Index" versus wavelength of related refractive index profile is

Silica Optical Fiber Technology for Devices and Components: Design, Fabrication, and International Standards, First Edition. By Kyunghwan Oh and Un-Chul Paek.
© 2012 John Wiley & Sons, Inc. Published 2012 by John Wiley & Sons, Inc.

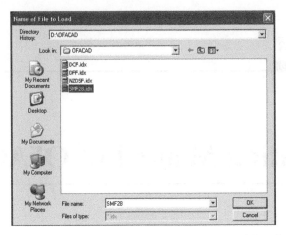

FIGURE B.1 Loading index file (SMF28.idx) for modal analysis.

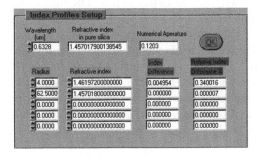

FIGURE B.2 Index profile setting in reference to pure silica glass.

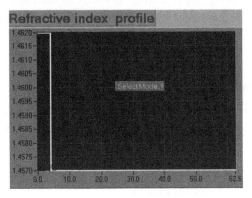

FIGURE B.3 Refractive index profile of the chosen file (SMF28.idx).

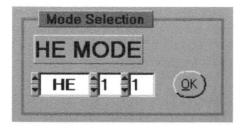

FIGURE B.4 Selecting the mode of interest, the fundamental HE_{11} mode.

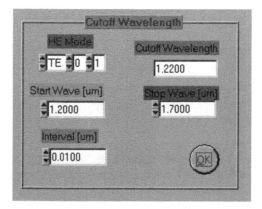

FIGURE B.5 Selecting the higher order mode to find its cut-off wavelength.

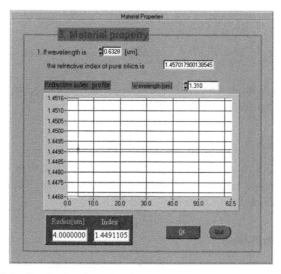

FIGURE B.6 Confirming the refractive index profile at different wavelengths.

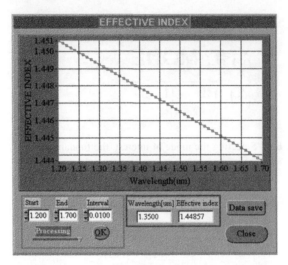

FIGURE B.7 Effective index of the chosen mode in the given spectral range.

plotted. By clicking on each point of the curve, its "wavelength" and also "Effective index" will be shown in the box located at the bottom of Figure B.7.

B.1.3 Group Delay versus Wavelength

Select "Group Delay" from "Viewer" in menu bar. Then click "OK." The "Group Delay" curve versus wavelength will appear. It is worthy to know group delay can be calculated by group index n_g:

$$n_g = n_e - \lambda \frac{dn_e}{d\lambda}, \qquad \text{Group}_{\text{delay}} = \frac{n_g}{c} \quad [\text{ns/km}]$$

By clicking on each point of Figure B.8, the properties of each point of the curve will be shown in the box located at the bottom.

B.1.4 Dispersion

Choose "Dispersion" from "Viewer" in menu bar. Then click "OK."

Dispersion (ps/(nm^2 km)) versus wavelength (μm) and Disp_Slope (ps/(nm^2 km)) versus wavelength (μm) will be plotted (Figure B.9).

B.1.5 Field Distribution

From "Viewer" in menu bar choose "Field." And then click "OK" (Figure B.10).

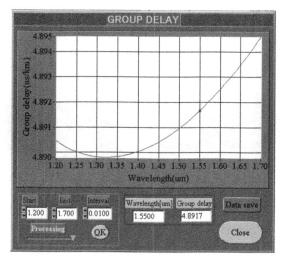

FIGURE B.8 Group delay of the chosen mode in the given spectral range.

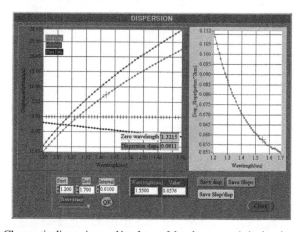

FIGURE B.9 Chromatic dispersion and its slope of the chosen mode in the given spectral range.

B.1.6   Intensity Profile

From "Viewer" in menu bar choose "Intensity" (Figure B.11). And then click "plot."
Press "3D Plot" button; then "3 Dimensional Mode field Pattern" figure will be plotted
that is shown in Figure B.12.

B.1.7   Mode Field Diameter

Mode field diameter (MFD) can be plotted by choosing "MFD and Effective Area"
from "Viewer" in menu tab. Then press "Calculate" button (Figure B.13).

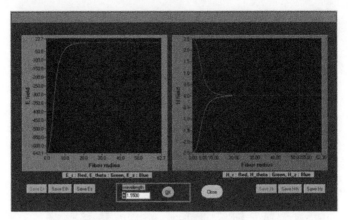

FIGURE B.10 Radial distribution of electric and magnetic field components at the given wavelength.

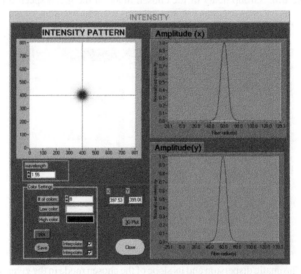

FIGURE B.11 1-D and 2-D Intensity pattern of the chosen mode at the given wavelength.

B.1.8 Optical Loss

To plot optical loss, select "Optical Loss" from "Loss" in menu bar. And click "Calculate" button (Figure B.14).

B.1.9 Bending Loss

Selecting "bending" from "Loss" in menu bar and then clicking "Calculate," Figure B.15 will be demonstrated.

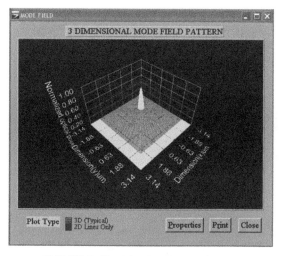

FIGURE B.12   3-D plot of Figure. B.11.

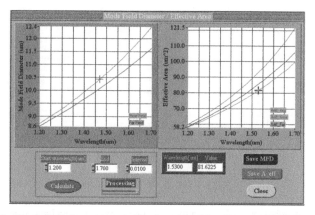

FIGURE B.13   Mode field diameter and effective area of the chosen mode in the given spectral range.

B.1.10   Splicing Loss

Selecting "splicing" from "Loss" in menu bar and then clicking "Calculate," Figure B.16 will be demonstrated.

B.2   NON-ZERO DISPERSION SHIFTED FIBER SIMULATION

As shown in Figure B.17, choose "NZDSF.idx" from the default index profiles and then click "OK." In OFACAD_EDU main page, click "OK" from "Index Profiles Setup" that is shown in Figure B.18.

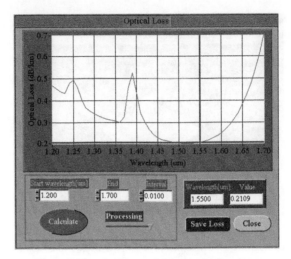

FIGURE B.14 Estimated optical loss of the fiber in the given spectral range.

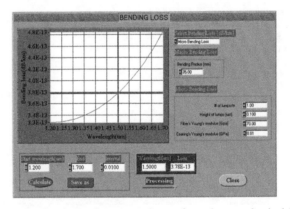

FIGURE B.15 Bending loss in the given spectral range for various mechanical data for the fiber.

Related refractive index profile will appear as shown in Figure B.19.

As indicated in Figure B.19, mode should be selected now. This program gives the facility to choose HE, TE, TM, and EH modes. Let us choose HE_11, which is the fundamental mode, as shown in Figure B.20, and then click "OK."

Now it is the appropriate time to check for the cutoff wavelength. The wavelength at which a mode ceases to propagate is called the cutoff wavelength for that mode. Usually just TE_01 mode is checked for cutoff wavelength (Figure B.21).

B.2.1 Material Property

Choose "Material property" from "Config" in the menu bar. Press "OK" button. The plot is demonstrated in Figure B.22. By putting wavelength in the upper part,

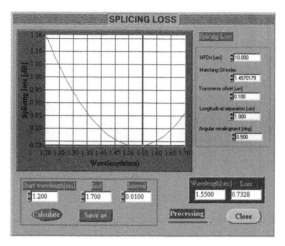

FIGURE B.16   Splice loss in the given spectral range for various alignment parameters.

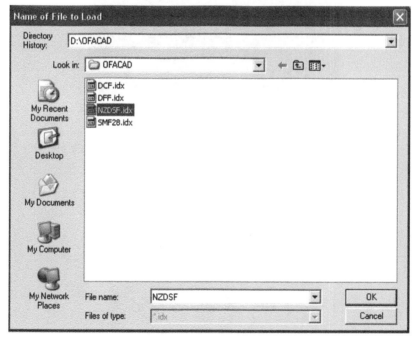

FIGURE B.17   Loading index file (NZDSF.idx) for modal analysis.

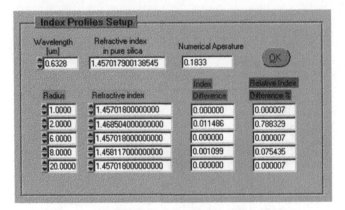

FIGURE B.18 Index profile setting in reference to pure silica glass.

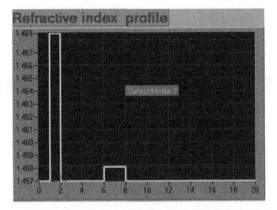

FIGURE B.19 Refractive index profile of the chosen file (NZDSF.idx).

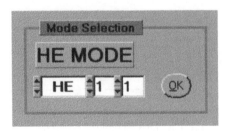

FIGURE B.20 Selecting the mode of interest, the fundamental HE_{11} mode.

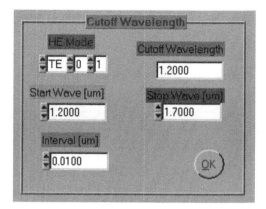

FIGURE B.21   Selecting the higher order mode to find its cut-off wavelength.

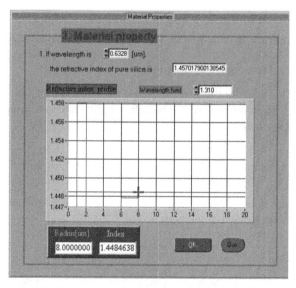

FIGURE B.22   Confirming the refractive index profile at different wavelengths.

refractive index profile of pure silica will be defined. Here for wavelength equal to 0.6328 μm, refractive index of pure silica is 1.4570179.

B.2.2   Effective Index versus Wavelength (μm)

Choose "Effective Index" from "Viewer" in menu tab. Then press "OK." And then curve of "Effective Index" versus wavelength of related refractive index profile is plotted. By clicking on each point of the curve, its "wavelength" and also "Effective index" will be shown in the box located at the bottom of Figure B.23.

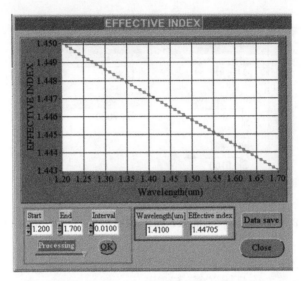

FIGURE B.23 Effective index of the chosen mode in the given spectral range.

B.2.3 Group Delay versus Wavelength

Select "Group Delay" from "Viewer" in menu bar. Then click "OK." The "Group Delay" curve versus wavelength will appear. It is worthy to know group delay can be calculated by group index n_g:

$$n_g = n_e - \lambda \frac{dn_e}{d\lambda}, \qquad \text{Group\_delay} = \frac{n_g}{c} \quad [\text{ns/km}]$$

By clicking on each point of Figure B.24, the properties of each point of the curve will be shown in the box located at the bottom.

B.2.4 Dispersion

Choose "Dispersion" from "Viewer" in menu bar. Then click "OK."

Dispersion (ps/(nm^2 km)) versus wavelength (μm) and Disp_Slope (ps/(nm^2 km)) versus wavelength (μm) will be plotted (Figure B.25).

B.2.5 Field Distribution

From "Viewer" in menu bar choose "Field." And then click "OK" (Figure B.26).

B.2.6 Intensity Profile

From "Viewer" in menu bar choose "Intensity" (Figure B.27). And then click "plot." Press "3D Plot" button; then "3 Dimensional Mode field Pattern" figure will be plotted that is shown in Figure B.28.

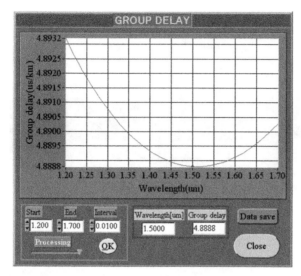

FIGURE B.24 Group delay of the chosen mode in the given spectral range.

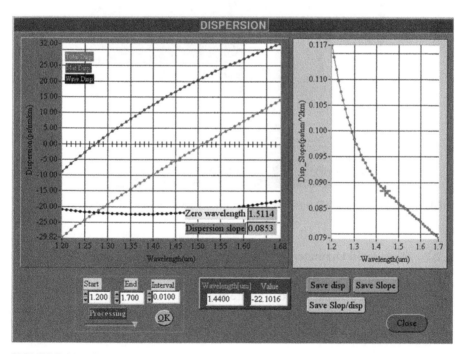

FIGURE B.25 Chromatic dispersion and its slope of the chosen mode in the given spectral range.

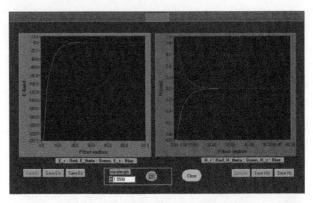

FIGURE B.26 Radial distribution of electric and magnetic field components at the given wavelength.

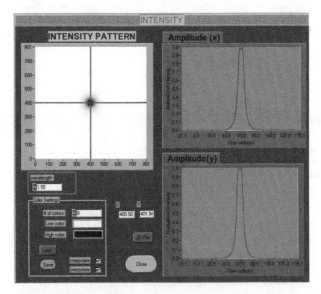

FIGURE B.27 1-D and 2-D Intensity pattern of the chosen mode at the given wavelength.

B.2.7 Mode Field Diameter

MFD can be plotted by choosing "MFD and Effective Area" from "Viewer" in menu tab. Then press "Calculate" button (Figure B.29).

B.2.8 Optical Loss

To plot optical loss, select "Optical Loss" from "Loss" in menu bar. And click "Calculate" button (Figure B.30).

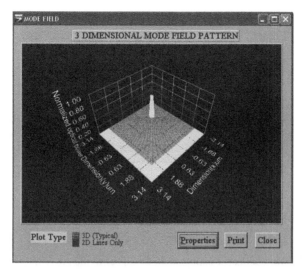

FIGURE B.28   3-D plot of Figure. B.27.

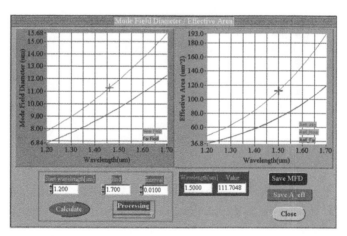

FIGURE B.29   Mode field diameter and effective area of the chosen mode in the given spectral range.

B.2.9   Bending Loss

Selecting "bending" from "Loss" in menu bar and then clicking "Calculate," Figure B.31 will be demonstrated.

B.2.10   Splicing Loss

Selecting "splicing" from "Loss" in menu bar and then clicking "Calculate," Figure B.32 will be demonstrated.

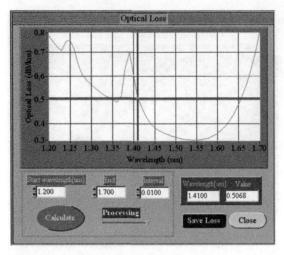

FIGURE B.30 Estimated optical loss of the fiber in the given spectral range.

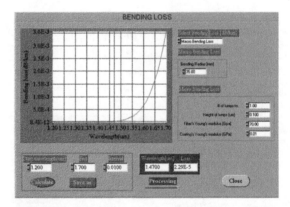

FIGURE B.31 Bending loss in the given spectral range for various mechanical data for the fiber.

B.3 DISPERSION FLATTENED FIBER SIMULATION

As shown in Figure B.33, choose "DFF.idx" from the default index profiles and then click "OK." In OFACAD_EDU main page, click "OK" from "Index Profiles Setup" that is shown in Figure B.34.

Related refractive index profile will appear as shown in Figure B.35.

As indicated in Figure B.35, mode should be selected now. This program gives the facility to choose HE, TE, TM, and EH modes. Let us choose HE_11, which is the fundamental mode, as shown in Figure B.36, and then click "OK."

Now it is the appropriate time to check for the cutoff wavelength. The wavelength at which a mode ceases to propagate is called the cutoff wavelength for that mode.

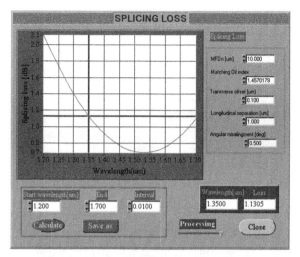

FIGURE B.32   Splice loss in the given spectral range for various alignment parameters.

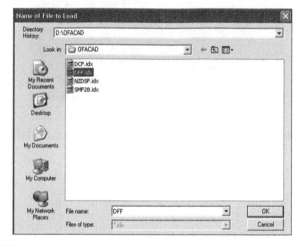

FIGURE B.33   Loading index file (DFF.idx) for modal analysis.

Usually just TE_01 mode is checked for cutoff wavelength. Click "OK" button (Figure B.37).

B.3.1   Material Property

Choose "Material property" from "Config" in the menu bar. Press "OK" button. The plot is demonstrated in Figure B.38. By putting wavelength in the upper part, refractive index profile of pure silica will be defined. Here for wavelength equal to $0.6328\,\mu m$, refractive index of pure silica is 1.4570179.

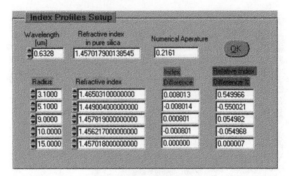

FIGURE B.34   Index profile setting in reference to pure silica glass.

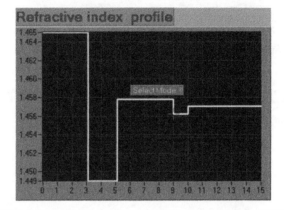

FIGURE B.35   Refractive index profile of the chosen file (DFF.idx).

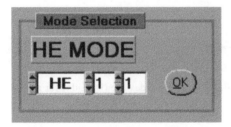

FIGURE B.36   Selecting the mode of interest, the fundamental HE_{11} mode.

B.3.2   Effective Index versus Wavelength (μm)

Choose "Effective Index" from "Viewer" in menu tab. Then press "OK." And then curve of "Effective Index" versus wavelength of related refractive index profile is plotted. By clicking on each point of the curve, its "wavelength" and also "Effective index" will be shown in the box located at the bottom of Figure B.39.

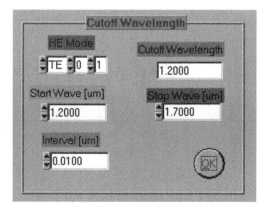

FIGURE B.37   Selecting the higher order mode to find its cut-off wavelength.

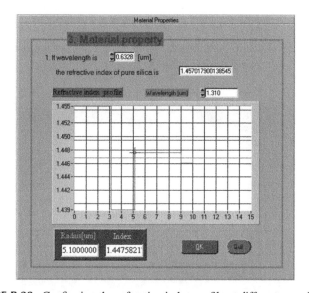

FIGURE B.38   Confirming the refractive index profile at different wavelengths.

B.3.3   Group Delay versus Wavelength

Select "Group Delay" from "Viewer" in menu bar. Then click "OK." The "Group Delay" curve versus wavelength will appear. It is worthy to know group delay can be calculated by group index n_g:

$$n_g = n_e - \lambda \frac{dn_e}{d\lambda}, \qquad \text{Group\_delay} = \frac{n_g}{c} \quad [\text{ns/km}]$$

By clicking on each point of Figure B.40, the properties of each point of the curve will be shown in the box located at the bottom.

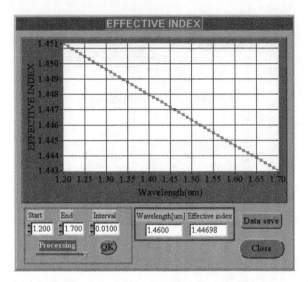

FIGURE B.39   Effective index of the chosen mode in the given spectral range.

FIGURE B.40   Group delay of the chosen mode in the given spectral range.

B.3.4   Dispersion

Choose "Dispersion" from "Viewer" in menu bar. Then click "OK."

Dispersion (ps/(nm^2 km)) versus wavelength (μm) and Disp_Slope (ps/(nm^2 km)) versus wavelength (μm) will be plotted (Figure B.41).

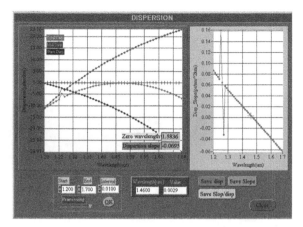

FIGURE B.41 Chromatic dispersion and its slope of the chosen mode in the given spectral range.

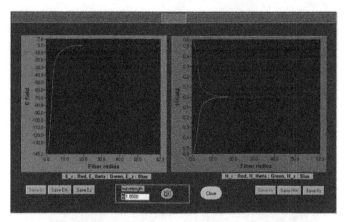

FIGURE B.42 Radial distribution of electric and magnetic field components at the given wavelength.

B.3.5 Field Distribution

From "Viewer" in menu bar choose "Field." And then click "OK" (Figure B.42).

B.3.6 Intensity Profile

From "Viewer" in menu bar choose "Intensity" (Figure B.43). And then click "plot." Press "3D Plot" button; then "3 Dimensional Mode field Pattern" figure will be plotted that is shown in Figure B.44.

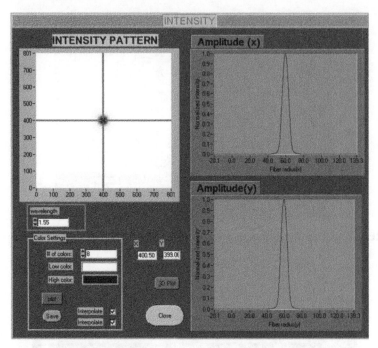

FIGURE B.43 1-D and 2-D Intensity pattern of the chosen mode at the given wavelength.

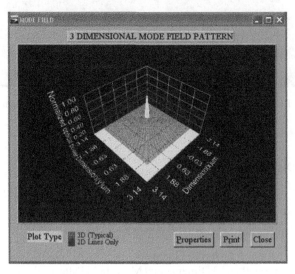

FIGURE B.44 3-D plot of Figure. B.43.

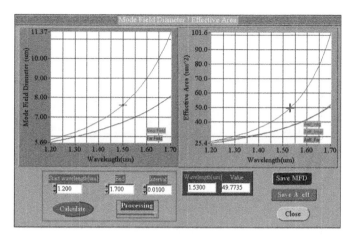

FIGURE B.45   Mode field diameter and effective area of the chosen mode in the given spectral range.

B.3.7   Mode Field Diameter

MFD can be plotted by choosing "MFD and Effective Area" from "Viewer" in menu tab. Then press "Calculate" button (Figure B.45).

B.3.8   Optical Loss

To plot optical loss, select "Optical Loss" from "Loss" in menu bar (Figure B.46).

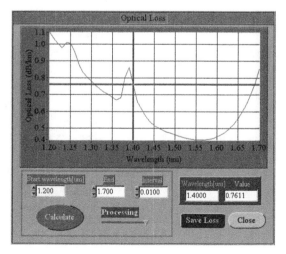

FIGURE B.46   Estimated optical loss of the fiber in the given spectral range.

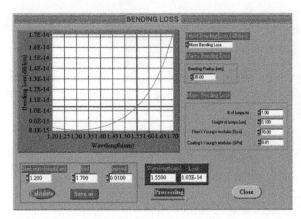

FIGURE B.47   Bending loss in the given spectral range for various mechanical data for the fiber.

B.3.9   Bending Loss

Selecting "bending" from "Loss" in menu bar and then clicking "Calculate," Figure B.47 will be demonstrated.

B.3.10   Splicing Loss

Selecting "splicing" from "Loss" in menu bar and then clicking "Calculate," Figure B.48 will be demonstrated.

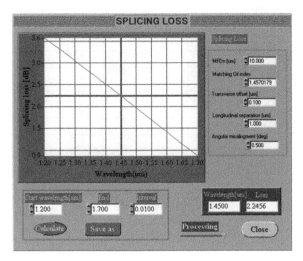

FIGURE B.48   Splice loss in the given spectral range for various alignment parameters.

Name of File to Load
Directory History:
D:\OFACAD
Look in:
OFACAD
DCF.idx
DFF.idx
NZDSF.idx
SMF28.idx
My Recent Documents
Desktop
My Documents
My Computer
My Network Places
File name:
DCF
Files of type:
OK
Cancel

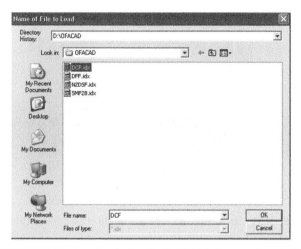

FIGURE B.49 Loading index file (DCF.idx) for modal analysis.

B.4 DISPERSION COMPENSATING FIBER SIMULATION

As shown in Figure B.49, choose "DCF.idx" from the default index profiles and then click "OK." In OFACAD_EDU main page, click "OK" from "Index Profiles Setup" that is shown in Figure B.50.

Related refractive index profile will appear as shown in Figure B.51.

As indicated in Figure B.51, mode should be selected now. This program gives the facility to choose HE, TE, TM, and EH modes. Let us choose HE_11, which is the fundamental mode, as shown in Figure B.52, and then click "OK."

Now it is the appropriate time to check for the cutoff wavelength. The wavelength at which a mode ceases to propagate is called the cutoff wavelength for that mode. Usually just TE_01 mode is checked for cutoff wavelength (Figure B.53).

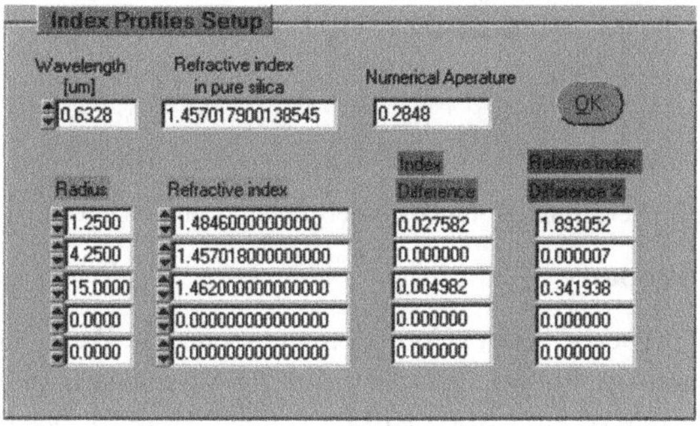

Index Profiles Setup
Wavelength [um]
0.6328
Refractive index in pure silica
1.457017900138545
Numerical Aperature
0.2848
OK
Radius
Refractive index
Index Difference
Relative Index Difference %
1.2500
1.48460000000000
0.027582
1.893052
4.2500
1.457018000000000
0.000000
0.000007
15.0000
1.462000000000000
0.004982
0.341938
0.0000
0.000000000000000
0.000000
0.000000
0.0000
0.000000000000000
0.000000
0.000000

FIGURE B.50 Index profile setting in reference to pure silica glass.

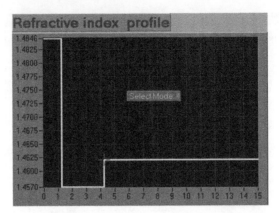

FIGURE B.51 Refractive index profile of the chosen file (DFF.idx).

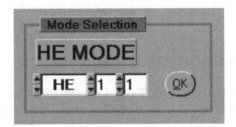

FIGURE B.52 Selecting the mode of interest, the fundamental HE_{11} mode.

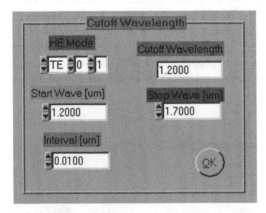

FIGURE B.53 Selecting the higher order mode to find its cut-off wavelength.

B.4.1 Material Property

Choose "Material property" from "Config" in the menu bar. Press "OK" button. The plot is demonstrated in Figure B.54. By putting wavelength in the upper part, refractive index profile of pure silica will be defined. Here for wavelength equal to $0.6328\,\mu m$, refractive index of pure silica is 1.4570179.

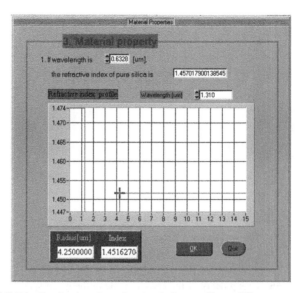

FIGURE B.54   Confirming the refractive index profile at different wavelengths.

B.4.2   Effective Index versus Wavelength (μm)

Choose "Effective Index" from "Viewer" in menu tab. Then press "OK." And then curve of "Effective Index" versus wavelength of related refractive index profile is plotted. By clicking on each point of the curve, its "wavelength" and also "Effective index" will be shown in the box located at the bottom of Figure B.55.

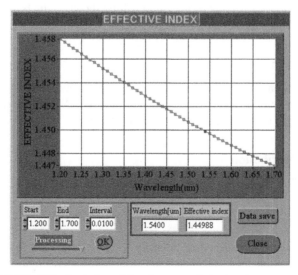

FIGURE B.55   Effective index of the chosen mode in the given spectral range.

FIGURE B.56 Group delay of the chosen mode in the given spectral range.

B.4.3 Group Delay versus Wavelength

Select "Group Delay" from "Viewer" in menu bar. Then click "OK." The "Group Delay" curve versus wavelength will appear. It is worthy to know group delay can be calculated by group index n_g:

$$n_g = n_e - \lambda \frac{dn_e}{d\lambda}, \qquad \text{Group\_delay} = \frac{n_g}{c} \quad [\text{ns/km}]$$

The properties of each point of the curve will be shown in the box located at the bottom, by clicking on each point of Figure B.56.

B.4.4 Dispersion

Choose "Dispersion" from "Viewer" in menu bar. Then click "OK."
Dispersion (ps/(nm² km)) versus wavelength (μm) and Disp_Slope (ps/(nm² km)) versus wavelength (μm) will be plotted (Figure B.57).

B.4.5 Field Distribution

From "Viewer" in menu bar choose "Field." And then click "OK" (Figure B.58).

B.4.6 Intensity Profile

From "Viewer" in menu bar choose "Intensity" (Figure B.59). And then click "plot." Press "3D Plot" button; then "3 Dimensional Mode field Pattern" figure will be plotted that is shown in Figure B.60.

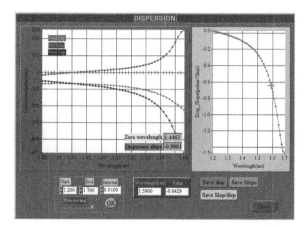

FIGURE B.57 Chromatic dispersion and its slope of the chosen mode in the given spectral range.

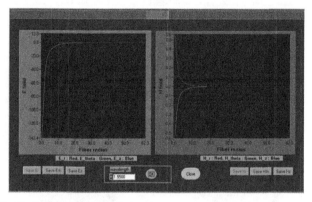

FIGURE B.58 Radial distribution of electric and magnetic field components at the given wavelength.

B.4.7 Mode Field Diameter

MFD can be plotted by choosing "MFD and Effective Area" from "Viewer" in menu tab. Then press "Calculate" button (Figure B.61).

B.4.8 Optical Loss

To plot optical loss, select "Optical Loss" from "Loss" in menu bar. And click "Calculate" button (Figure B.62).

B.4.9 Bending Loss

Selecting "bending" from "Loss" in menu bar and then clicking "Calculate," Figure B.63 will be demonstrated.

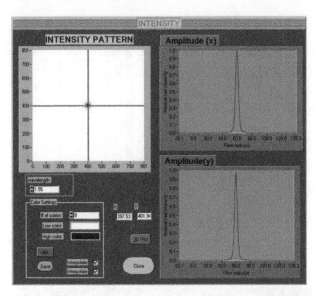

FIGURE B.59 1-D and 2-D Intensity pattern of the chosen mode at the given wavelength.

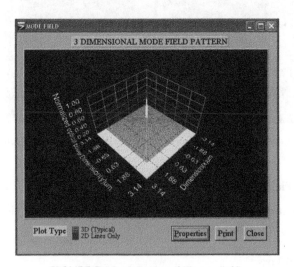

FIGURE B.60 3-D plot of Figure. B.27.

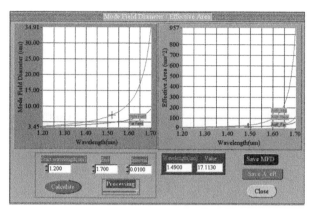

FIGURE B.61   Mode field diameter and effective area of the chosen mode in the given spectral range.

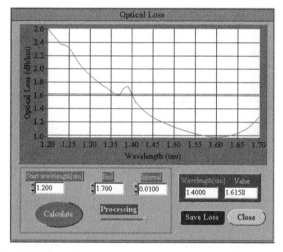

FIGURE B.62   Estimated optical loss of the fiber in the given spectral range.

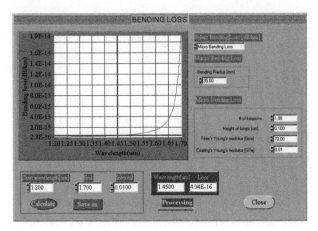

FIGURE B.63   Bending loss in the given spectral range for various mechanical data for the fiber.

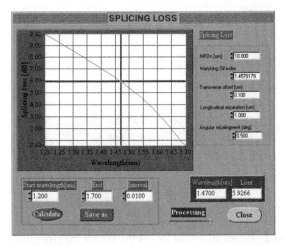

FIGURE B.64   Splice loss in the given spectral range for various alignment parameters.

B.4.10   Splicing Loss

Selecting "splicing" from "Loss" in menu bar and then clicking "Calculate," Figure B.64 will be demonstrated.

Index

air silica holey fiber, ASHF   15, 321
  bend loss edge wavelength, λ_{ble} 334
  bending loss, α_{bend} 334
  effective index guiding, EIG   324
  endless single mode fiber   325
  fabrication   323
  fundamental space filling mode,
    FSM   327
  large mode area, LMA   331
  normalized frequency   328
anti Stokes shift   235
attenuation coefficient, $a(\lambda)$ 2, 54, 65,
  243

beat length, L_p 283
bend insensitive single mode fiber   66, 68,
  71, 78
Bessel function, $Jm(x)$, $Km(x)$ 24, 25
birefringence, B 283
boundary conditions   26
bow-tie fiber   292, 301
Brillouin scattering   235, 253
  bandwidth, $\Delta\nu_B$ 236
  frequency shift, ν_B 236, 254
  gain spectra   256, 260, 264
  longitudinal acoustic veolocity, V_L 254
  peak gain, g_B 236, 254
  threshold power, P_B^{th} 236

bubbler   86

cascaded boundary method, CBM   378
characteristic equation   29
cladding   24, 25
coarse wavelength division multiplexing,
  CWDM   133, 135, 136
collapse   99, 102
communication bands, O, E, S, C, L, U   10,
  66
confinement loss   345
conventional single mode fiber see standard
  single mode fiber
core   24
cross phase modulation, XPM   145
cut-off shifted single mode fiber   66, 68,
  137
cut-off wavelength, λ_c 52

3dB bandwidth   207
defect engineering   363
dense wavelength division multiplexing,
  DWDM   133, 134, 135
deposition efficiency   95
deposition rate   96
differential Group Delay, DGD   62
dispersion compensating fiber, DCF   133,
  139, 161, 338

*Silica Optical Fiber Technology for Devices and Components: Design, Fabrication, and
International Standards*, First Edition. By Kyunghwan Oh and Un-Chul Paek.
© 2012 John Wiley & Sons, Inc. Published 2012 by John Wiley & Sons, Inc.

dispersion compensating module, DCM 133
dispersion compensation using higher order
     mode 167, 174, 178, 181
dispersion flattened fiber, DFF 137, 153, 337
dispersion in optical fiber 4
   chromatic dispersion boundary 72, 77
   chromatic dispersion in single mode
     fiber, D 6, 7, 390
   dispersion penalty 142, 144
   modal dispersion in multi mode fiber 5
   polarization mode dispersion 7, 8, 38, 56,
     57, 59
dispersion penalty 142
dispersion shifted fiber, DSF 65, 66, 137,
     139, 151

effective area, A_{eff} 51
effective index, n_{eff} 35, 388
effective modal bandwidth, EMB 214
effective nonlinearity, γ 229, 347
elliptic clad fiber 292, 299
elliptic core fiber 291, 297
elongation 101, 102
etching 90
ethernet 186, 220

fast axis 285
F-doping 89, 90
fiber Brillouin sensor 267
fiber channel 220
fiber drawing 83, 105
   coating zone 120
   cooling zone 114, 118
   heating zone 106, 108
   neck down 107, 114
   photopolymerization 124
   tension 116
fiber Raman laser, FRL 249
fiber spinning 311
fiber twist 311
figure of merit, FOM 159, 347
flat clad fiber 292
four wave mixing, FWM 145

group delay, t_g 36, 37, 57, 388
group velocity, v_g 35
   optical glasses 44

Helmholtz equation in cylindrical
     coordinate 23

high-birefringence fiber, HBF 294, 305
   fabrication 295
highly nonlinear fiber, HNLF 232
hollow optical fiber, HOF 355

IEC (International Electrotechnical
     Commission) standard 9, 10, 11
   IEC 60793-1-1 9
   IEC 60793-1-20 9, 11
   IEC 60793-1-30 9, 11
   IEC 60793-1-40 9, 11
   IEC 60793-1-42 9, 11
   IEC 60793-1-44 9, 11
   IEC 60793-1-45 9, 11
   IEC 60793-1-47 9, 11
   IEC 60793-1-48 9, 11
   IEC 60793-2 9
   IEC 60793-2-10 A1 10, 215, 216, 217
   IEC 60793-2-20 A2 215, 217
   IEC 60793-2-30 A3 215, 217
   IEC 60793-2-40 A4 215, 217
   IEC 60793-2-50 B1.1 10, 66
   IEC 60793-2-50 B1.2 10, 66
   IEC 60793-2-50 B1.3 10, 66
   IEC 60793-2-50 B2 10, 66
   IEC 60793-2-50 B4 10, 66
   IEC 60793-2-50 B5 10, 66
   IEC 60793-2-50 B6 10, 66
   IEC 60794-1-1 9
   IEC 60794-2 9
   IEC 60794-3 9
IR loss, α_{IR} 54
ISO/IEC 11810 216, 217
ITU-T (International Telecommunication
     Union – Telecommunication)
     standard 9, 10, 11
   G.650 9, 11, 67
   G.651 9, 10
   G.652 9, 10
   G.652.A, B, C, D 66, 68
   G.653 9, 10
   G.653.A, B 66, 71, 72
   G.654 9, 10
   G.654.A, B, C 66, 68
   G.655 9, 10
   G.655.A, B, C, D, E 66, 68, 73, 74,
     76, 77
   G.656 9, 10, 66, 73, 74
   G.657 9, 10, 66, 68

Kramers-Kronig relation   41

low OH fiber   see low water peak fiber
low water peak fiber   65, 66, 67, 68
low-birefringence fiber, LBF   294, 310

material dispersion   41, 42
  pure silica and GeO$_2$-doped glass   45
Maxwell's equations   18
Miller's rule   228, 229
mode converter   360
mode field diameter, MFD, $2w$ 49, 50, 51
modified chemical vapor deposition, MCVD
    86, 87, 89
  chemical reactions   89, 90
  precursors   88
multimode fiber
  differential modal delay, DMD   203, 209
  graded index, GI   190
  laser optimized   212
  modal delay optimized   12
  optimal α, α_{opt} 195, 198
  α-profile   190
  technology evolution   3

nonlinear refractive index, n_2 144, 228,
    230, 231
non-zero dispersion shifted fiber, NZDSF
    65, 66, 67, 137, 139
  Alcatel TeraLight Ultra^{TM} 138, 160
  Corning LEAF^{TM} 138, 160
  Lucent TrueWave REACH^{TM} 138, 160
  Lucent TrueWave RS^{TM} 138, 160
  Sumitomo Pure Guide^{TM} 138, 160
normalized frequency, V 29, 328

OH loss   54, 343
optical fiber analysis computer-aided design,
    OFACAD   378
optical fiber history   1
optical fiber modes   30, 32
  HE, EH, TE, TM mode   30, 37, 176
  LP modes   32, 33, 35, 170
  LP$_{01}$ mode   33, 34
  LP$_{02}$ mode   177
  LP$_{11}$ mode   168, 169
optical power in the core and clad   34
outside vapor deposition, OVD   86, 87, 89

overfilled launch, OFL   187, 188, 204

PANDA fiber   292, 302
phase mismatch, $\Delta\beta$ 147
phase velocity, v_p 35
plasma chemical vapor deposition, PCVD
    86, 87, 89
Pointcaré sphere   286
polarization maintaining fiber, PMF   281,
    294, 352
polarization mode delay, PM delay   59, 60, 61
polarization   280
power ration in the core, P$_{core}$ 349
Poynting vector   34
precursors in MCVD   88
preform   83
preform spinning   311
propagation constant
  β_1, β_2, β_3 140
  longitudinal, β 22, 25, 33, 34, 139
  α-profile multimode fiber   193
  transverse, in the clad, w 24, 29
  transverse, in the core, u 24, 29
propagation equation   145
pulse distortion   140

Raman amplifier   246
Raman scattering   235, 237
  bandwidth, $\Delta\nu_R$ 236
  frequency shift, ν_R 236
  gain spectra   238, 239, 240
  peak gain, g_R 236, 238, 242
  threshold power, P_R^{th} 236
Rayleigh scattering loss, α_R 54, 343
refractive index, n 20, 41, 44
  pure silica and doped silica glass   44
relative dispersion slope, RDS   159, 339
relative index difference, Δ,   29, 46,
  binary silica glasses   48
  GeO$_2$ doped silica glass   47
rod in tube, RIT   103

sealing   99
selective mode launch   187, 188, 204
self phase modulation, SPM   145
Sellmeier coefficients   43
Sellmeier equation   40, 41, 42, 46
side tunnel fiber   291

side-pit fiber 291
silica glass
  density 112
  elastic modulus 112
  linear expansion coefficient 112, 296
  nonlinear refractive index, n_2 144, 230, 231
  refractive index, n 44
  specific heat 112
  surface tension 112
  thermal conductivity 112
simulation examples 386
  conventional single mode fiber 386
  dispersion compensating fiber 397
  dispersion flattened fiber 391
  non-zero dispersion shifted fiber 394
single mode fiber
  birefringence controlled 14
  dispersion tailored 13
  optical loss spectrum 2, 65, 244
  technology evolution 3
single polarization fiber, SPF 293, 304
sintering 96
slow axis 285

standard single mode fiber, 65, 66, 67, 137
stimulated Brillouin scattering
  suppression 264
Stokes shift 235
Stokes's vector, $\vec{S}$ 287
stress applying part, SAP 291, 303, 354

thermoporesis 94
transition metal impurities 84

UV loss, α_{UV} 54

vapor axial deposition, VAD 86, 87, 89
vapor pressure of chloride precursors 84
very short reach, VSR 220
viscosity of glasses 98

wave equation 18
wavelength division multiplexing,
  WDM 133

zero dispersion wavelength, λ_0 57
zero-dispersion slope, S_{0max} 57

WILEY SERIES IN MICROWAVE AND OPTICAL ENGINEERING

KAI CHANG, Editor
Texas A&M University

FIBER-OPTIC COMMUNICATION SYSTEMS, Fourth Edition • *Govind P. Agrawal*

ASYMMETRIC PASSIVE COMPONENTS IN MICROWAVE INTEGRATED CIRCUITS • *Hee-Ran Ahn*

COHERENT OPTICAL COMMUNICATIONS SYSTEMS • *Silvello Betti, Giancarlo De Marchis, and Eugenio Iannone*

PHASED ARRAY ANTENNAS: FLOQUET ANALYSIS, SYNTHESIS, BFNs, AND ACTIVE ARRAY SYSTEMS • *Arun K. Bhattacharyya*

HIGH-FREQUENCY ELECTROMAGNETIC TECHNIQUES: RECENT ADVANCES AND APPLICATIONS • *Asoke K. Bhattacharyya*

RADIO PROPAGATION AND ADAPTIVE ANTENNAS FOR WIRELESS COMMUNICATION LINKS: TERRESTRIAL, ATMOSPHERIC, AND IONOSPHERIC • *Nathan Blaunstein and Christos G. Christodoulou*

COMPUTATIONAL METHODS FOR ELECTROMAGNETICS AND MICROWAVES • *Richard C. Booton, Jr.*

ELECTROMAGNETIC SHIELDING • *Salvatore Celozzi, Rodolfo Araneo, and Giampiero Lovat*

MICROWAVE RING CIRCUITS AND ANTENNAS • *Kai Chang*

MICROWAVE SOLID-STATE CIRCUITS AND APPLICATIONS • *Kai Chang*

RF AND MICROWAVE WIRELESS SYSTEMS • *Kai Chang*

RF AND MICROWAVE CIRCUIT AND COMPONENT DESIGN FOR WIRELESS SYSTEMS • *Kai Chang, Inder Bahl, and Vijay Nair*

MICROWAVE RING CIRCUITS AND RELATED STRUCTURES, Second Edition • *Kai Chang and Lung-Hwa Hsieh*

MULTIRESOLUTION TIME DOMAIN SCHEME FOR ELECTROMAGNETIC ENGINEERING • *Yinchao Chen, Qunsheng Cao, and Raj Mittra*

DIODE LASERS AND PHOTONIC INTEGRATED CIRCUITS • *Larry Coldren and Scott Corzine*

EM DETECTION OF CONCEALED TARGETS • *David J. Daniels*

RADIO FREQUENCY CIRCUIT DESIGN • *W. Alan Davis and Krishna Agarwal*

RADIO FREQUENCY CIRCUIT DESIGN, Second Edition • *W. Alan Davis*

MULTICONDUCTOR TRANSMISSION-LINE STRUCTURES: MODAL ANALYSIS TECHNIQUES • *J. A. Brandão Faria*

PHASED ARRAY-BASED SYSTEMS AND APPLICATIONS • *Nick Fourikis*

SOLAR CELLS AND THEIR APPLICATIONS, Second Edition • *Lewis M. Fraas and Larry D. Partain*

FUNDAMENTALS OF MICROWAVE TRANSMISSION LINES • *Jon C. Freeman*

OPTICAL SEMICONDUCTOR DEVICES • *Mitsuo Fukuda*

MICROSTRIP CIRCUITS • *Fred Gardiol*

HIGH-SPEED VLSI INTERCONNECTIONS, Second Edition • *Ashok K. Goel*

FUNDAMENTALS OF WAVELETS: THEORY, ALGORITHMS, AND APPLICATIONS, Second Edition • *Jaideva C. Goswami and Andrew K. Chan*

HIGH-FREQUENCY ANALOG INTEGRATED CIRCUIT DESIGN • *Ravender Goyal (ed.)*

RF AND MICROWAVE TRANSMITTER DESIGN • *Andrei Grebennikov*

ANALYSIS AND DESIGN OF INTEGRATED CIRCUIT ANTENNA MODULES • *K. C. Gupta and Peter S. Hall*

PHASED ARRAY ANTENNAS, Second Edition • *R. C. Hansen*

STRIPLINE CIRCULATORS • *Joseph Helszajn*

THE STRIPLINE CIRCULATOR: THEORY AND PRACTICE • *Joseph Helszajn*

LOCALIZED WAVES • *Hugo E. Hernández-Figueroa, Michel Zamboni-Rached, and Erasmo Recami (eds.)*

MICROSTRIP FILTERS FOR RF/MICROWAVE APPLICATIONS, Second Edition • *Jia-Sheng Hong*

MICROWAVE APPROACH TO HIGHLY IRREGULAR FIBER OPTICS • *Huang Hung-Chia*

NONLINEAR OPTICAL COMMUNICATION NETWORKS • *Eugenio Iannone, Francesco Matera, Antonio Mecozzi, and Marina Settembre*

FINITE ELEMENT SOFTWARE FOR MICROWAVE ENGINEERING • *Tatsuo Itoh, Giuseppe Pelosi, and Peter P. Silvester (eds.)*

INFRARED TECHNOLOGY: APPLICATIONS TO ELECTROOPTICS, PHOTONIC DEVICES, AND SENSORS • *A. R. Jha*

SUPERCONDUCTOR TECHNOLOGY: APPLICATIONS TO MICROWAVE, ELECTRO-OPTICS, ELECTRICAL MACHINES, AND PROPULSION SYSTEMS • *A. R. Jha*

TIME AND FREQUENCY DOMAIN SOLUTIONS OF EM PROBLEMS USING INTEGTRAL EQUATIONS AND A HYBRID METHODOLOGY • *B. H. Jung, T. K. Sarkar, S. W. Ting, Y. Zhang, Z. Mei, Z. Ji, M. Yuan, A. De, M. Salazar-Palma, and S. M. Rao*

OPTICAL COMPUTING: AN INTRODUCTION • *M. A. Karim and A. S. S. Awwal*

INTRODUCTION TO ELECTROMAGNETIC AND MICROWAVE ENGINEERING • *Paul R. Karmel, Gabriel D. Colef, and Raymond L. Camisa*

MILLIMETER WAVE OPTICAL DIELECTRIC INTEGRATED GUIDES AND CIRCUITS • *Shiban K. Koul*

ADVANCED INTEGRATED COMMUNICATION MICROSYSTEMS • *Joy Laskar, Sudipto Chakraborty, Manos Tentzeris, Franklin Bien, and Anh-Vu Pham*

MICROWAVE DEVICES, CIRCUITS AND THEIR INTERACTION • *Charles A. Lee and G. Conrad Dalman*

ADVANCES IN MICROSTRIP AND PRINTED ANTENNAS • *Kai-Fong Lee and Wei Chen (eds.)*

SPHEROIDAL WAVE FUNCTIONS IN ELECTROMAGNETIC THEORY • *Le-Wei Li, Xiao-Kang Kang, and Mook-Seng Leong*

ARITHMETIC AND LOGIC IN COMPUTER SYSTEMS • *Mi Lu*

OPTICAL FILTER DESIGN AND ANALYSIS: A SIGNAL PROCESSING APPROACH • *Christi K. Madsen and Jian H. Zhao*

THEORY AND PRACTICE OF INFRARED TECHNOLOGY FOR NONDESTRUCTIVE TESTING • *Xavier P. V. Maldague*

METAMATERIALS WITH NEGATIVE PARAMETERS: THEORY, DESIGN, AND MICROWAVE APPLICATIONS • Ricardo Marqués, Ferran Martín, and Mario Sorolla

OPTOELECTRONIC PACKAGING • *A. R. Mickelson, N. R. Basavanhally, and Y. C. Lee (eds.)*

OPTICAL CHARACTER RECOGNITION • *Shunji Mori, Hirobumi Nishida, and Hiromitsu Yamada*

ANTENNAS FOR RADAR AND COMMUNICATIONS: A POLARIMETRIC APPROACH • *Harold Mott*

INTEGRATED ACTIVE ANTENNAS AND SPATIAL POWER COMBINING • *Julio A. Navarro and Kai Chang*

ANALYSIS METHODS FOR RF, MICROWAVE, AND MILLIMETER-WAVE PLANAR TRANSMISSION LINE STRUCTURES • *Cam Nguyen*

LASER DIODES AND THEIR APPLICATIONS TO COMMUNICATIONS AND INFORMATION PROCESSING • *Takahiro Numai*

FREQUENCY CONTROL OF SEMICONDUCTOR LASERS • *Motoichi Ohtsu (ed.)*

SILICA OPTICAL FIBER TECHNOLOGY FOR DEVICE AND COMPONENTS: DESIGN, FABRICATION, AND INTERNATIONAL STANDARDS • *Un-Chul Paek and Kyunghwan Oh*

WAVELETS IN ELECTROMAGNETICS AND DEVICE MODELING • *George W. Pan*

OPTICAL SWITCHING • Georgios Papadimitriou, Chrisoula Papazoglou, and Andreas S. Pomportsis

MICROWAVE IMAGING • *Matteo Pastorino*

ANALYSIS OF MULTICONDUCTOR TRANSMISSION LINES • *Clayton R. Paul*

INTRODUCTION TO ELECTROMAGNETIC COMPATIBILITY, Second Edition • *Clayton R. Paul*

ADAPTIVE OPTICS FOR VISION SCIENCE: PRINCIPLES, PRACTICES, DESIGN AND APPLICATIONS • *Jason Porter, Hope Queener, Julianna Lin, Karen Thorn, and Abdul Awwal (eds.)*

ELECTROMAGNETIC OPTIMIZATION BY GENETIC ALGORITHMS • *Yahya Rahmat-Samii and Eric Michielssen (eds.)*

INTRODUCTION TO HIGH-SPEED ELECTRONICS AND OPTOELECTRONICS • *Leonard M. Riaziat*

NEW FRONTIERS IN MEDICAL DEVICE TECHNOLOGY • *Arye Rosen and Harel Rosen (eds.)*

ELECTROMAGNETIC PROPAGATION IN MULTI-MODE RANDOM MEDIA • *Harrison E. Rowe*

ELECTROMAGNETIC PROPAGATION IN ONE-DIMENSIONAL RANDOM MEDIA • *Harrison E. Rowe*

HISTORY OF WIRELESS • *Tapan K. Sarkar, Robert J. Mailloux, Arthur A. Oliner, Magdalena Salazar-Palma, and Dipak L. Sengupta*

PHYSICS OF MULTIANTENNA SYSTEMS AND BROADBAND PROCESSING • *Tapan K. Sarkar, Magdalena Salazar-Palma, and Eric L. Mokole*

SMART ANTENNAS • *Tapan K. Sarkar, Michael C. Wicks, Magdalena Salazar-Palma, and Robert J. Bonneau*

NONLINEAR OPTICS • *E. G. Sauter*

APPLIED ELECTROMAGNETICS AND ELECTROMAGNETIC COMPATIBILITY • *Dipak L. Sengupta and Valdis V. Liepa*

COPLANAR WAVEGUIDE CIRCUITS, COMPONENTS, AND SYSTEMS • *Rainee N. Simons*

ELECTROMAGNETIC FIELDS IN UNCONVENTIONAL MATERIALS AND STRUCTURES • *Onkar N. Singh and Akhlesh Lakhtakia (eds.)*

ANALYSIS AND DESIGN OF AUTONOMOUS MICROWAVE CIRCUITS • *Almudena Suárez*

ELECTRON BEAMS AND MICROWAVE VACUUM ELECTRONICS • *Shulim E. Tsimring*

FUNDAMENTALS OF GLOBAL POSITIONING SYSTEM RECEIVERS: A SOFTWARE APPROACH, Second Edition • *James Bao-yen Tsui*

SUBSURFACE SENSING • *Ahmet S. Turk, A. Koksal Hocaoglu, and Alexey A. Vertiy (eds.)*

RF/MICROWAVE INTERACTION WITH BIOLOGICAL TISSUES • *André Vander Vorst, Arye Rosen, and Youji Kotsuka*

InP-BASED MATERIALS AND DEVICES: PHYSICS AND TECHNOLOGY • *Osamu Wada and Hideki Hasegawa (eds.)*

COMPACT AND BROADBAND MICROSTRIP ANTENNAS • *Kin-Lu Wong*

DESIGN OF NONPLANAR MICROSTRIP ANTENNAS AND TRANSMISSION LINES • *Kin-Lu Wong*

PLANAR ANTENNAS FOR WIRELESS COMMUNICATIONS • *Kin-Lu Wong*

FREQUENCY SELECTIVE SURFACE AND GRID ARRAY • *T. K. Wu (ed.)*

ACTIVE AND QUASI-OPTICAL ARRAYS FOR SOLID-STATE POWER COMBINING • *Robert A. York and Zoya B. Popovic (eds.)*

OPTICAL SIGNAL PROCESSING, COMPUTING AND NEURAL NETWORKS • *Francis T. S. Yu and Suganda Jutamulia*

ELECTROMAGNETIC SIMULATION TECHNIQUES BASED ON THE FDTD METHOD • *Wenhua Yu, Xiaoling Yang, Yongjun Liu, and Raj Mittra*

SiGe, GaAs, AND InP HETEROJUNCTION BIPOLAR TRANSISTORS • *Jiann Yuan*

PARALLEL SOLUTION OF INTEGRAL EQUATION-BASED EM PROBLEMS • *Yu Zhang and Tapan K. Sarkar*

ELECTRODYNAMICS OF SOLIDS AND MICROWAVE SUPERCONDUCTIVITY • *Shu-Ang Zhou*

Printed and bound by CPI Group (UK) Ltd, Croydon, CR0 4YY